METHODS IN MOLECULAR BIOLOGY™

Series Editor
John M. Walker
School of Life Sciences
University of Hertfordshire
Hatfield, Hertfordshire, AL10 9AB, UK

For further volumes:
http://www.springer.com/series/7651

Protein Dynamics

Methods and Protocols

Edited by

Dennis R. Livesay

Department of Bioinformatics and Genomic, University of North Carolina at Charlotte
Charlotte, North Carolina, USA

Humana Press

Editor
Dennis R. Livesay
Department of Bioinformatics
and Genomic
University of North Carolina at Charlotte
Charlotte, North Carolina, USA

ISSN 1064-3745 ISSN 1940-6029 (electronic)
ISBN 978-1-4939-6307-2 ISBN 978-1-62703-658-0 (eBook)
DOI 10.1007/978-1-62703-658-0
Springer New York Heidelberg Dordrecht London

Printed on acid-free paper

Humana Press is a brand of Springer
Springer is part of Springer Science+Business Media (www.springer.com)

To my teachers
Mr. Jack Young, Prof. Terry Kruger, and Prof. Shankar Subramaniam.
Thank you for making science wonderful.

Preface

From the hand-drawn images of Irving Geis to modern computer-generated representations, pictures of protein structures are both artful and revealing. In fact, the foundation of structural biology is built upon these images. So much so that "structure gazing" is a commonly used verb in our discipline, leading to enormous increases in our understanding of stability mechanisms and how proteins actually work. Nevertheless, it has been recognized for over five decades that these static snapshots do not—in fact, they cannot!—tell the whole story. Like all molecules, proteins undergo significant conformational fluctuations, and these fluctuations are intimately tied to stability, function, and regulation. Feynman said it best when he famously wrote, "*Everything that living things do can be understood in terms of the jigglings and wigglings of atoms*" [1].

The pioneering works of Koshland [2], Straub [3], Karush [4], and others emphasized the importance of protein dynamics. However, our understanding of protein dynamics is just now starting to fully blossom. This is due to the fact that the "*jigglings and wigglings*" occur over a wide range of amplitudes and timescales, making it impossible to characterize the entire ballet [5] by viewing through a single prism. Rather, our modern understanding is built upon a series of methods where each respectively interrogates the motions that occur on specific timescales, including various NMR techniques and a wide range of spectroscopies. A second barrier to understanding has been the fact that experimental methods tend to be both time-consuming and expensive. As a result, many computational stand-ins have been developed and utilized to speed up our interrogations of protein dynamics. Theoretically, brute-force molecular dynamics simulations could be used to simultaneously characterize all timescales. However, practical limits on computing times prevent this, leading to a number of coarse-grained techniques that tend to be optimized for a given set of dynamical timescales.

This volume is roughly divided into two halves, experimental and computational methods. In each chapter, a method is introduced and detailed best-practice recipes are provided, covering most of the experimental and computational "prisms" available. The decision to combine both experiment and computation into a single book reflects modern protein research. Investigators must be fluent in both, and they regularly integrate results, either through collaboration or increasingly by utilizing both in a single lab. Consequently, this book is an important resource for anyone studying protein dynamics because it describes the primary methods used to characterize the molecular-level fluctuations that underlie "*Everything that living things do.*"

Charlotte, North Carolina, USA | ***Dennis R. Livesay***

References

1. Jiao RJ, Simpson TW, Siddique Z (2007) Product family design and platform-based product development: a state-of-the-art review. J Intell Manuf 18(1):5–29
2. Martin MV, Ishii K (2002) Design for variety: developing standardized and modularized product platform architectures. Res Eng Des 13(4):213–235

3. Simpson TW, Marion TJ, de Weck O, Holtta-Otto K, Kokkolaras M, Shooter SB (2006) Platform-based design and development: current trends and needs in industry. In: ASME design engineering technical conferences – design automation conference ASME, Philadelphia, Pennsylvania, Paper No. DETC2006/DAC-99229
4. Simpson TW, Siddique Z, Jiao J (2005) Product platform and product family design: methods and applications. Springer, New York, NY

Acknowledgment

This work was supported in part by NIH grants GM101570 and GM073082.

Contents

Contributors

GREGORY R. BOWMAN • *Departments of Molecular & Cell Biology and Chemistry, University of California, Berkeley, Berkeley, CA, USA*

MATTHEW C. BROWN • *Department of Bioinformatics and Genomics, University of North Carolina, Charlotte, NC, USA*

LAURA S. BUSENLEHNER • *Department of Chemistry, The University of Alabama, Tuscaloosa, AL, USA*

JHIH-WEI CHU • *Department of Chemical and Biomolecular Engineering, University of California, Berkeley, Berkeley, CA, USA*

CHARLES C. DAVID • *Department of Physics and Optical Science, University of North Carolina, Charlotte, NC, USA*

JUN-TAO GUO • *Department of Bioinformatics and Genomics, University of North Carolina, Charlotte, NC, USA*

JEFFREY A. HANSON • *Department of Chemical and Biomolecular Engineering, University of California, Berkeley, Berkeley, CA, USA*

RONALD D. HILLS JR. • *Department of Pharmaceutical Sciences, University of New England, Portland, ME, USA*

VINCENT J. HILSER • *Departments of Biology and Biophysics, Johns Hopkins University, Baltimore, MD, USA*

RIEKO ISHIMA • *Department of Structural Biology, University of Pittsburgh, Pittsburgh, PA, USA*

DONALD J. JACOBS • *Department of Physics and Optical Science, University of North Carolina at Charlotte, Charlotte, NC, USA*

PATRICE KOEHL • *Department of Computer Science and Genome Center, University of California, Davis, Davis, CA, USA*

ANDREW L. LEE • *Division of Chemical Biology and Medicinal Chemistry, University of North Carolina at Chapel Hill, Chapel Hill, NC, USA*

DENNIS R. LIVESAY • *Department of Bioinformatics and Genomics, University of North Carolina, Charlotte, NC, USA*

YURI E. NESMELOV • *Department of Physics and Optical Science, University of North Carolina at Charlotte, Charlotte, NC, USA*

CHAD M. PETIT • *Division of Chemical Biology and Medicinal Chemistry, University of North Carolina at Chapel Hill, Chapel Hill, NC, USA*

FLOYD E. ROMESBERG • *Department of Chemistry, The Scripps Research Institute, La Jolla, CA, USA*

CHRISTIAN RUSSELL • *Department of Bioinformatics and Genomics, University of North Carolina, Charlotte, NC, USA*

HARSIMRAN SINGH • *Department of Chemistry, The University of Alabama, Tuscaloosa, AL, USA*

YAN-WEN TAN • *Department of Physics, Fudan University, Shanghai, P. R. China*

MUSTAFA TEKPINAR • *Department of Physics, University at Buffalo, Buffalo, NY, USA*

DEEPTAK VERMA • *Department of Bioinformatics and Genomics, University of North Carolina, Charlotte, NC, USA*

DAHLIA R. WEISS • *Department of Pharmaceutical Chemistry, University of California San Francisco, San Francisco, CA, USA*
STEPHEN A. WELLS • *Department of Physics, University of Bath, Bath, UK*
STEVEN T. WHITTEN • *Department of Chemistry and Biochemistry, Texas State University, San Marcos, TX, USA*
HAW YANG • *Department of Chemistry, Princeton University, Princeton, NJ, USA*
WENJUN ZHENG • *Department of Physics, University at Buffalo, Buffalo, NY, USA*
JÖRG ZIMMERMANN • *Department of Chemistry, The Scripps Research Institute, La Jolla, CA, USA*

Part I

Experimental Methods for Characterizing Protein Dynamics

Chapter 1

Monitoring Side-Chain Dynamics of Proteins Using ^{2}H Relaxation

Chad M. Petit and Andrew L. Lee

Abstract

Nuclear magnetic resonance (NMR) is a powerful technique capable of monitoring a wide range of motions in proteins on a per residue basis. A variety of ^{2}H relaxation experiments have been developed for monitoring side-chain methyl group motions on the picosecond–nanosecond timescale. These experiments enable determination of the order parameter, S^2_{axis}, which reports on the rigidity of the C-CH_3 bond for side-chain methyl groups. The application of a commonly used subset of these experiments is described in this chapter. It is intended to serve as a practical guide to investigators interested in monitoring side-chain motions.

Key words Nuclear magnetic resonance (NMR), Protein dynamics, Methyl dynamics, Spin relaxation, Model-free analysis, Order parameters, Deuterium relaxation

1 Introduction

Proteins are dynamic molecules that depend on the coordination of atomic fluctuations to function properly. These motions occur over a large range of timescales [1, 2], from picosecond bond librations to folding and unfolding transitions which may take seconds or longer (Fig. 1). Many computational and experimental techniques have been developed to explore these motions and their roles in biological processes. These techniques include molecular dynamics simulations, crystallographic B-factor analysis, time-resolved crystallography, fluorescence spectroscopy, and nuclear magnetic resonance (NMR). NMR offers a distinct advantage by allowing experimental investigation into site-specific dynamics over a wide range of timescales. Biological processes such as catalysis and conformational switching have been shown by NMR to be dependent on motions occurring on the microsecond to millisecond (μs–ms), or "slow," timescale [3–5]. Some proteins, however, do not undergo significant motions on this slow timescale. In contrast, motions on the picosecond to nanosecond (ps–ns), or "fast,"

Dennis R. Livesay (ed.), *Protein Dynamics: Methods and Protocols*, Methods in Molecular Biology, vol. 1084,
DOI 10.1007/978-1-62703-658-0_1,

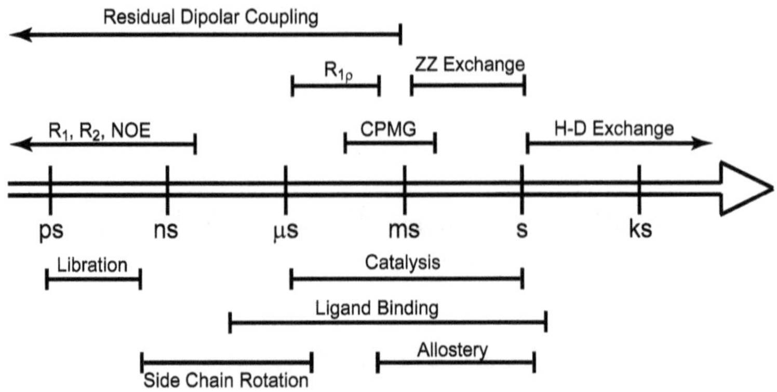

Fig. 1 Protein motions and their approximate timescales

timescale are ubiquitous throughout the proteome. Many of these motions result from local fluctuations of backbone and side-chain dihedral angles. This review focuses on NMR characterization of side-chain motions of methyl-bearing residues using ^{2}H relaxation experiments, while the analysis of backbone motions using ^{15}N relaxation experiments is covered in a separate chapter.

The order parameter, S^2, is a measure of the internal reorientational freedom of a given bond vector on the ps–ns timescale. It ranges from 0, corresponding to no favored position of the bond vector in the molecular frame, to 1, indicating complete rigidity of the bond. Typically, order parameters are measured for backbone amide and side-chain methyl bonds of individual residues in a protein using NMR spin relaxation experiments. Backbone amide order parameters provide information on site-specific motions involving the main chain of the protein and are therefore primarily dictated by secondary structure. Side-chain order parameters provide a measure of amplitude of motions largely independent of secondary structure and are thus excellent probes for monitoring the internal motions of folded proteins. While the range for backbone order parameters is relatively narrow, methyl side-chain order parameters vary considerably throughout the protein (Fig. 2). It is this variability in the amplitude of side-chain motions that has captured the interest of NMR spectroscopists since the early stages of investigations into protein dynamics. What underlies this variability is unclear, as side-chain motions appear not to be correlated with depth of burial, packing density, or solvent accessible surface area [6]. What is abundantly clear, however, is that the majority of ps–ns motions in a protein occur in the side chains, making them indispensable to anyone interested in the motions of proteins.

Recent studies have established the functional relevance of side-chain motions on the ps–ns time regime [7]. The potential functional relevance of these motions was first realized when atomic fluctuations, as measured by S^2, were shown to be related to the

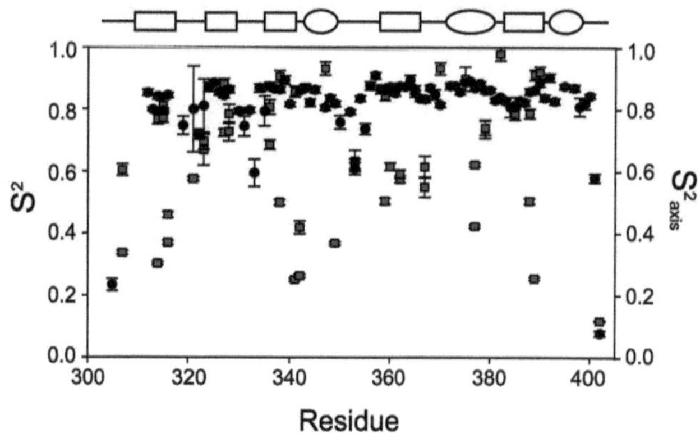

Fig. 2 Distribution of PDZ3 order parameters. Backbone order parameters, S^2, are indicated by *black circles* while side-chain methyl order parameters, S^2_{axis}, are indicated by *red squares*. The secondary structure of PDZ3 is represented on top of figure with ovals depicting α-helices and rectangles depicting β-sheets. Note that backbone order parameters are tightly clustered around 0.9 except in loops. This highlights that values of S^2 are dependent on secondary structure whereas S^2_{axis} has no such dependence, as evidenced by their heterogeneous distribution throughout the protein (Color figure online)

Gibbs free energy equation [8]. An analytical relationship between order parameters measured by NMR and the local residual entropy of proteins was subsequently derived [9, 10]. While these derivations did not provide a reliable method for calculating absolute entropies associated from order parameters, they did provide a more reliable assessment of *relative* entropic differences (e.g., differences between free and bound states). It was therefore predicted that differences in order parameters could serve as a proxy for changes in conformational entropy. This prediction was ultimately verified experimentally by Wand and coworkers using calmodulin as a model system [11]. In addition to serving as sources of conformational entropy, side chains also facilitate intramolecular communication between sites that would not be expected to be energetically linked based on their three dimensional structure [12–16]. Taken together, the notion that side-chain motions can facilitate intramolecular communication, along with their ability to serve as a source of conformational entropy, suggests that these motions may offer a basis for allosteric communication without detectable conformational change [17–19]. The idea that side-chain motions are able to facilitate this type of "dynamic allostery" was demonstrated by Petit et al. [20]. This work will be detailed later in this review.

Given the recent number of reports on the role of side-chain motions in protein function, this review seeks to provide a resource for those interested in investigating side-chain motions in other systems. This review will primarily focus on the practical aspects of measuring ps–ns side-chain motions using ^{2}H relaxation on

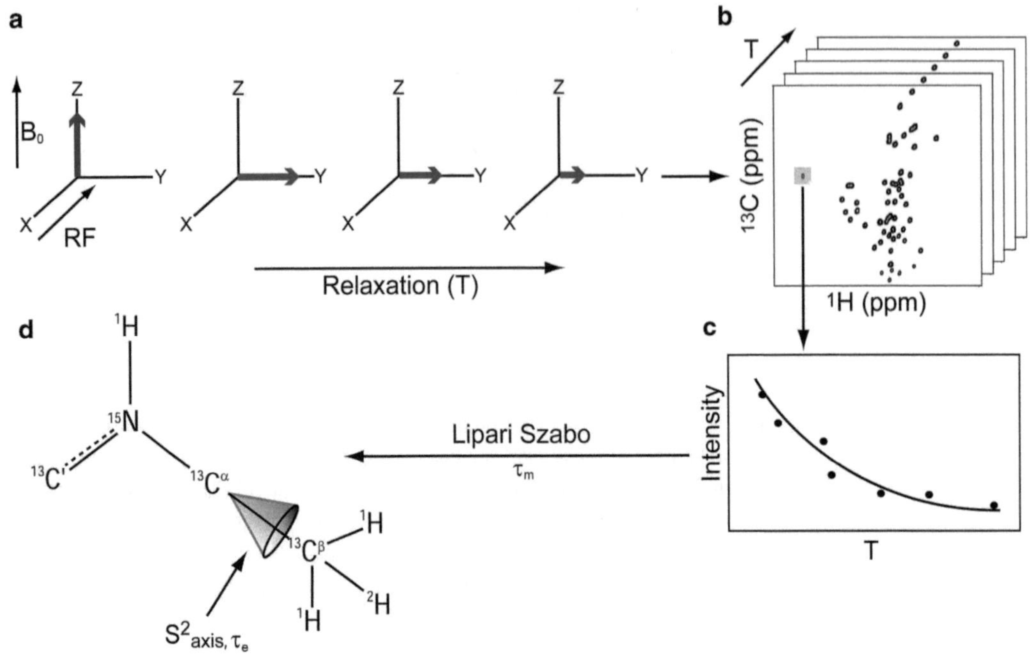

Fig. 3 General outline of process for obtaining methyl order parameters. (**a**) A graphical representation of D_y, or transverse, relaxation. The direction of the static magnetic field is shown as a *vertical arrow*-labeled B_0, and the RF pulse applied to rotate bulk magnetization into the *xy* plane is shown as a an arrow pointing towards -x. Following the RF pulse, the coherent magnetization precesses around B_0 in the *x–y* plane and immediately begins to dephase. This loss of coherence is a function of the relaxation time (*T*) and is depicted by the *gray arrow*. (**b**) ^{2}H Relaxation is monitored by acquiring multiple 2D ^{1}H–^{13}C HSQC type spectra with different relaxation times. (**c**) The typical experiments used to determine methyl order parameters measure ^{2}H D_z and D_y relaxation rates. To obtain these rates, peak intensities are measured as a function of relaxation time and fit to single exponential equations. (**d**) The relaxation parameters are fit to the Lipari–Szabo model-free equations to obtain methyl dynamic parameters which describe the amplitude (S^2_{axis}) and the characteristic time (τ_e) of the C–CH_3 symmetry axis (Color figure online)

proteins less than 25–30 kDa. It is our intention to guide the reader through the general process of measuring side-chain order parameters as outlined in Fig. 3. For proteins larger than 30 kDa, the reader is referred to a review by Tugarinov [21] that details experimental approaches for studying side-chain dynamics in high-molecular-weight proteins (*see* **Note 1**). For more in-depth theoretical considerations regarding ^{15}N (backbone) and ^{2}H (side-chain) relaxation theory and analysis, readers are referred to reviews by Stone [22] and Wand [6], respectively.

1.1 Spin Relaxation

Dynamics on the ps–ns time regime are measured by NMR using spin relaxation experiments. In these experiments, nuclear spins are perturbed away from their ground state by one or more radio frequency pulses (Fig. 3a). Their subsequent return to equilibrium, or "relaxation," is then monitored as a function of time to obtain relaxation rates. Excited nuclei are subject to two types of relaxation processes, longitudinal and transverse. Longitudinal relaxation involves the return of bulk magnetization along the static magnetic field, thereby regenerating the equilibrium Boltzmann population. Transverse relaxation involves the dephasing of coherent magnetization as it precesses around the static magnetic field in the *x–y* plane. These processes do not occur spontaneously but are stimulated by local magnetic field fluctuations (near the relaxing spin) at or near the Larmor frequency of the target nucleus. Since these fluctuations arise from the movement of nuclei relative to one another in the presence of a static magnetic field (B_0), they are extremely sensitive to molecular motion. It is this relationship between motion and relaxation that enables spin relaxation measurements to report on the motions of bond vectors.

For biomolecules, spin relaxation rates are commonly measured using a series of 2D heteronuclear single-quantum coherence (HSQC)-type experiments (Fig. 3b). Each HSQC in the series utilizes a different relaxation time delay (T) during which molecular motions stimulate spin relaxation. Because peak intensities are modulated as a function of this delay, relaxation rates can be measured by fitting the observed decay in intensity to a single exponential (Fig. 3c). There are multiple mechanisms by which ps–ns motions can relax excited nuclear spins back to equilibrium. Each mechanism's total contribution to the overall relaxation process is dependent on the type of nucleus being analyzed along with the strength of the static magnetic field. For ^{15}N and ^{13}C, the dominant mechanisms are dipole–dipole interactions and chemical shift anisotropy (CSA). Dipole–dipole-induced relaxation occurs when the magnetic field of a nuclear spin affects the local magnetic field of another. Relaxation due to CSA occurs because of local variations in shielding from the static magnetic field that result from the inhomogeneity of electron density surrounding the observed nucleus. This causes the magnetic field at the nucleus to fluctuate as the molecule rotates which, in turn, induces relaxation.

Although dipole–dipole and CSA relaxation mechanisms are active when analyzing side-chain motions, they are not the dominant mechanism for ^{2}H relaxation. Nuclei with $I > 1/2$ possess nuclear electric quadrupole moments whose nuclear charge distribution is not spherically symmetric. Interactions between the quadrupole moment and local oscillations in the surrounding electric field gradients provide a relaxation mechanism for excited spins. These interactions are strong and are extremely efficient at promoting relaxation, exceeding relaxation caused by dipole–dipole and CSA by 1–2 orders of magnitude. Consequently, quadrupolar relaxation is the dominant relaxation mechanism for ^{2}H spins, even in fractionally deuterated side chains. Because they are not contaminated by other relaxation mechanisms and are not significantly affected by cross-correlated relaxation from neighboring dipoles [23], ^{2}H relaxation rates are monoexponential and straightforward to interpret. However, it is important to note that these relaxation rates are for the axially symmetric electric field gradient of the ^{2}H nucleus and not the C–D bond vector of the side-chain methyl (*see* Subheading 1.3).

The deuterium nucleus has five independent operators, or relaxation modes, with distinct relaxation rates that can be linearly combined to describe the density matrix. The equations describing these five relaxation modes are as follows [24]:

$$R^{Q}(D_z) = \frac{3}{16}\left(\frac{e^2qQ}{\hbar}\right)^2 [J(\omega_D) + 4J(2\omega_D)] \quad (1a)$$

$$R^{Q}(3D_z^2 - 2) = \frac{3}{16}\left(\frac{e^2qQ}{\hbar}\right)^2 [3J(\omega_D)] \quad (1b)$$

$$R^{Q}(D_y) = \frac{1}{32}\left(\frac{e^2qQ}{\hbar}\right)^2 [9J(0) + 15J(\omega_D) + 6J(2\omega_D)] \quad (1c)$$

$$R^{Q}(D_+D_z + D_zD_+) = \frac{1}{32}\left(\frac{e^2qQ}{\hbar}\right)^2 [9J(0) + 3J(\omega_D) + 6J(2\omega_D)] \quad (1d)$$

$$R^{Q}(D_+^2) = R^{Q}(D_X^2 + D_Y^2) = \frac{3}{16}\left(\frac{e^2qQ}{\hbar}\right)^2 [J(\omega_D) + 2J(2\omega_D)] \quad (1e)$$

The quadrupolar coupling constant $\left(\frac{e^2qQ}{\hbar}\right)$ is approximately 167 kHz [25], and $J(\omega)$ is the value of the spectral density, or frequency of motion, evaluated at the indicated frequency (0, ω_D, $2\omega_D$). D_z (longitudinal relaxation) and $3D_z^2 - 2$ relaxation rates (quadrupolar order) are related to the populations of spin states, i.e., longitudinal relaxation, while D_y (in-phase transverse magnetization), $D_+D_z + D_zD_+$ (antiphase transverse magnetization), and D_+^2 relaxation rates (double-quantum magnetization) involve transitions between spin states, i.e., transverse relaxation. These equations relate measured relaxation rates to the spectral density function (*see* Analysis), thereby providing a link between relaxation measurements and the frequency of molecular motions.

1.2 ^{2}H Relaxation Pulse Sequences

The original relaxation experiments used to determine methyl order parameters were designed to measure the D_z and D_y relaxation rates of ^{2}H nuclei in CH_2D isotopomers (Fig. 4, *see* **Note 2**) [23]. It should be noted that the pulse sequence used to measure ^{2}H D_y relaxation is a $T_{1\rho}$ experiment that utilizes a spin lock pulse (Fig. 4b). This effectively removes contributions to the relaxation rate from chemical exchange (R_{ex}), thus eliminating the need for alternate models that account for R_{ex} such as those used in ^{15}N relaxation analysis. The pulse sequences are based on constant-time

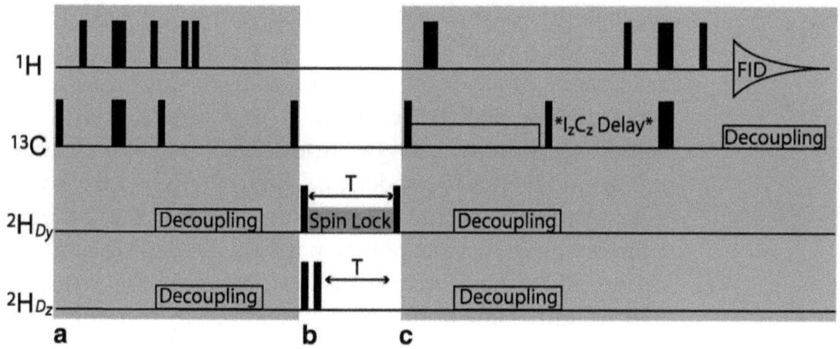

Fig. 4 General schematic of pulse sequences used to measure ^{2}H D_y and D_z relaxation rates for CH_2D isotopomers. (**a**) Magnetization is transferred to ^{2}H nuclei in the CH_2D isotopomers. (**b**) Relaxation is allowed to occur for period *T*. (**c**) Magnetization is transferred back to ^{1}H for detection. Position for optional delay that allows "on the fly" subtraction of the I_zC_z component is noted

^{1}H–^{13}C heteronuclear single-quantum coherence (HSQC) experiments and are fully detailed in Muhandiram et al. [23]. Briefly, magnetization begins on proton and is transferred via INEPT [26] to the attached ^{13}C nuclei. The magnetization is then allowed to evolve for a period of $1/(2J_{CH})$ at which point the selection of signals originating from $^{13}CH_2D$ isotopomers is achieved by editing pulses (Fig. 4a). Next, relaxation of the triple spin term $I_zC_zD_z$ (longitudinal) or $I_zC_zD_y$ (transverse) is allowed to occur for a specified relaxation period (T) (Fig. 4b). I_z, C_z, and D_z denote the z components of the ^{1}H, ^{13}C, and ^{2}H nuclei, respectively, that constitute the $^{13}CH_2D$ isotopomers. After the relaxation period, ^{13}C chemical shift is recorded in a constant-time manner with magnetization subsequently being returned to proton for detection via reverse INEPT [26] transfer (Fig. 4c). The pulse sequences in Muhandiram et al. [23] measure triple spin coherences and therefore require an additional experiment that measures I_zC_z relaxation rates in order to obtain pure D_z or D_y relaxation rates. Kay and coworkers have since updated these sequences to internally subtract the I_zC_z component "on the fly," making the additional experiment unnecessary [27]. This "on-the-fly" subtraction is accomplished by the insertion of a delay in the pulse sequence (Fig. 4c) that accounts for the I_zC_z component of relaxation, thereby allowing the effective relaxation rate, D_z or D_y, to be measured. Finally, an additional three experiments that measure D_+^2 (double-quantum magnetization), $3D_z^2 - 2$ (quadrupolar order), and $D_+D_z + D_zD_+$ (transverse antiphase magnetization) relaxation rates can also be used if fitting of the LS3 model is desired (*see* Analysis). These additional experiments are based on the Muhandiram et al. [23] pulse sequences and are described in Millet et al. [27].

The pulse sequences from Muhandiram et al. [23] are used to measure the D_z and D_y relaxation rates needed for characterizing methyl side-chain motions. They consist of a series of HSQC experiments with each HSQC in the series utilizing a different relaxation time delay (*T*). Relaxation rates can therefore be measured by fitting the decay in peak intensity as a function of relaxation delay to a single exponential. To obtain accurate rates, it is good practice to properly calibrate all pulses (*see* **Note 3**) and to adjust the relaxation times to ensure proper sampling of the relaxation curve. These time points may vary depending on the relaxation properties of the protein of interest, which are influenced by molecular weight. Time points may be selected by running a series of 1D D_z and D_y experiments, each with a varied relaxation time point. This will allow the overall intensity of the signal to be monitored as a function of relaxation delay. The delays should be incrementally increased until signal can no longer be detected. At this point, the appropriate spacing of time points can be calculated with the knowledge that the decay in peak intensity is a single exponential. Typically, at least ~10 relaxation time points (including a few duplicates) are used to fit the relaxation curve. Duplicate points are collected to determine the error associated with peak intensities and are critical for determining error in D_z and D_y relaxation rates. It is good practice to perform these experiments at two different magnetic field strengths, as this allows for high-quality fits of the spectral density function (*see* Analysis).

1.3 Spectral Density Analysis

NMR spin relaxation measurements allow extraction of dynamics in the form of bond vector motions. This is carried out using the "model-free" theory of Lipari and Szabo [28, 29] which is a form of the spectral density function that allows interpretation of how experimentally measured relaxation rates (Eqs. 1a–1e) relate to the frequency of motions in proteins. The theoretical framework necessary for this bridge between experimental measurement and molecular motions will now be briefly detailed.

To describe the reorientational motions of a bond vector, a time-dependent rotational autocorrelation function $C(t)$ must be defined. This correlation function provides the basis for simple parametric models of angular bond vector motion and is particularly well suited to separate motions occurring on different timescales [22, 30]. The Fourier transform of $C(t)$ is the spectral density function, $J(\omega)$, which describes the probability that a bond vector has molecular motion at a given frequency. The spectral density is defined by the equation

$$J(\omega) = \frac{\tau_m}{1 + (\omega\tau_m)^2}, \tag{2}$$

where the rotational correlation time, τ_{m}, is the time required for the molecule to rotate 1 rad in the laboratory frame. This form of the spectral density function assumes a motional model with no internal motions and isotropic tumbling. However, this model is not appropriate to use for proteins since (1) proteins are not internally static and (2) this equation cannot distinguish between molecular tumbling motions and internal protein dynamics.

To alleviate these limitations, a modified spectral density function was proposed by Lipari and Szabo, known as the "model-free" formalism, as it assumes no specific physical motional model [28, 29]. The fundamental assumption made in the model-free formalism is that the internal dynamics of a protein are faster than, and consequentially independent of, overall molecular tumbling. This assumption allows the isolation of internal motion from overall tumbling. The original model-free spectral density function is comprised of the three dynamical parameters τ_m, S^2, and τ_e and can be expressed as

$$J(\omega) = \frac{2}{5}\left[\frac{S^2\tau_m}{1+(\omega\tau_m)^2} + \frac{\tau(1-S^2)}{1+(\omega\tau)^2}\right], \tag{3}$$

where $\tau^{-1} = \tau_m{}^{-1} + \tau_e{}^{-1}$. This form of the model-free spectral density is referred to as LS2 because it yields two internal dynamics parameters, S^2 and τ_e. S^2 is the order parameter describing the degree of angular restriction for a given bond vector, while τ_e represents the characteristic time of the bond vector motion. Because the rotational correlation time, τ_m, cannot be reliably obtained using ^{2}H relaxation experiments, it is necessary to perform the standard set of ^{15}N relaxation measurements [31] to determine the value of τ_m to use when fitting methyl order parameters.

Residue-specific methyl order parameters are fit using a nonlinear least-squares minimization of the error function

$$\chi^2 - \sum_{j}^{M}\left(\frac{\text{obs}_j - \text{calc}_j}{\lambda_j^{\text{obs}}}\right)^2, \tag{4}$$

in which M is the number of relaxation measurements for a given spin, obs_j is the jth measured relaxation rate, calc_j is the jth calculated relaxation rate, and λ_j^{obs} is the estimated uncertainty in obs_j. The measured relaxation rate, obs_j, is obtained using the pulse sequences previously described. Briefly, relaxation rates are measured by fitting the observed decay in peak intensity to a single exponential using a nonlinear least-squares two-parameter fit algorithm (e.g., Levenberg–Marquardt). Duplicate relaxation points allow uncertainties in the measured relaxation rates, λ_j^{obs}, to be estimated by either Monte Carlo simulations, or they can be taken from the covariation matrix when fitting with the Levenberg–Marquardt algorithm. Because these rate uncertainties are the only source of empirical error when fitting methyl order parameters, they are fundamental to obtaining accurate order parameters and must be determined as reliably as possible. The calculated relaxation rate, calc_j, is determined using the equations that define relaxation parameters in terms of spectral densities (Eqs. 1a–1e and 3). To ensure that the minimization successfully locates the global minimum, an initial grid search of the relevant parameter space is performed prior to final minimization. Errors associated with internal dynamics parameters are estimated using Monte Carlo simulations. Although the fitting described above has typically been performed using in-house software, software for fitting methyl order parameters from ^{2}H relaxation data is expected to be freely available from the Wand lab (A. Joshua Wand, personal communication).

As noted above, ^{2}H relaxation measurements report on motions involving the principle axis of the electric field gradient tensor and not the bond vector of interest (Fig. 3d). Therefore, fitting ^{2}H relaxation data to the model-free formalism actually yields S^2 and τ_e for the principle axis of the electric field gradient tensor. It is therefore necessary to make a series of assumptions when establishing how the measured ^{2}H relaxation rates relate to biologically relevant motions of side chains. For S^2, the first assumption made is that the principle axis of the electric field gradient tensor is collinear with the C–D bond vector [25, 32]. The second assumption pertains to the nature of motions experienced by the C–D bond. These motions involve rapid rotations *about* the methyl threefold symmetry axis and the actual motions *of* the symmetry axis itself. Rapid rotation about the axis is the primary motion experienced by the C–D bond, yet is not particularly biologically relevant. Thus, it is desirable to remove this component from the S^2 parameter. By assuming that the rapidly rotating methyl group has tetrahedral geometry, the effect of this motion can be removed using Eq. 5, in which θ is the angle between the C–D bond and the averaging axis (109.5°) [23, 25].

$$S = S_{\text{axis}}\left[(3\cos^2\theta - 1)/2\right] \tag{5}$$

Therefore, $S^2{}_{\text{axis}}$ reports on the motions of the methyl symmetry axis only (Fig. 3d). For τ_e, it is not possible to isolate the characteristic time of motions *about* the methyl symmetry axis from the characteristic time of motions *of* the symmetry axis using ^{2}H relaxation. τ_e is therefore a weighted average of the characteristic times of both dynamic processes. While it is presently unclear how to best interpret this dynamics parameter, τ_e does provide a sensitive probe when monitoring changes in protein dynamics upon perturbation [12, 14].

For the majority of methyl groups, a single order parameter and correlation time is sufficient to describe local dynamics. However, a more detailed analysis of side-chain motions has been developed for residues that have additional, slower motions (i.e., low nanosecond) that are incompatible with the two-parameter model [27, 33]. This alternate model, termed LS3, replaces τ_m with the adjustable parameter, $\tau_{c,\text{eff}}$, which is the correlation time for the combined motions of overall molecular rotation and "slow" side-chain reorientation (on a timescale slightly slower than tumbling). This three parameter model of the spectral density function can be expressed as [33]

$$J(\omega) = \frac{2}{5}\left[\frac{S^2\tau_{c,\text{eff}}}{1+(\omega\tau_{c,\text{eff}})^2} + \frac{\tau(1-S^2)}{1+(\omega\tau)^2}\right], \tag{6}$$

in which $\tau^{-1} = \tau_{c,\text{eff}}{}^{-1} + \tau_e{}^{-1}$. To successfully utilize the LS3 model, it is necessary to measure at least three ^{2}H relaxation rates at a single field with one rate being from transverse relaxation to evaluate the spectral density at $J(0)$ [33]. However, it has been shown that determining all five ^{2}H relaxation rates at a single field greatly improves the accuracy of analysis when compared to obtaining only the minimum number necessary [33]. An *F*-test statistical analysis is typically used to justify the use of the more complex LS3 model over the standard LS2 model.

One caveat to using the Lipari–Szabo model-free formalism is that it makes certain assumptions by presupposing a functional form of the spectral density. A method pioneered by Peng and Wagner was developed to circumvent this limitation by directly solving for the spectral density [34, 35]. The calculated spectral density value can then, in turn, be used to evaluate the accuracy of "model" spectral density functions that are currently used to determine site-specific order parameters (e.g., Lipari–Szabo model-free formalism). However, spectral density mapping requires that an expanded set of relaxation experiments be performed to enable discrete frequencies of the spectral density to be uniquely determined. For ^{2}H, the equations describing the dominant quadrupolar relaxation mechanism indicate that their decay rates depend on the spectral density function evaluated at three distinct frequencies (Eqs. 1a–1e) [24]. These frequencies are 0, ω_D, and $2\omega_D$, where ω_D is the ^{2}H Larmor frequency. From measurement of the five modes of relaxation for each ^{2}H nuclei at a single field, it is possible to explicitly back calculate the three spectral density values (Eqs. 1a–1e). Spectral density mapping for ^{2}H generally agrees with the Lipari–Szabo model-free formalism [6, 33]. However, in addition to requiring an expanded number of experiments, spectral density mapping does not allow for the separation of internal and global motions. As a result, in many cases, the Lipari–Szabo model-free formalism is preferred when assessing the ps–ns motions of bond vectors.

1.4 Example Application

We now present a specific example that may serve as a "training system" for investigators interested in using ^{2}H relaxation experiments to monitor side-chain motions in proteins. This example is taken from the study by Petit et al. in which side-chain dynamics were shown to play an integral role in the function of a PDZ domain [20]. PDZ domains are small monomeric proteins that typically bind the carboxyl terminal tails of their target proteins. They are often found in scaffolding proteins that contain modular structures in the context of larger multidomained proteins. The third PDZ domain from the neuronal scaffolding protein PSD-95/SAP90 (PDZ3, residues 303–402) is commonly thought of as the archetype of PDZ domains. Although the PDZ family has a highly conserved fold consisting of a 6-stranded half β-barrel and 2 α-helices [36], some members exhibit additional secondary structural elements or differences in lengths of helices, β-strands, or loops [37–41]. Although the archetype of the PDZ family, PDZ3 contains an atypical third α-helix that has been shown to be phosphorylated in vivo at position Y397 [42]. We sought to investigate the effects of this modification through structural and dynamic characterization of a phosphorylation mimic [43] that was constructed by deleting the 7 carboxyl terminal residues, referred to here as Δ7ct. Isothermal titration calorimetry measurements indicate that the binding of Δ7ct is reduced by 21-fold from PDZ3 and that, interestingly, the differences between the two binding energies appear to be entirely entropic. Backbone dynamics measured using ^{15}N relaxation experiments reveals essentially no differences between the two constructs. A comparison of the side-chain dynamics, however, shows that Δ7ct undergoes global decreases in S^2_{axis}, indicating that the side chains become more flexible upon deletion of α3. However, the enhanced motions are quenched upon binding to CRIPT peptide ligand, indicating that binding of CRIPT peptide "snaps" the side-chain dynamics back to that of CRIPT-bound PDZ3. As there is no evidence of gross structural changes that could account for the observed dynamic changes, we conclude that the additional entropic penalty paid upon peptide binding observed for Δ7ct is due to increased motions (i.e., conformation entropy) in the side chains of the unbound protein. In addition, dynamic changes occur throughout the protein with a significant number being distal to the peptide binding pocket. Taken together, we demonstrate that the changes in conformational entropy associated with side-chain dynamics is the driving force behind the allosteric behavior observed in PDZ3. This study is an excellent example of how the knowledge of side-chain motions can help to understand biological phenomena. Without this knowledge, the source of the reduction in binding affinity between the two proteins would have remained a mystery.

2 Materials

1. Expression Vector.
   - PSD-95 PDZ3 (AddGene: [Plasmid 31229: PDZ3-pET28a]).

2. Purification Columns.
   - Source 30Q (GE Biosciences, Inc.)
   - Fast-Flow Q Sepharose (GE Biosciences, Inc.)
   - G50-Sephadex (GE Biosciences, Inc.)
3. Stock of 10× M9 Salts 1 L, pH 7.4.
   - 67.6 g Na_2HPO_4.
   - 30.0 g KH_2PO_4.
   - 5.0 g NaCl.
4. 1 L M9 Growth Media.
   - 100 mL 10× M9 salts, pH 7.4.
   - 600 mL D_2O.
   - 300 mL ddH_2O.
   - 1.0 g ^{15}N NH_4Cl.
   - 2.0 g ^{13}C-labeled glucose.
   - 2.0 mL 1 M $MgSO_4$.
   - 0.1 mL 1 M $CaCl_2$.
   - 10.0 mg thiamine hydrochloride.
   - 50–100 mg relevant antibiotic.
   - 10.0 mg $FeSO_4$.
5. Lysis Buffer.
   - 50 mM Tris-HCl pH 7.5.
   - 150 mM NaCl.
   - 1 mM EDTA.
6. Q Sepharose Loading Buffer.
   - 50 mM Tris-Hcl pH 7.5.
   - 1 mM EDTA.
7. Q Sepharose Elution Buffer.
   - 50 mM Tris-Hcl pH 7.5.
   - 1 M NaCl.
   - 1.0 mM EDTA.
8. NMR Buffer
   - 22 mM sodium phosphate pH 6.8.
   - 55 mM NaCl.
   - 1.1 mM EDTA.
   - 0.02 % NaN_3.

3 Methods

The following protocol is meant to guide the reader through the entire process of obtaining methyl order parameters for the PDZ3 domain used in a previous study [20]. A PDZ3 expression construct can be obtained from AddGene, which can be used to express and purify PDZ3 from *Escherichia coli.* This protein can be used as a training vehicle to gain experience in ^{2}H methyl relaxation and analysis. Table 1 contains all of the relevant relaxation

Table 1
PDZ3 chemical shift assignments, ^{2}H relaxation data, and S^2_{axis} order parameters

PDZ3 ^{2}H Relaxation Data													
Residue	Atom	^{1}H(ppm)	^{13}C(ppm)	D_y (500)	Λ_y (500)	D_y (600)	Λ_y (600)	D_z (500)	Λ_z (500)	D_z (600)	Λ_z (600)	S^2_{axis}	Λ
I307	cδ1	0.50	12.57	21.0	0.5	23.1	0.3	54.4	1.4	62.7	1.0	0.34	0.01
I307	cγ2	0.52	16.84	10.7	0.4	11.5	0.2	25.0	0.9	26.5	0.6	0.61	0.02
I314	cδ1	0.66	16.63	25.9	0.6	28.6	0.4	79.3	2.1	95.1	1.6	0.30	0.01
I314	cγ2	0.75	19.38	10.6	0.4	11.6	0.2	36.0	1.2	41.5	0.8	0.77	0.02
V315	cγ1	0.53	21.07	10.2	0.3	10.9	0.2	35.1	1.0	38.9	0.7	0.81	0.02
V315	cγ2	0.85	20.92	11.1	0.3	12.1	0.2	42.1	1.0	49.4	0.7	0.77	0.01
I316	cδ1	0.83	13.70	19.6	0.5	21.5	0.3	51.3	1.4	60.8	0.9	0.37	0.01
I316	cγ2	0.84	18.38	13.8	0.3	15.3	0.2	31.4	0.7	35.0	0.4	0.46	0.01
T321	cγ2	1.19	21.53	13.4	0.2	14.0	0.2	37.6	0.7	41.6	0.5	0.58	0.01
L323	cδ1	0.84	26.73	11.9	1.1	11.5	0.6	27.6	2.5	33.3	1.7	0.67	0.05
L323	cδ2	0.70	22.63	12.1	0.7	12.8	0.4	42.8	2.5	47.5	1.6	0.70	0.03
I327	cδ1	0.46	14.23	13.2	0.4	14.4	0.2	79.0	3.1	97.7	2.5	0.73	0.01
I327	cγ2	0.86	19.43	9.9	0.4	11.0	0.2	42.5	1.6	49.9	1.0	0.88	0.02
V328	cγ1	0.87	21.23	9.0	0.4	8.8	0.2	21.4	1.1	20.8	0.5	0.79	0.03
V328	cγ2	0.60	19.65	9.7	0.5	10.2	0.3	24.2	1.1	25.9	0.7	0.73	0.03
I336	cδ1	0.49	9.11	13.1	0.4	14.2	0.3	55.8	2.0	67.3	1.6	0.69	0.02
I336	cγ2	0.69	18.21	10.5	0.4	11.8	0.3	42.9	1.7	48.7	1.1	0.81	0.02
I338	cδ1	0.14	13.40	16.2	0.4	18.6	0.3	59.1	1.8	70.6	1.4	0.50	0.01
I338	cγ2	0.65	17.64	10.2	0.4	10.8	0.2	44.7	1.7	54.5	1.3	0.91	0.02
I341	cδ1	0.45	12.36	27.6	0.5	30.1	0.4	68.8	1.4	76.4	1.1	0.25	0.01
I341	cγ2	0.67	17.34	10.6	0.4	11.5	0.2	48.9	1.8	58.6	1.3	0.86	0.02
L342	cδ1	1.06	23.65	23.3	0.5	24.7	0.3	47.3	0.9	53.0	0.6	0.26	0.01

(continued)

Table 1
(continued)

PDZ3 ^{2}H Relaxation Data													
Residue	Atom	$^1H_{(ppm)}$	$^{13}C_{(ppm)}$	D_y (500)	λ_y (500)	D_y (600)	λ_y (600)	D_z (500)	λ_z (500)	D_z (600)	λ_z (600)	S^2_{axis}	λ
L342	cδ2	0.95	25.97	17.3	1.3	18.5	0.6	45.3	3.3	50.4	1.7	0.42	0.02
A343	cβ	1.39	17.77	9.7	0.2	10.3	0.1	33.9	0.7	38.9	0.5	0.87	0.01
A347	cβ	1.53	19.09	9.2	0.4	10.1	0.2	38.5	1.6	42.6	1.0	0.93	0.02
L349	cδ1	0.82	25.24	ND	ND	ND	ND	ND	ND	ND	ND	ND	ND
L349	cδ2	0.84	22.52	19.2	0.4	20.4	0.2	47.0	1.0	53.3	0.6	0.37	0.01
L353	cδ1	0.48	25.40	13.8	1.5	13.3	0.5	34.2	3.6	44.0	1.7	0.63	0.03
L353	cδ2	0.38	23.95	13.3	0.5	14.4	0.3	45.5	1.9	53.7	1.3	0.62	0.02
I359	cδ1	0.61	13.16	16.8	0.5	18.8	0.3	66.7	2.5	80.0	1.9	0.51	0.01
I359	cγ2	0.65	18.81	10.8	0.3	11.7	0.2	48.1	1.7	59.1	1.2	0.84	0.02
L360	cδ1	0.81	25.24	ND	ND	ND	ND	ND	ND	ND	ND	ND	ND
L360	cδ2	0.63	21.99	13.1	0.4	13.8	0.2	41.5	1.2	45.5	0.8	0.62	0.01
V362	cγ1	0.75	20.91	12.9	0.3	14.3	0.2	35.3	0.9	42.6	0.6	0.58	0.01
V362	cγ2	0.78	22.84	12.0	0.3	13.4	0.2	32.3	0.9	36.7	0.6	0.59	0.01
V365	cγ1	0.82	20.70	10.0	0.3	10.7	0.2	38.0	1.0	42.2	0.6	0.86	0.02
V365	cγ2	0.97	21.20	ND	ND	ND	ND	ND	ND	ND	ND	ND	ND
L367	cδ1	0.69	26.07	14.5	1.0	14.2	0.6	36.4	2.6	41.9	1.7	0.55	0.03
L367	cδ2	0.61	23.02	10.4	0.7	11.5	0.4	23.3	1.6	26.8	1.0	0.62	0.04
A370	cβ	1.41	21.00	9.4	0.3	10.2	0.2	37.3	1.2	44.5	0.9	0.93	0.02
A375	cβ	1.31	19.42	8.4	0.4	8.9	0.3	23.6	1.1	25.5	0.7	0.90	0.04
A376	cβ	1.43	17.78	ND	ND	ND	ND	ND	ND	ND	ND	ND	ND
I377	cδ1	0.83	13.22	19.2	0.4	21.0	0.2	64.8	1.4	74.2	0.9	0.42	0.01
I377	cγ2	0.90	17.51	12.9	0.3	14.0	0.2	43.0	0.9	48.8	0.6	0.62	0.01
A378	cβ	1.43	17.78	ND	ND	ND	ND	ND	ND	ND	ND	ND	ND
L379	cδ1	0.80	25.79	11.2	1.1	11.3	0.4	34.1	3.3	36.0	1.1	0.74	0.03
L379	cδ2	0.82	24.00	11.3	0.5	11.9	0.3	39.2	1.7	42.5	0.9	0.74	0.02
A382	cβ	1.48	20.26	9.3	0.3	9.9	0.2	40.0	1.4	47.5	1.0	0.98	0.02
T385	cγ2	0.91	22.17	10.3	0.3	10.9	0.2	34.4	1.1	35.7	0.6	0.78	0.02
V386	cγ1	0.78	21.93	ND	ND	ND	ND	ND	ND	ND	ND	ND	ND
V386	cγ2	0.77	22.00	ND	ND	ND	ND	ND	ND	ND	ND	ND	ND
T387	cγ2	0.97	21.28	ND	ND	ND	ND	ND	ND	ND	ND	ND	ND

(continued)

Table 1
(continued)

PDZ3 ^{2}H Relaxation Data													
Residue	Atom	$^1H_{(ppm)}$	$^{13}C_{(ppm)}$	D_y (500)	Λ_y (500)	D_y (600)	Λ_y (600)	D_z (500)	Λ_z (500)	D_z (600)	Λ_z (600)	S^2_{axis}	Λ
I388	cδ1	0.74	13.99	17.5	0.4	18.8	0.3	68.3	2.1	83.1	1.6	0.51	0.01
I388	cγ2	0.78	17.63	10.8	0.3	11.8	0.2	38.9	1.2	47.3	0.8	0.79	0.02
I389	cδ1	0.67	10.44	26.6	0.6	28.5	0.4	59.7	1.3	70.4	1.0	0.26	0.01
I389	cγ2	0.74	17.73	10.3	0.3	11.2	0.2	53.9	1.9	64.9	1.4	0.91	0.02
A390	cβ	0.96	23.11	9.7	0.3	10.4	0.2	38.5	1.2	46.9	0.8	0.92	0.02
A402	cβ	1.30	20.15	21.0	0.5	42.1	0.3	65.0	0.6	71.0	0.4	0.12	0.00

Data was collected at 25 °C using Varian Inova spectrometers operating at 500 and 600 MHz ^{1}H frequencies. S^2_{axis} order parameters were fit using a τ_m of 5.89 ns. All times are in milliseconds

measurements and spectral information for assessing the accuracy of the reader's experimental measurements. If an alternate protein is desired, please *see* **Notes 1**, **4**, **5**, **6**, and **7** for suggestions and preparations to consider.

3.1 Protein Purification and Sample Preparation

1. For purification of PDZ3, it is first necessary to subclone the PDZ3 gene into a T7 expression vector or obtain the expression plasmid used in Petit et al. [20] from AddGene.
2. Transform the expression plasmid into DE3 Star *Escherichia coli* cells.
3. Inoculate a 3 mL starter culture of LB broth with a single colony and allow the culture to grow for 8 h at 37 °C.
4. While growing, prepare a liter of M9 growth media supplemented with $^{15}NH_4Cl$, U-$^{13}C_6$ D-glucose, and 60 % D_2O (*see* Subheading 2).
5. After 8 h of growth, inoculate a 50 mL aliquot of M9 growth media with 200 μL of the LB starter culture and incubate at 37 °C for 18 h.
6. Use the 50 mL culture to inoculate the remaining 950 mL of M9 growth media.
7. Allow the 1 L culture to grow at 37 °C until the culture reaches an OD_{600} of 0.6–0.8.
8. Induce protein expression by adding isopropyl β-D-1-thiogalactopyranoside (IPTG) to a final concentration of 1 mM.
9. Allow the protein to express for 4 h at 37 °C.
10. Pellet the bacteria by centrifugation and then resuspend in Lysis Buffer.

11. Lyse the bacteria using two freeze–thaw cycles followed by ultrasonication in an ice/water bath.
12. Pellet the cellular debris by centrifugation and discard the pellet.
13. Add 250 μL of a 5 % (w/v) polyethyleneimine solution drop-wise to the bacterial lysate stirring on ice for 5 min.
14. Pellet the precipitated DNA and discard.
15. Load the lysate onto a Source 30Q column using Lysis Buffer.
16. Collect the flow-through and dialyze overnight at 4 °C in 2 L of Q Sepharose Loading Buffer.
17. The next morning, replace dialysis buffer with an additional 2 L of Q Sepharose Loading Buffer and allow to dialyze for another 4 h at 4 °C.
18. Load the equilibrated lysate onto a Fast-Flow Q Sepharose column and elute using a shallow gradient of Q Sepharose Elution Buffer. PDZ3 typically begins to elute at ~10 % of Q Sepharose Elution Buffer, however, an SDS/PAGE gel should be used for verification as PDZ3 has a low extinction coefficient at 280 nm ($\varepsilon_{280} = 2{,}980\ \text{M}^{-1}\ \text{cm}^{-1}$).
19. Pool fractions containing PDZ3 together and concentrate to 5 mL.
20. Load the concentrated lysate onto a G50-Sephadex column pre equilibrated with NMR buffer. Typical protein yields are 30 mg mL^{-1} at approximately 95 % purity as verified by SDS/PAGE analysis.
21. Concentrate fractions containing PDZ3 to allow for preparation of an NMR sample that contains 1 mM PDZ3 with the addition of 5–10 % D_2O.

3.2 Data Collection

Temperature should be calibrated to 25 °C using a methanol standard (*see* **Note 8**). Pulse sequences that internally subtract the I_zC_z component were not in use at the time of data collection so it was necessary to manually subtract out this component to obtain pure D_z and D_y relaxation rates. The relaxation delays are as follows with delays used for duplicate points underlined: <u>12.15</u>, 76.15, <u>44.15</u>, <u>20.15</u>, 68.15, 52.15, 28.15, 60.15, and <u>36.15</u> ms for I_zC_z; <u>1.1</u>, 30.0, 12.4, <u>3.2</u>, <u>25.1</u>, 16.3, 5.8, 20.5, and 8.9 ms for $I_zC_zD_y$; <u>2.95</u>, 65.55, 27.45, <u>7.35</u>, <u>54.95</u>, 35.95, 13.05, 45.15, 19.85 ms for $I_zC_zD_z$. Relaxation delays should be collected non-sequentially while the FIDs are collected interleaved (*see* **Note 9**).

3.3 Data Processing

^{2}H relaxation experiments are processed using NMRPipe, an extensive software system for processing and analyzing NMR spectroscopic data [44]. It is first necessary to convert the raw data to a file or files capable of being processed by NMRPipe. If the data are

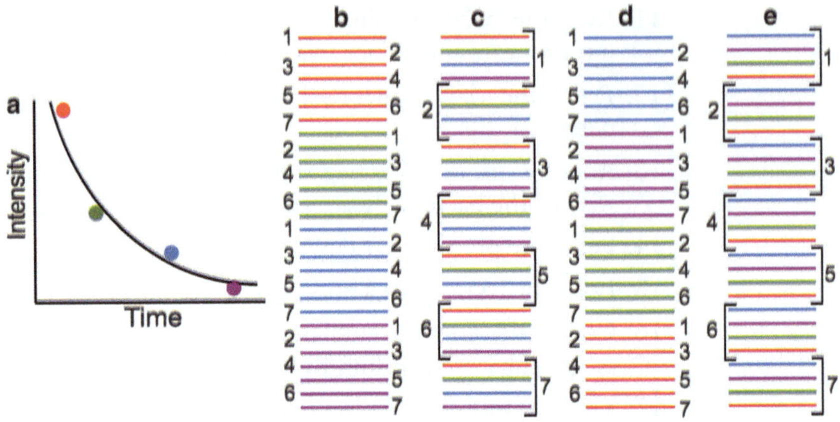

Fig. 5 Options for ordering of relaxation data collection. Hypothetical t_1 values for a given FID or group of FIDs are labeled *1–7*. (**a**) Schematic of a relaxation curve indicating the relationship between peak intensity and relaxation time. Relaxation time points can be collected sequentially (**b**) or non-sequentially (**d**). FIDs associated with a particular relaxation time can be collected sequentially (**b**, **d**) or interleaved (**c**, **e**)

collected in an interleaved manner (Fig. 5c, e), the FIDs need to be reordered using the NMRPipe command QMIX. Once converted, the FIDs are read in using the xyz2pipe command for further processing and Fourier Transformation. The processed data are then written out using the xyz2pipe command which prepares the processed data for analysis using a number of available software packages (NMRView [45], SPARKY [46], etc.).

In this case, NMRView was used to extract peak intensities for each relaxation delay which were then used in subsequent fitting algorithms (*see* Analysis). Briefly, the extracted intensities are fit to a single exponential to obtain ^{2}H relaxation rates. These relaxation rates are then fit to the standard model-free formalism [28] using a τ_m of 5.89 ns. An initial grid search of parameter space followed by Powell minimization of the error function (Eq. 4) was done using the in-house software program, RVI. RVI is a front-end interface for relxn2.2[32], another in-house program written for fitting order parameters. The obtained S^2 is then divided by a factor of 0.111 to correct for the methyl rotation symmetry axis [23] thereby yielding S^2_{axis} order parameters. Errors in S^2_{axis} were estimated using 150 Monte Carlo simulations.

4 Notes

1. Dynamic investigation of proteins using NMR has, in the past, been restricted to smaller systems with molecular weights typically less than 30 kDa [21]. Indeed, this chapter is intended for proteins in this molecular weight range. However, recent combinations of advances in isotopic labeling schemes and pulse

sequence development has allowed for dynamics applications to be applied to larger systems such as malate synthase G [55] (82 kDa) and the proteasome core particle [56] (670 kDa) [21, 57]. Robust and effective techniques are now available that allow the specific labeling of isoleucine, leucine, valine, alanine, and threonine methyl groups [58–60]. These techniques use isotopically labeled biosynthetic precursors, to selectively label the methyl sites in proteins expressed in D_2O-based minimal media. In addition, these precursors are available in multiple isotopic labeling schemes, making them suitable for a number of NMR applications. The use of transverse relaxation optimized spectroscopy (TROSY) [61] in pulse sequence development has also facilitated larger proteins to be dynamically and structurally characterized using NMR. TROSY-based pulse sequences allow the selection and isolation of slower relaxing coherences from their shorter lived counterparts by preventing the intermixing of these pathways throughout the NMR experiment. These sequences therefore enable enhanced sensitivity and resolution for proteins larger than ~30 kDa.

2. The NMR experiments described in this review use isotope-edited pulse sequences and therefore require appropriate spin isotopes in protein samples to function properly. For biomolecular NMR, the most commonly used nuclear spins are ^{1}H, ^{2}H, ^{15}N, and ^{13}C. As the natural abundance of ^{2}H (0.015 %), ^{15}N (0.37 %), and ^{13}C (1.1 %) are extremely low, it is necessary to isotopically enrich the protein being studied. This is accomplished by expressing the protein in M9 minimal media supplemented with isotopically labeled NH_4Cl (^{15}N, 99 %) and/or D-glucose (U-$^{13}C_6$, 99 %) depending on the experiments being performed. For ^{2}H relaxation experiments, it is necessary to use M9 supplemented with both ^{15}N-labeled NH_4Cl and ^{13}C-labeled glucose as well as to express the protein in 60 % D_2O in order to generate the CH_2D isotopomers needed by the pulse sequence [23]. A protocol for expressing and purifying an isotopically labeled protein (PDZ3) is included in this chapter.

3. Each pulse used in the ^{2}H relaxation experiments must be properly calibrated to ensure accurate and consistent relaxation measurements. ^{1}H pulses should be calibrated on the protein sample, while a ^{13}C-labeled methyl iodide standard may be used to calibrate ^{13}C pulses. Recent spectrometer software updates allow automated calibrations for ^{1}H, ^{13}C, and ^{15}N rectangular pulses on the sample of interest. Unfortunately, this automated option is not available for ^{2}H pulse calibrations on many older instruments (especially Varian), necessitating manual calibration of ^{2}H pulses. If the channel used for ^{2}H pulsing cannot be directly observed, it is necessary to re-cable the instrument to allow observation of the ^{2}H signal to

calibrate the necessary pulses on the pulsing channel. As the re-cabling is hardware dependent, there is no universal protocol for calibrating ^{2}H pulses. However, direct observation of the deuterium channel has become a standard feature of newer instruments, thereby making ^{2}H pulse calibration much more straightforward.

4. NMR is an inherently insensitive method and thus requires relatively concentrated protein samples to obtain high-quality data. For relaxation studies, typical samples are composed of ~550 μL (for Varian probes) of protein solution ranging in concentration from 0.5 to 1.0 mM. Depending on the molecular weight, a single sample may require tens of milligrams of expressed, purified protein. Recent advances in NMR technology, however, have allowed the development of higher field magnets as well as cryogenically cooled probes, both of which serve to improve sensitivity. Shigemi tubes (Allison Park, PA) use susceptibility matched plugs to reduce the required volume of sample (~300 μL) which, in turn, reduces the amount of expressed protein needed. In addition to these technological advances, there are several further steps that can be taken to improve sensitivity (i.e., signal to noise). Increasing the protein concentration of the sample is one way to improve sensitivity. However, as protein solubility may impede this option, it may be necessary to assess different combinations of pH, storage temperature, buffers, and salt concentration to improve protein solubility [47]. It should be noted that a buffer's ionic strength negatively impacts probe sensitivity, particularly for cryogenically cooled probes. Therefore, a balance between the ionic strength of a sample and the improvement in solubility must be made to get the optimal benefit of increased solubility without sacrificing more sensitivity than needed. Another option to achieve greater sensitivity is to signal average more by increasing the number of scans performed in an experiment. This option is limited by the fact that the signal to noise ratio only increases as the square root of the number of transients collected.

5. When studying a new protein, it may be necessary to determine if the protein will remain natively folded under the conditions and for the duration of any experiments that are performed. To determine this, it is advisable to store the sample under the desired experimental conditions while obtaining several periodic ^{1}H–^{15}N HSQC spectra over the course of time that you plan on running the experiments. This will ensure that a protein will be able to remain stable throughout the experiment(s) without wasting the time and expenses of running a multiday experiment on a denatured protein. Another advisable practice is to obtain a quick ^{1}H–^{15}N HSQC spectra before and after any experiments of considerable length to determine

if there were any changes in the sample throughout the course of the experiment(s).

6. Before adding the sample, the NMR tube should be washed and dried to ensure no contamination. Typically, NMR tubes are washed with 2 L of ddH_2O, rinsed with 100 % ethanol, and are then allowed to dry thoroughly. A solvent jet washer/cleaner (Wilmad Glass, Inc.) attached to a vacuum pump can be used to expedite both washing and drying of the tube. Drying NMR tubes by laying them in an oven should be avoided as this can warp the tubes over time. Dirty tubes should be filled with a 50 % solution of nitric acid and allowed to incubate overnight. Special care must then be taken to ensure no residual nitric acid remains, as it may potentially alter the pH of your sample or unfold the protein of interest.
7. Assignments of the protein's methyl groups must be made if a site-specific analysis of side-chain dynamics is to be undertaken. Methyl assignments can be made by using a number of experiments including HCCH-TOCSY [50], (H)C(CO)NH-TOCSY [51], and the (H)CCH_3-TOCSY [52]. We find that, for small- and medium-sized proteins, an efficient way to assign a protein's methyl groups is to first assign the C^α and C^β of each residue using standard HNCACB and CBCA(CO)NH experiments [53]. Once the C^α and C^β are assigned, the (H)CCH_3-TOCSY[52] can then be used to assign the chemical shifts of side-chain carbons and protons. Methyl assignments are made by correlating the assigned C^α and C^β chemical shifts with spin systems from the (H)CCH_3-TOCSY that contain the complete spectra of side-chain resonances. Stereospecific assignments of prochiral methyl groups are conveniently determined using a fractionally ^{13}C-labeled sample [54].
8. Temperature calibration before each series of relaxation experiments is critical to ensure consistent and reproducible measurements. Temperature standards composed of neat methanol (for $T < 25$ °C) or ethylene glycol (for $T > 25$ °C) are typically used to obtain accurate temperature calibrations. Higher (lower) temperatures will decrease (increase) the τ_m by altering the viscosity of the protein solution. If an incorrect τ_m is used for fitting methyl order parameters, the fit S^2_{axis} values will not be accurate. It is therefore vital to perform the ^2H relaxation experiments at the same temperature as the ^{15}N relaxation experiments used to determine the protein's τ_m. In addition to tumbling, changes in temperature have also been shown to affect the value of the methyl order parameters themselves [48, 49].
9. ^2H relaxation experiments are composed of several 2D experiments ($^1H/^{13}C$) each collected with a different relaxation delay

and are thus commonly referred to as "pseudo-3D" experiments. Relaxation rates for individual residues can be determined by fitting its decay in peak intensity to a single exponential (Fig. 1.5a). The "pseudo-3D" can be contrasted with an actual 3D experiment, such as the HNCA in which each dimension corresponds to time evolution of oscillatory signals. Consequently, ^{2}H relaxation experiments can be collected in the following ways: (1) Collect the relaxation time points and the individual FIDs for each time point sequentially (Fig. 5b). (2) Collect the relaxation time points sequentially and interleave the individual FIDs for each relaxation time point (Fig. 5c). (3) Collect the relaxation time points non-sequentially and the individual FIDs for each relaxation time point sequentially (Fig. 5d). (4) Collect the relaxation time points non-sequentially and interleave the individual FIDs for each relaxation time point (Fig. 5e). Each of the collection methods has unique advantages in terms of the quality of the exponential decay curves that yield the D_z or D_y relaxation rates. By collecting the relaxation points non-sequentially, potential problems with sample heating can be mitigated. The advantage to interleaving the individual FIDs for each relaxation time point is that any systematic error that develops over the course of the experiment will be spread over all of the relaxation points so that no one point contains more error than any other. If each relaxation time point is acquired in its entirety before moving on to the next, each relaxation point's error will be independent of all the other points. However, a tradeoff of interleaving the FIDs is that if a problem arises during a defined segment of the data acquisition, the entire relaxation data set is compromised. To minimize the error in each relaxation point, we recommend collecting the relaxation time non-sequentially and to interleave the individual FIDs (Fig. 5e).

References

1. Frauenfelder H, Sligar SG, Wolynes PG (1991) The energy landscapes and motions of proteins. Science 254(5038):1598–1603
2. Onuchic JN, LutheySchulten Z, Wolynes PG (1997) Theory of protein folding: the energy landscape perspective. Annu Rev Phys Chem 48:545–600
3. Eisenmesser EZ, Bosco DA, Akke M, Kern D (2002) Enzyme dynamics during catalysis. Science 295(5559):1520–1523
4. Eisenmesser EZ, Millet O, Labeikovsky W, Korzhnev DM, Wolf-Watz M, Bosco DA, Skalicky JJ, Kay LE, Kern D (2005) Intrinsic dynamics of an enzyme underlies catalysis. Nature 438(7064):117–121
5. Boehr DD, McElheny D, Dyson HJ, Wright PE (2006) The dynamic energy landscape of dihydrofolate reductase catalysis. Science 313 (5793):1638–1642
6. Igumenova TI, Frederick KK, Wand AJ (2006) Characterization of the fast dynamics of protein amino acid side chains using NMR relaxation in solution. Chem Rev 106(5): 1672–1699
7. Sapienza PJ, Lee AL (2010) Using NMR to study fast dynamics in proteins: methods and

applications. Curr Opin Pharmacol 10 (6):723–730

8. Akke M, Bruschweiler R, Palmer AG (1993) Nmr order parameters and free-energy - an analytical approach and its application to cooperative Ca2+ binding by calbindin-D(9k). J Am Chem Soc 115(21):9832–9833
9. Li ZG, Raychaudhuri S, Wand AJ (1996) Insights into the local residual entropy of proteins provided by NMR relaxation. Protein Sci 5(12):2647–2650
10. Yang DW, Kay LE (1996) Contributions to conformational entropy arising from bond vector fluctuations measured from NMR-derived order parameters: application to protein folding. J Mol Biol 263(2):369–382
11. Frederick KK, Marlow MS, Valentine KG, Wand AJ (2007) Conformational entropy in molecular recognition by proteins. Nature 448(7151):325–329
12. Clarkson MW, Lee AL (2004) Long-range dynamic effects of point mutations propagate through side chains in the serine protease inhibitor eglin c. Biochemistry 43 (39):12448–12458
13. Fuentes EJ, Der CJ, Lee AL (2004) Ligand-dependent dynamics and intramolecular signaling in a PDZ domain. J Mol Biol 335 (4):1105–1115
14. Clarkson MW, Gilmore SA, Edgell MH, Lee AL (2006) Dynamic coupling and allosteric behavior in a nonallosteric protein. Biochemistry 45(25):7693–7699
15. Fuentes EJ, Gilmore SA, Mauldin RV, Lee AL (2006) Evaluation of energetic and dynamic coupling networks in a PDZ domain protein. J Mol Biol 364(3):337–351
16. Namanja AT, Peng T, Zintsmaster JS, Elson AC, Shakour MG, Peng JW (2007) Substrate recognition reduces side-chain flexibility for conserved hydrophobic residues in human Pin1. Structure 15(3):313–327
17. Cooper A, Dryden DT (1984) Allostery without conformational change. A plausible model. Eur Biophys J 11(2):103–109
18. Wand AJ (2001) Dynamic activation of protein function: a view emerging from NMR spectroscopy. Nat Struct Biol 8(11):926–931
19. Tsai CJ, del Sol A, Nussinov R (2008) Allostery: absence of a change in shape does not imply that allostery is not at play. J Mol Biol 378(1):1–11
20. Petit CM, Zhang J, Sapienza PJ, Fuentes EJ, Lee AL (2009) Hidden dynamic allostery in a PDZ domain. Proc Natl Acad Sci USA 106 (43):18249–18254
21. Sheppard D, Sprangers R, Tugarinov V (2010) Experimental approaches for NMR studies of side-chain dynamics in high-molecular-weight proteins. Prog Nucl Magn Reson Spectrosc 56 (1):1–45
22. Jarymowycz VA, Stone MJ (2006) Fast time scale dynamics of protein backbones: NMR relaxation methods, applications, and functional consequences. Chem Rev 106(5): 1624–1671
23. Muhandiram DR, Yamazaki T, Sykes BD, Kay LE (1995) Measurement of H-2 T-1 and T-1p relaxation-times in uniformly C-13-labeled and fractionally H-2-labeled proteins in solution. J Am Chem Soc 117(46):11536–11544
24. Jacobsen JP, Bildsoe HK, Schaumburg K (1976) Application of density matrix formalism in Nmr-spectroscopy.2. One-spin-1 case in anisotropic phase. J Magn Reson 23(1): 153–164
25. Mittermaier A, Kay LE (1999) Measurement of methyl H-2 quadrupolar couplings in oriented proteins. How uniform is the quadrupolar coupling constant? J Am Chem Soc 121 (45):10608–10613
26. Morris GA, Freeman R (1979) Enhancement of nuclear magnetic-resonance signals by polarization transfer. J Am Chem Soc 101(3): 760–762
27. Millet O, Muhandiram DR, Skrynnikov NR, Kay LE (2002) Deuterium spin probes of side-chain dynamics in proteins. 1. Measurement of five relaxation rates per deuteron in (13)C-labeled and fractionally (2)H-enriched proteins in solution. J Am Chem Soc 124 (22):6439–6448
28. Lipari G, Szabo A (1982) Model-free approach to the interpretation of nuclear magnetic-resonance relaxation in macromolecules.1. Theory and range of validity. J Am Chem Soc 104(17):4546–4559
29. Lipari G, Szabo A (1982) Model-free approach to the interpretation of nuclear magnetic-resonance relaxation in macromolecules.2. Analysis of experimental results. J Am Chem Soc 104(17):4559–4570
30. Lee AL, Wand AJ (2001) Nuclear magnetic resonance (NMR) spectroscopy for monitoring molecular dynamics in solution. In: José María Valpuesta, eLS. John Wiley & Sons, Ltd. doi:10.1038/npg.els.0003104
31. Farrow NA, Muhandiram R, Singer AU, Pascal SM, Kay CM, Gish G, Shoelson SE, Pawson T, Formankay JD, Kay LE (1994) Backbone dynamics of a free and a phosphopeptide-complexed Src homology-2 domain studied

by N-15 Nmr relaxation. Biochemistry 33 (19):5984–6003

32. Lee AL, Flynn PF, Wand AJ (1999) Comparison of H-2 and C-13 NMR relaxation techniques for the study of protein methyl group dynamics in solution. J Am Chem Soc 121 (12):2891–2902
33. Skrynnikov NR, Millet O, Kay LE (2002) Deuterium spin probes of side-chain dynamics in proteins. 2. Spectral density mapping and identification of nanosecond time-scale side-chain motions. J Am Chem Soc 124(22):6449–6460
34. Peng JW, Wagner G (1992) Mapping of spectral density-functions using heteronuclear Nmr relaxation measurements. J Magn Reson 98 (2):308–332
35. Peng JW, Wagner G (1992) Mapping of the spectral densities of N-H bond motions in Eglin-C using heteronuclear relaxation experiments. Biochemistry 31(36):8571–8586
36. Doyle DA, Lee A, Lewis J, Kim E, Sheng M, MacKinnon R (1996) Crystal structures of a complexed and peptide-free membrane protein-binding domain: Molecular basis of peptide recognition by PDZ. Cell 85 (7):1067–1076
37. Morais Cabral JH, Petosa C, Sutcliffe MJ, Raza S, Byron O, Poy F, Marfatia SM, Chishti AH, Liddington RC (1996) Crystal structure of a PDZ domain. Nature 382(6592):649–652
38. Birrane G, Chung J, Ladias JA (2003) Novel mode of ligand recognition by the Erbin PDZ domain. J Biol Chem 278(3):1399–1402
39. Peterson FC, Penkert RR, Volkman BF, Prehoda KE (2004) Cdc42 regulates the Par-6 PDZ domain through an allosteric CRIB-PDZ transition. Mol Cell 13(5):665–676
40. Mishra P, Socolich M, Wall MA, Graves J, Wang Z, Ranganathan R (2007) Dynamic scaffolding in a G protein-coupled signaling system. Cell 131(1):80–92
41. Bhattacharya S, Dai Z, Li J, Baxter S, Callaway DJ, Cowburn D, Bu Z (2010) A conformational switch in the scaffolding protein NHERF1 controls autoinhibition and complex formation. J Biol Chem 285(13):9981–9994
42. Ballif BA, Carey GR, Sunyaev SR, Gygi SP (2008) Large-scale identification and evolution indexing of tyrosine phosphorylation sites from murine brain. J Proteome Res 7(1):311–318
43. Zhang J, Petit CM, King DS, Lee AL (2011) Phosphorylation of a PDZ domain extension modulates binding affinity and interdomain interactions in postsynaptic density-95 (PSD-95) protein, a membrane-associated guanylate kinase (MAGUK). J Biol Chem 286 (48):41776–41785
44. Delaglio F, Grzesiek S, Vuister GW, Zhu G, Pfeifer J, Bax A (1995) Nmrpipe - a multidimensional spectral processing system based on Unix pipes. J Biomol NMR 6(3):277–293
45. Johnson BA, Blevins RA (1994) Nmr view - a computer-program for the visualization and analysis of Nmr data. J Biomol NMR 4 (5):603–614
46. Goddard TD, Kneller DG. SPARKY 3. University of California, San Francisco
47. Bagby S, Tong KI, Ikura M (2001) Optimization of protein solubility and stability for protein nuclear magnetic resonance. Methods Enzymol 339:20–41
48. Lee AL, Wand AJ (2001) Microscopic origins of entropy, heat capacity and the glass transition in proteins. Nature 411(6836):501–504
49. Lee AL, Sharp KA, Kranz JK, Song XJ, Wand AJ (2002) Temperature dependence of the internal dynamics of a calmodulin-peptide complex. Biochemistry 41(46):13814–13825
50. Kay LE, Xu GY, Singer AU, Muhandiram DR, Formankay JD (1993) A gradient-enhanced hcch tocsy experiment for recording side-chain H-1 and C-13 correlations in H2O samples of proteins. J Magn Reson Ser B 101 (3):333–337
51. Montelione GT, Lyons BA, Emerson SD, Tashiro M (1992) An efficient triple resonance experiment using C-13 isotropic mixing for determining sequence-specific resonance assignments of isotopically-enriched proteins. J Am Chem Soc 114(27):10974–10975
52. Uhrin D, Uhrinova S, Leadbeater C, Nairn J, Price NC, Barlow PN (2000) 3D HCCH3-TOCSY for resonance assignment of methyl-containing side chains in C-13-labeled proteins. J Magn Reson 142(2):288–293
53. Muhandiram DR, Kay LE (1994) Gradient-enhanced triple-resonance 3-dimensional Nmr experiments with improved sensitivity. J Magn Reson Ser B 103(3):203–216
54. Neri D, Szyperski T, Otting G, Senn H, Wuthrich K (1989) Stereospecific nuclear magnetic-resonance assignments of the methyl-groups of valine and leucine in the DNA-binding domain of the 434-repressor by biosynthetically directed fractional C-13 labeling. Biochemistry 28(19):7510–7516
55. Tugarinov V, Muhandiram R, Ayed A, Kay LE (2002) Four-dimensional NMR spectroscopy of a 723-residue protein: chemical shift assignments and secondary structure of malate synthase G. J Am Chem Soc 124(34): 10025–10035
56. Sprangers R, Kay LE (2007) Quantitative dynamics and binding studies of the 20S

proteasome by NMR. Nature 445 (7128):618–622

57. Tugarinov V, Hwang PM, Kay LE (2004) Nuclear magnetic resonance spectroscopy of high-molecular-weight proteins. Annu Rev Biochem 73:107–146
58. Tugarinov V, Kanelis V, Kay LE (2006) Isotope labeling strategies for the study of high-molecular-weight proteins by solution NMR spectroscopy. Nat Protoc 1(2):749–754
59. Ayala I, Sounier R, Use N, Gans P, Boisbouvier J (2009) An efficient protocol for the complete incorporation of methyl-protonated alanine in perdeuterated protein. J Biomol NMR 43 (2):111–119
60. Sinha K, Jen-Jacobson L, Rule GS (2011) Specific labeling of threonine methyl groups for NMR studies of protein-nucleic acid complexes. Biochemistry 50(47):10189–10191
61. Pervushin K, Riek R, Wider G, Wuthrich K (1997) Attenuated T-2 relaxation by mutual cancellation of dipole-dipole coupling and chemical shift anisotropy indicates an avenue to NMR structures of very large biological macromolecules in solution. Proc Natl Acad Sci USA 94(23):12366–12371

Chapter 2

CPMG Relaxation Dispersion

Rieko Ishima

Abstract

NMR relaxation is sensitive to molecular and internal motion of proteins. ^{15}N longitudinal relaxation rate (R_1), transverse relaxation rate (R_2), and {^{1}H}-^{15}N Nuclear Overhauser Effect (NOE) experiments are often performed to globally elucidate protein dynamics, primarily on the sub-nanosecond timescale. In contrast, constant relaxation time R_2 dispersion experiments are applied to characterize protein equilibrium conformations that interconvert on the millisecond timescale. Information on local conformational equilibria of proteins provides important insights about protein energy landscapes and is useful to interpret molecular recognition mechanisms as well. Here, we describe a protocol for performing ^{15}N Carr–Purcell–Meiboom–Gill (CPMG) R_2 dispersion measurements in solution, including protein preparation, step-by-step experimental parameter settings, and the first step of data analysis.

Key words NMR, Relaxation, CPMG, Protein, Dynamics, Conformation, Equilibrium

1 Introduction

1.1 Relaxation Dispersion in General

Recently, relaxation dispersion experiments have been extensively used for high-resolution protein NMR studies in solution [1–15]. These experiments detect the transverse relaxation rate, R_2, as a function of the effective Carr–Purcell–Meiboom–Gill (CPMG) or spin-lock field strength, namely, B_1-dependent relaxation dispersion. In contrast to ^{15}N R_1, R_2, and ^{15}N-{^{1}H} NOE experiments that characterize sub-nanosecond motions and milli-microsecond motions by applying the model-free analysis [16–18], the CPMG relaxation dispersion experiment provides quantitative information about milli-microsecond timescale motions that affect the chemical exchange term in R_2 and have been used to characterize protein conformational equilibria and kinetics on the milli-microsecond timescale. For example, although it is difficult to detect milli-microsecond dynamics in the model-free analysis when the site undergoes fast internal motion or experiences effects of anisotropic molecular diffusion [19, 20], the CPMG relaxation dispersion

Dennis R. Livesay (ed.), *Protein Dynamics: Methods and Protocols*, Methods in Molecular Biology, vol. 1084,
DOI 10.1007/978-1-62703-658-0_2, © Springer Science+Business Media New York 2014

experiment readily detects conformational exchange in the presence of these phenomena.

In a related experiment, relaxation dispersion is recorded as a function of the static magnetic field strength, B_0. In this B_0-dependent relaxation dispersion experiment that is called "relaxometry," "field cycling," or "nuclear magnetic relaxation dispersion (NMRD)," the longitudinal relaxation rate, R_1, is measured as a function of B_0 [21–30]. This B_0-dependent dispersion experiment determines a spectral density $J(\omega)$ as a function of B_0 ($= \omega_i/\gamma_i$, in which ω_i and γ_i are resonance angular frequency and the gyromagnetic ratio of the observed nuclear spins, respectively) and has been applied to provide insights into the dynamic processes of biomolecules. However, in this review, we exclusively describe the protocol for the B_1-dependent relaxation dispersion for backbone studies, ^{15}N constant-time (CT) CPMG relaxation (or sometimes called the CPMG R_2 dispersion) experiment.

In the CT-CPMG relaxation dispersion experiment, a reference signal intensity of ^{15}N transverse magnetization at time zero, $I(0)$, and a series of intensities, $I^i(T_{CP})$, at time T_{CP} at different effective field strength, ν_{eff}, are recorded. Here, T_{CP} is a constant period of CPMG relaxation. The effective field strength is varied ($i = 1$ to n): the ith effective field is defined by $\nu_{eff} = 1/(4\tau_{CP})$ with τ_{CP} as a half duration from the center of one CPMG pulse to the center of the subsequent CPMG pulse, i.e., inter-pulse delay. For $I^i(T_{CP})$ at each ith ν_{eff}, $R_2{}^i$ is numerically calculated using the following equations:

$$I^i(T_{CP}) = I^0 \exp(-R_2{}^i\, T_{CP}) \tag{1.1}$$

$$R_2{}^i = -(1/T_{CP})\, \ln(I^i(T_{CP})/I^0) \tag{1.2}$$

When the experimental noise is ΔI^i, the uncertainty of $R_2{}^i$, $\Delta R_2{}^i$, is calculated by Eq. 2 [31, 32].

$$\Delta R^i/R_2{}^i = (\Delta I^i/I^0)\left[1 + \exp(2\, R_2{}^i\, T)\right]^{1/2}/(R_2{}^i\, T) \tag{2}$$

Here, experimental noise ΔI^i of $I^i(T_{cp})$ is the same as that of I^0. When the exchange rate, k_{ex}, is much greater than $R_2{}^0$, $R_2{}^i$ can be written as the sum of $R_{ex}(\nu_{eff})^i$, the chemical exchange contribution, and $R_2{}^0$, the intrinsic relaxation rate, that is determined by relaxation due to dipolar coupling and chemical shift anisotropy:

$$R_2{}^i = R_2{}^0 + R_{ex}(\nu_{eff})^i \tag{3}$$

There have been two important developments in CT-CPMG experiments for protein backbone dynamics studies. One is a pulse scheme that provides uniform values of $R_2{}^0$ either (a) by using an rc-INEPT that averages the contributions of both inphase and the antiphase (N_{XY} and $2N_{XY}H_Z$, respectively) components to the intrinsic relaxation rate, $R_2{}^0$ [7], or (b) by applying a continuous wave (CW) ^1H pulse during a CPMG relaxation period, T_{CP}, in which the antiphase $2N_{XY}H_Z$ component is decoupled [33]. The second development is the use of a constant relaxation time T_{CP} that allows a two-exponential fit using the signal intensity at time zero and T_{CP} [8, 9].

1.2 Information Obtained by CPMG Relaxation Dispersion

Analysis of CPMG relaxation dispersion data provides information on the parameters that determine the chemical exchange phenomenon: assuming a two-site exchange system with states A and B, the exchange rate, k_{ex}, the fractional populations, p_A and p_B, where $p_A + p_B = 1$, the difference in chemical shifts of the two states, $\delta\omega$, ($= |\omega_A - \omega_B|$), and their $R_2{}^0$ values. Although $R_2{}^0$ in the two states, $R_2{}^{0A}$ and $R_2{}^{0B}$, can differ, they are typically assumed to be the same [34]. The exchange rate is related to the individual rates, k_{AB} and k_{BA}, by $k_{AB} = p_B k_{ex}$ and $k_{BA} = p_A k_{ex}$, respectively. Relative populations provide the Gibbs free-energy difference between the two states. In the current commercial NMR software, the upper limit of accessible k_{ex} is limited by instrumentation instability and heating at short τ_{CP} (i.e., high ν_{eff}) that are required to characterize the dispersion at high k_{ex}. On the other hand, the lower limit of the observed k_{ex} is determined by R_2 because a longer T_{CP} is required to record smaller k_{ex}. Thus, in recent 15N CPMG experiments, ν_{eff} is often varied approximately from 50 Hz to 1 kHz, which detects k_{ex} in the approximate range of 100 s^{-1} to 10^4 s^{-1}.

1.3 Models Used to Analyze CPMG Relaxation Dispersion

There are three steps in the analysis of CPMG R_2 dispersion data. One is to obtain a residue profile of chemical exchange, the second is optimization of the exchange parameters (p_A, k_{ex}, $\delta\omega$, $R_2{}^0$) for each site, and the last is optimization of the exchange parameters (p_A and k_{ex}) for a group of residues. The residue profile is provided by calculating a root-mean-square deviation of $R_2{}^i$ values in one dispersion profile (i.e., for $i = 1$ to n), $R_2{}^{RMSD}$,

$$R_2{}^{RMSD} = \left\{\Sigma\left(R_2{}^{i,\exp} - R_2{}^{ave}\right)^2/n\right\}^{1/2} \tag{4}$$

or $R_2{}^{rmsd}$, or by minimizing the following χ^2, $\chi_{R20}{}^2$ (Eqs. 5 and 6):

$$R_2{}^{i,\,cal} = R_2{}^0 \tag{5}$$

$$\chi_{R20}{}^2 = \Sigma\left(R_2{}^{i,\exp} - R_2{}^{i,\,cal}\right)^2/\left(\Delta R^i\right)^2 \tag{6}$$

Here, $R_2{}^{ave}$ is an average of $R_2{}^i$ for $i = 1$ to n. The χ^2 in Eq. 6 describes whether the dispersion curve can fit to a uniform $R_2{}^0$, i.e., to a case of no significant dispersion. Large values of $R_2{}^{rmsd}$ or $\chi_{R20}{}^2$ both indicate deviation of $R_2{}^i$ from $R_2{}^0$. A plot of $R_2{}^{rmsd}$ or $\chi_{R20}{}^2$ as a function of residue number provides an overview of the conformational flexibility of the protein. $R_2{}^{RMSD}$ is useful to indicate which sites undergo conformational exchange when the uncertainty of $R_2{}^{i,\,\exp}$ is not quantitatively determined. In contrast,

χ_{R20}^2 sensitively detects conformational exchange when changes in $R_2^{i,\ \exp}$ are small and the uncertainty of $R_2^{i,\ \exp}$ is accurately determined.

In the second step, parameters for exchange are calculated by solving the Bloch–McConnell equation, including the effects of 180° pulses, iteratively or by using its analytical solutions [2, 35–37]. A typical analytical equation applied for CPMG R_2 dispersion may be the Carver–Richards equation. This equation is suitable to analyze intermediate exchange and fast exchange [2, 37]. Currently, there are several programs that determine the exchange parameters [14, 38–41]. If needed, an additional step, involving a group fit, may be performed when a region of interest or a whole protein is expected to undergo a cooperative dynamic process characterized by a unique set of p_A and k_{ex}.

2 Materials

2.1 NMR Samples

1. TSP sample: 3-(trimethylsilyl)propionic-2,2,3,3-d4 acid sodium salt (TSP) is dissolved in 0.5 ml of 99.9 % deuterium oxide (D_2O) at 60 mM (~10 mg/ml) concentration.
2. ^{15}N-labeled urea sample: ^{15}N urea is dissolved in deuterated DMSO, 0.5 ml at 200 mM (~12 mg/ml) concentration.
3. Uniformly ^{15}N-labeled and perdeuterated ubiquitin as a standard sample (Isotec, in Sigma-Aldrich).
4. Uniformly ^{15}N-labeled and perdeuterated protein to be studied at >0.4 mM concentration. For overexpression of the protein in *E. coli*, ^{15}N ammonium chloride and D_2O are used.

2.2 NMR Tubes

1. Regular NMR tubes for TSP and urea.
2. Shigemi NMR tubes for protein samples for a Bruker Avance NMR spectrometer (Shigemi Inc, Allison Park, PA.)

3 Methods

Here, we describe the protocol used by our group to record ^{15}N CT-CPMG relaxation dispersion data for a perdeuterated, uniformly ^{15}N-labeled protein using a Bruker Avance 600 NMR instrument (14.09 T) operated by Topspin software. Based on the fact that groups often share high-field NMR instruments, the protocol includes initial calibration steps before starting the CPMG experiment. ^{15}N CT-CPMG relaxation dispersion experiment is sensitive to ^{15}N pulse calibration error and off-resonance error, and ^{1}H–^{1}H J-coupling effect because of additional ^{1}H pulses in the ^{15}N CPMG sequence. We typically check ^{15}N pulse width as well as ^1H pulse width before setting up the relaxation experiments. The protocol includes all these steps. Note that settings of the experimental parameters may vary by the slight differences of software and hardware: the following protocol may need to modify for actual application for your NMR instruments.

3.1 ^1H TSP Temperature Check

Since NMR instruments at high magnetic field strengths are often shared, it is important to check that the instrument is fully functional before running a protein experiment. For this purpose, in our group, a commonly applied sample with a strong deuterium lock, i.e., TSP in D_2O, is tested first. Since the chemical shift difference between the TSP and water proton signals depends on temperature, a one-dimensional proton NMR spectrum of the TSP sample in D_2O is also used to check whether the temperature of the sample is the same as that calibrated previously (Fig. 1).

1. Place the TSP sample in the magnet.
2. Set the sample temperature from the temperature regulation command EDTE in Bruker Topspin software.
3. Apply deuterium lock by selecting D_2O as a deuterium solvent.
4. Record one-dimensional proton NMR spectrum by employing the pulse sequence, (p1)-detection, with f1 channel for ^1H.
5. Ensure water and TSP signals are observed.
6. Measure the chemical shift differences between the water and TSP signals. If the value is different from the one used previously, calibrate the temperature using a standard method applicable to the NMR instrument (*see* **Note 1**).

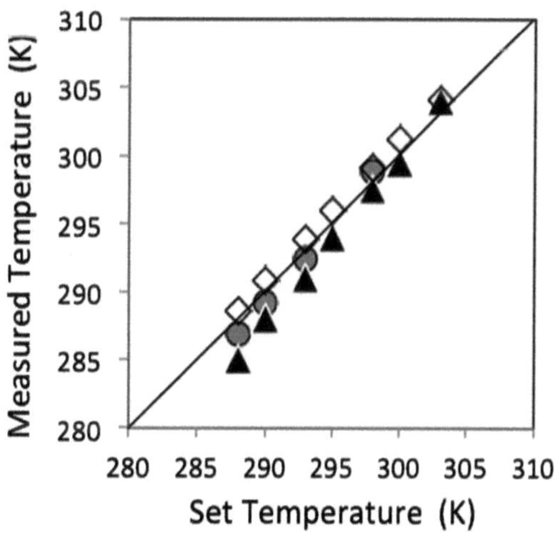

Fig. 1 Temperature calibration curve. The data were recorded using deuterated methanol (*diamond*), pure methanol (*circle*), and TSP in D_2O (*triangle*). *See* **Note 1**

3.2 ^{15}N Pulse Calibration

Using the ^{15}N urea sample in deuterated DMSO, ^{15}N pulse width is calibrated before the protein experiment.

1. Place the ^{15}N urea sample in the magnet.
2. Apply deuterium lock for DMSO as a deuterium solvent.
3. Run *decp90f3* in the Bruker Topspin software: d1-^{1}H(p_1)-d2-^{15}N(p_{21})-acquisition. Here, ^{1}H(p_1) is a tip pulse for ^{1}H. d1 is a pulse repetition delay (~5 s). d2 is given by $1/2J_{NH}$ = 5.5 ms. ^{15}N(p_{21}) is a ^{15}N pulse.
4. Repeat step 3 with a varying ^{15}N pulse width to find out the 90° degree rotation in which ^{1}H–^{15}N antiphase signals are minimized.

3.3 Heteronuclear Single-Quantum Coherence (HSQC) Experiment

^{1}H–^{15}N HSQC experiment provides a 2D (two-dimensional) heteronuclear chemical shift correlation map between directly bonded ^{1}H and ^{15}N in proteins. Typically, ^{1}H–^{15}N HSQC spectra are already available for any proteins upon which relaxation experiments are performed. However, it is useful to record the ^{1}H–^{15}N HSQC experiment again before and after the relaxation experiments to verify the sample stability. In addition, even though the protein concentration itself is measured by UV absorption or by other methods, it is critical to know the NMR sensitivity of the current sample and the NMR experimental setting are suitable in a shared use NMR instrument before staring the relaxation experiments. Thus, we typically record a short ^{1}H–^{15}N HSQC experiment before running the dispersion experiments.

1. Place the sample in the NMR magnet. Turn on deuterium lock, tune and match the probe, optimize shimming, and calibrate ^{1}H pulse width.
2. Choose the pulse sequence "*hsqcfpf3gpphwg*" (or choose an equivalent HSQC pulse sequence that was used previously) for a ^{1}H–^{15}N HSQC experiment. The following parameters are used to run the HSQC experiment in our 600 MHz NMR instrument: ^{1}H pulse width (10 μs), ^{15}N pulse width (45 μs), ^{15}N pulse width for acquisition decoupling (220 μs), 1,024 complex points for an acquisition dimension with a spectral width of 8,012 Hz (ca. 60 ms acquisition), 256 complex points for an indirect (t_1) dimension with a spectral width of 2,083 Hz, 8 scans, with a recycle delay of 1 s.
3. Once the acquisition is started, check the deuterium lock window, and confirm that there is not a severe reduction of deuterium lock due to heating of the probe or the sample.
4. Fourier transform the first free-induction decay of the HSQC, and overlay it with one that was recorded previously. Confirm that the pattern of the 1D spectrum is the same as that recorded previously; any intensity difference between the two spectra

should mostly reflect the estimated sample concentration ratio. If not, there is likely to be a mistake in the protein sample preparation or the NMR instrumentation.

3.4 ^{15}N CPMG Dispersion Experiment at 600 MHz

There are two versions of ^{15}N CPMG dispersion experiments: (1) with an rc INEPT period and (2) with ^{1}H continuous wave (CW) pulse scheme [9, 33] (*see* **Note 2**). The experiment with ^{1}H CW pulse has the advantage that achieves a twofold smaller value of ν_{eff} at the same T_{CP} [33]. However, strong ^{1}H CW decoupling may cause sample heating in high salt samples [42–45]. In general, the ^{1}H CW version is recommended whereas the rc-INEPT version is suggested for high salt samples.

First, the dispersion experiment will be run on a 600 MHz instrument that corresponds to 61 MHz ^{15}N resonance frequency. This is a set of 2D relaxation data at a reference point ($T_{CP} = 0$) and a varying ν_{eff} (and therefore, τ_{CP}) point. In our sequence, the experiments are recorded in an interleaved manner, in which free-induction decay signals are first accumulated with a small number of scans (NS = 4) for the complete set of 2D data. Then the experiments are repeated (l5 = 4), and the data is added to the previously acquired data, so that the total number of data acquisitions is NS × l5 = 16 (see below). In this way, the heating effect at different values of τ_{CP} is more distributed than is the case by recording an entire individual 2D data set sequentially.

1. The pulse sequence we used is shown in **Note 3**. Set the acquisition parameter window for a 2D data collection of ^{1}H direct and ^{15}N indirect dimension similar to that of ^{1}H–^{15}N HSQC experiment. Set ^{1}H and ^{15}N carrier frequencies and the ^{1}H spectral width. The experimental parameters are listed in (*see* **Note 4**).
2. ^{1}H pulses are set: a hard rectangular 90° pulse (p1 @ pl1), a hard 180° pulse (p2 @ pl1), a soft water 90° pulse (p11 @ sp1), and a CW power (pl8). In our protocol, these are p1 = 10.3 μs, p2 = 20.6 μs, p11 = 1 ms at corresponding pulse powers, respectively. pl8 was set to be 11 kHz as a B_{SL} field strength (22.7 μs as a rectangular 90° pulse).
3. ^{15}N pulses are set: a hard 90° pulse (p21 @ pl3), a hard 180° pulse (p22 @ pl3), and an acquisition composite-decoupling 90° pulse (pcpd3 @ pl16): p21 = 45 μs, p22 = 90 μs, pcpd3 = 240 μs (*see* **Note 5**).
4. Delays to adjust the starting of the CPMG, d13 and d14, are calculated based on the ^{1}H and ^{15}N pulse widths: $d13 = 1/(2\pi B_{SL}) - (4/\pi)p1$ and $d14 = p21 - (2/\pi)p1 + 8$ μs, respectively.
5. Manually calculate vc and vp to make the total T_{cp} uniform, 32 ms: calculate vc and vp to satisfy $T_{cp}/2 = (2{*}vp + p22){*}vc$. Use even number of vc (vc = 0, 2, 4, 6, 8, ...).

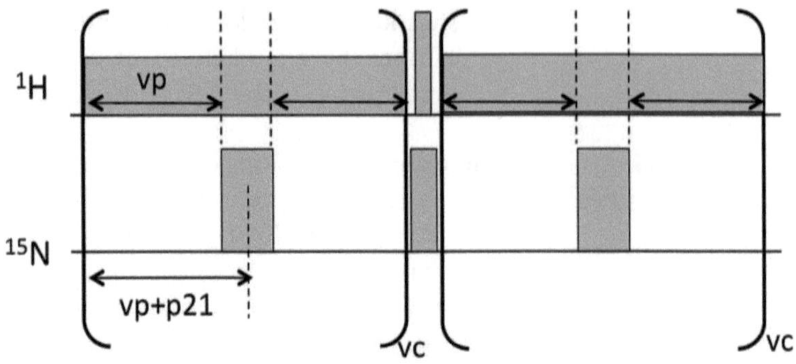

Fig. 2 Setting of the T_{CP} at variable vp and vc. To acquire seven data points for $T_{CP} = 32$ ms, the vclist (0, 32, 1, 16, 2, 8, and 4) and the vplist (4 u, 205 u, 7955 u, 455 u, 3955 u, 955 u, and 1955 u) were used. Here, "u" is μs unit in Bruker vplist. The total ν_{eff} calculated from 0.25/(vp + p21)

The starting vc = 0 corresponds to the reference ($T_{CP} = 0$) data point whereas other vc values are used to provide the delay at $T_{CP} = 32$ ms. Then, save the values to the vc and vp lists, respectively (Fig. 2). At each data set, the effective field strength is given by $\nu_{eff} = 0.25/(\text{vp} + \text{p21})$.

6. Delays for t_1 evolution is set. Semi-constant time evolution is used for the indirect dimension in this pulse sequence. Here, twice of l3 is the total number of t_1 increment points (i.e., indirect data point) and sw_1 is the spectral width of the indirect dimension:
   (a) in0 = d0/(l3 + 1)
   (b) in10 = 1/(2*sw_1)
   (c) in9 = in10 − in0
7. Set parameters for interleaved acquisition to produce uniform the sample heating effect at different τ_{CP}. The total number of 2D data sets (the number of dispersion points, n), l4 = n, and the number of total scans, ns*l5. Note that in this pulse sequence (*see* **Note 3**), free-induction decay (FID) data are acquired in the following manner. First, FID data of a referent point and a series of the different τ_{CP} values are recorded (the loop for l4). Next, quadrature data sets in the t_1 dimension are recorded. Third, this acquisition protocol is repeated with t_1 increments (the loop for l3). Finally, the interleave loop acquires the entire data sets again and adds the data to that already acquired (the loop for l5).
8. Set total number of indirect complex points, 1TD, given by l4*2*l3 in the acquisition parameter window. This experiment stores the entire data set as one large 2D data.
9. Calculate the total experiment time using the "expt" command and also by hand. Total time should be approximately (d1 + 0.1 s)*(ns*l5)*(l4*2*l3)/60/60 h.

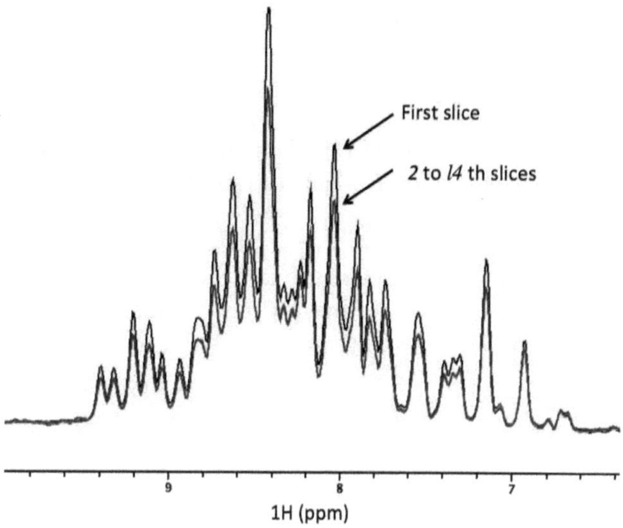

Fig. 3 Overlay of the first seven slices of the dispersion data of ubiquitin. Signal intensity of the first 1D slice, a reference point at $T_{CP} = 0$, is large whereas others are almost equivalent to each other, from 2nd to the 14th slices, because there is not much chemical exchange in ubiquitin

10. Run the experiments firstly by setting vc = 0 for all just to check the performance. Then, increase vc values step by step if there is any particular heating (e.g., no change of the deuterium lock level) detected.
11. Once the experiment runs without heating, read the 1D slice by "*rser 1*" for the first *n* (= 14) data sets. If most of the protein does not undergo chemical exchange, the 1D slices from second FID to the *n*th FID (14 = *n*) should be almost equivalent to each other (Fig. 3, *see* **Note 6**).

3.5 ^{15}N CPMG Dispersion Experiment at 900 MHz

CPMG R_2 dispersion experiments should be recorded at multiple (at least two) magnetic field strengths for better optimization of parameters derived from the analysis of exchange profiles. This is critical in order to obtain a robust value of the exchange rates: when chemical exchange is in the fast exchange regime, significant increase in the exchange rate, $R_{ex}(\nu_{eff})^i$ (*see* Eq. 3), should be observed. Thus, the above experiment in Subheading 3.4 is repeated on an 800 or 900 MHz instrument, on which ^{15}N resonance frequency is 81 or 91 MHz, respectively. Here, an example at 900 MHz is described. Regarding the dispersion experiments at high magnetic field strength, there are two important points. (1) Although shorter pulse widths may be used to excite the same spectral width (in ppm) at higher magnetic field strength, the shorter pulses should not be used for the relaxation dispersion experiment as they cause sample heating. Instead, signals that are located far from the carrier frequency are discarded. (2) Intrinsically, signal-to-noise ratio increases with increase in the static magnetic field strength. Because of this advantage, a shorter T_{CP} than that at 600 MHz may be used to achieve $\Delta R^i/R_2{}^i$ (Eq. 2) at 900 MHz data similar to that at 600 MHz [32, 46]. Such reduction of T_{CP} helps to suppress (or reduce) sample heating.

1. Set ^{15}N carrier frequency at 108 ppm.
2. Use similar ^{1}H and ^{15}N pulse widths to those used at 600 MHz if the instrument allows (corresponding to **steps 1–4** in Subheading 3.4).
3. Set $T_{CP} = 24$ ms, and calculate vc and vp accordingly.
4. Set semi-constant time evolution. Here, sw_1 should be 900/600 wider (i.e., in10 is 600/900 smaller) than that used in the 600 MHz experiments. in9 is set consistent with the change of in10.
5. Set l3, l4, and l5 and the number of indirect dimension points. Check the total experimental time.
6. Test the experiments by using a short vc first. Then, run the experiment.
7. Once the data have been successfully collected, repeat the experiment with a ^{15}N carrier frequency at 122 ppm.

3.6 Data Analysis: Determination of R_2

At the end of the experiments, one obtains a dispersion data set at 600 MHz and two data sets, recorded at two ^{15}N carrier frequencies, at 900 MHz. The same data processing protocol is used for all the data sets. For the two at 900 MHz, the data are sorted, which is described below:

1. The acquired data can be processed using nmrPipe and nmrDraw [47], in which fid.com and all.com are run (*see* **Note** 7).
2. Once the spectra are processed, signal assignments are added to the nmrDraw assignment table, assign.tab, using the first 2D data set that was acquired without $T_{cp} = 0$ (i.e., vc = 0).
3. Using SeriesTab command, the peak height under each assigned chemical shift is taken. This is done with averaging of 1 or 2 data points (dx, dy = 1~3) so that only the peak top heights are used (*see* **Note 8**).
   seriesTab –in assign.tab –out out.tab –list relax.list –dx 2 –dy 2
4. Intensity of the noise for each spectrum is found in the "estimate noise" command in nmrDraw. The noise should be similar within all the experiments. R_2 and the uncertainty are numerically calculated using Eqs. (1.1, 1.2) and (2) with the peak height ratio and the noise.

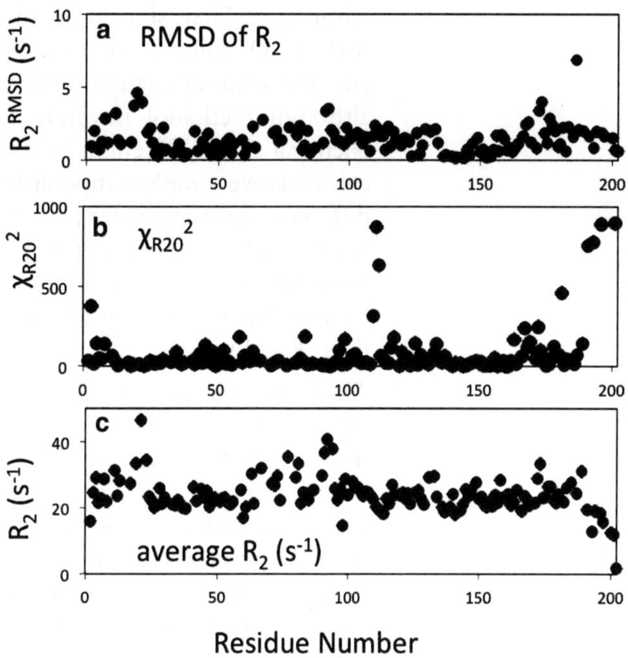

Fig. 4 (**a**) R_2^{RMSD} and (**b**) χ_{R20}^2 plot of ^{15}N CPMG dispersion data, calculated using Eq. 4 or 5. The average R_2 of each dispersion curve is plotted in (**c**). Increase in the χ_{R20}^2 value indicates away from the uniform R_2 values. Note that χ_{R20}^2 may represent the dynamics at the C-terminus well whereas R_2^{RMSD} may represent the dynamics in a region between residues 60 and 90 well

5. Repeat the above analysis for the data sets acquired at 900 MHz. This results in two R_2 dispersion data sets for each amide signal. For the signals with chemical shift below 115 ppm, take the R_2 values acquired at 108 ppm ^{15}N carrier frequency. On the other hand, for the signals above 115 ppm, take the R_2 values acquired at 122 ppm ^{15}N carrier frequency. For the signals around 115 ppm, R_2 values should be similar in the two data sets.
6. Once R_2 values are obtained for the series of the dispersion data (for $i = 1$ to n), R_2^{RMSD} or χ_{R20}^2 are calculated using Eq. 4 or 5. The figure provides an overall profile of the conformational exchange in milli-micro second timescale of the protein (Fig. 4). Further determination of exchange parameters is done using programs that are publically available [14, 38–41].

4 Notes

1. At the start of NMR machine time, it is better to check the instrument performance when multiple users share the NMR instrument. Thus, we firstly record ^{1}H one-dimensional spectrum of TSP [48]. Measurement of the TSP spectrum has two

advantages. First, since most of the previous users use samples with D_2O solvent for deuterium lock, ^1H experiment is checked without changing deuterium lock frequency. Second, although methanol is often used to calibrate temperature, chemical shift differences between water and TSP signal indicates relative sample temperature. Note that the chemical shift difference between water and TSP also depends on salt concentration and pH of the TSP sample. The application of TSP for temperature calibration is possible only when the calibration curve is already made under a quantitatively calibrated condition (Fig. 1).

2. There are two versions of ^{15}N CPMG dispersion experiments: (1) with an rc-INEPT period and (2) with a strong ^1H continuous wave (CW) decoupling scheme [7, 9, 33]. The former, originally developed version uses less ^1H power compared to the latter and therefore is appropriate to avoid heating of protein samples in high salt solution at high magnetic field strengths. The latter version is more suitable for protonated and ^{15}N-labeled large proteins in which the ^1H–^1H dipolar coupling network may disturb sufficient cancellation of the ^1H effect by a single ^1H 180° pulse. $R_2{}^0$ obtained by using ^1H CW pulse is smaller than that obtained using the rc-INEPT and is therefore advantageous also for larger proteins. This is because $R_2{}^0$ obtained using rc-INEPT contains a ^1H relaxation effect in the ^1H–^{15}N antiphase term. A drawback of the ^1H CW version is that because of the relatively strong ^1H pulse, the constant-time CPMG period, T_{CP}, is limited to be relatively short. Although, technically, protonated protein could be used for ^{15}N CPMG dispersion experiments, we often use a perdeuterated ^{15}N sample to avoid potential artifacts caused by the ^1H–^1H dipolar coupling network.

3. An example of a pulse sequence for ^{15}N CT-CPMG relaxation dispersion. Depending on the NMR instruments, delays and powers may need to be adjusted. The authors are not responsible for any loss or damage occurred by the use of this sequence. Also, note that some of 4 μs delay may be deleted depending on spectrometers. We kept them to avoid potential I/O errors.

Pulse sequences

```
#include <Grad.incl>
#include <Avance.incl>
#define TWO_D

;null power pl0 = 120
;p1 proton 90 at pl1, 9u
;p2  proton 180 at pl1, 18u
;p11 1ms proton 90  at sp1
```

```
;15N f3
;p21 90 pulse @ pl3 n15 90
;p22 180 pulse @ pl3
;pcpd3 low power n15 90 (240us) on f3 at pl16
;  Set minimum acquisition time (~40-50 ms)

;nitrogen evolution:
;in0=d0/(l3+1)
;in9=in10-in0
;in10=1/(2sw)
"d10=2.7m"
"d9=4u"
"d0=d9+d10"

;other parameters to be set
;vc  variable counter list for number of loop
;vp variable counter list for slock pulse @ pl8
;l4 total number of T2 data points
;  Set Tcp/2  = (vp+180pwn+vp)*vc
; Tcp = 32 ms, p22=90us, p21=45us, p8(sl)=22 us @ pl8,
; vc = 0      2        4      8   12         16      25     32
;    vp = 2u  3955u  1955u  955u  621.6u  455u  275u  205u

;l5 A loop to set total number of scan is NS*l5.
; Do not increase NS,
;  instead, increase l5 in order to suppress heating.
;
;  Please be carcful when you set vp < 0.3ms or Tcp > 32ms
; Increase acquisition time once no heating is checked.

;NEED ADJUSTED AT DIFFERENT PL8, PL1, and PL3.
; In d14, we have additional 8 us for power switch time.

"d13=1.21u"; ;1/wSL - (4/pi)pw
"d14=46.4u" ; pwn-(2/pi)pwh+8u


"d4=2.25m"
"d6=d4-p12-8u"
"d7=2.7m"
"d5=d7-p12-10u"

"d11=50m"
"d12=10m"
"d26=p21-p1"
"d27=p21-p8"
"d15=p1*4.22+4u+0.318*p21"

;Gradient pulses
"p12=0.4m"  ; gp1 = +10%
"p13=1.0m"  ; gp0 = +50%
"p14=0.9m"  ; gp0 = +50%
```

```
"p15=0.6m"  ; gp0 = +50%
"p16=0.6m"  ; gp2 = +36%

#define ON
#undef OFF

1  ze
2  d11 do:f3 BLKGRAD
  d12
3  d12
21 d12*5.0
4  d12*8.0
25 10u do:f2 do:f3
  10u pl3:f3

#ifdef ON
  d1 fq1:f1
  1m UNBLKGRAD
  10u pl0:f1
  10u pl3:f3
  (p21 ph6):f3
  10u
    (p11:sp1 ph14):f1
    10u
  p13:gp0
  0.5m pl1:f1
;***** start 90-degree on h-n *****
 (p1 ph0):f1
 4u
 p12:gp1
 4u
 d6
 (p22 ph6):f3 (d26 p2 ph0):f1
 4u
 p12:gp1
 4u
 d6
;***** hsqc to nitrogen *****
 (p1 ph4):f1
 4u
 p16:gp2
 0.5m
 (p21 ph6):f3
   4u
   p12:gp1
   4u
 d5
 (p22 ph6):f3 (d26 p2 ph0):f1
   4u
```

```
   p12:gp1
   4u
 d5
 (p21 ph9):f3 (d26 p2 ph0):f1
 2.5m  ;
 p13:gp0 ; Total 4.5 ~ 5 ms
 1m fq1:f1 ;

;***** first R2 disp delay ******

 (p1 ph1):f1
 d13 ; 1/wSL - (4/pi)pw
 (p1 ph0):f1
 d14 ; pwn-(2/pi)pwh+8u
 (p2 ph1):f1
 (p21 ph3):f3
 4u
 4u pl8:f1


69          (vp ph8):f1
 (p22 ph9):f3 (p22 ph8):f1
 (vp ph8):f1
 lo to 69 times c

 4u
  (p22 ph5):f3 (d27 p8*2 ph0):f1
 4u

70       (vp ph8):f1
         (p22 ph9):f3 (p22 ph8):f1
         (vp ph8):f1
         lo to 70 times c
 4u
 4u pl1:f1
 (p21 ph6):f3
 (p2 ph2):f1
   d14  ; pwn-(2/pi)pwh+8u
  (p1 ph0):f1
  d13  ; 1/wSL - (4/pi)pw
 (p1 ph2):f1

 2.5m
 p13:gp0 ;  Total 4.5 ~ 5 ms
 1m  fq1:f1

;***** n15 evolution delay *******
 (p21 ph10):f3
 d0
 d15
 (p22 ph9):f3
 d9
 (p1 ph0 2u p1*2.22 ph4 2u p1 ph0):f1
```

```
d10
(p21 ph7):f3
2u
p14:gp0
1.0m
          4u pl0:fl
        (pl1:spl ph13):fl
4u
          4u pl1:fl
(pl ph0):fl

4u
p15:gp0
610u pl0:fl
(pl1:spl ph13):fl
4u
5u pl1:fl

(pl*2 ph15):fl
5u pl0:fl
(p22 ph6):f3 (pl1:spl ph13):fl
4u
p15:gp0

610u pl16:f3

#endif
#ifdef TWO_D
 go=2 ph31 cpd3:f3
 100u do:f3
     1m BLKGRAD
 d11 do:f3 wr #0 if #0 zd
#endif
#ifdef TWO_D
d12*0.5 ivc
d12*0.5 ivp
lo to 3 times l4 ;adjust for # of T2
d12 ip10
lo to 21 times 2 ;QUAD
d12 dd0
d12 id9
d12 id10
d12 ip31
d12 ip31
lo to 4 times l3 ;N t1 pts
d12 rd0
d12 rd9
d12 rd10
d12 rf #0
```

```
d12 ip10
d12 ip10
d12 ip31
d12 ip31
lo to 25 times l5

#endif
d11 do:f3
d11 do:f2
exit

ph0=0
ph1=1 1 3 3
ph2=3 3 1 1
ph3=0 2
ph4=1
ph5=0 0 2 2
ph6=0
ph7=0
ph8=0
ph9=1
ph10=1 1 1 1
ph13=2
ph14=3
ph15=0
ph16=0
ph31=0 2 0 2 ;
```

4. The following parameters were used to acquire the dispersion data:

Parameter	Value	Parameter	Value	Parameter	Value
p1 @ pl1	10.3 μs	d14	38.6 μs	in10	256.85 μs
p11 @ sp1	1 ms	d0	2.704 ms	l3	100
pl8 for CW	11 kHz[a]	d9	4 μs	l4	7
p21 @ pl3	45 μs	d10	2.7 ms	l5	2
pcpd3 @ pl16	240 μs	in0	15 μs	1TD	1,400
d13	1.27 μs	in9	241.85 μs	d1	2.4 s

[a]pl8 needs to be stronger than this at higher magnetic field strength and for protonated samples that have multiple ^{1}H–^{1}H coupling effects

5. It is important to avoid sample heating and probe heating. Thus, the ^{15}N composite-decoupling pulse power and the acquisition time are minimized. Since GARP decoupling at 1 kHz (250 μs as a 90° pulse) covers ~5 kHz bandwidth, we typically employ a 220–240 μs pulse. To avoid heating, we typically set the acquisition time to 50–60 ms at 600 MHz

and to 30–40 ms at 900 MHz, respectively. This results in similar resolution in both 600 and 900 MHz data.

6. In the first real data point correction, the experiment records an FID of the reference spectrum at $T_{CP} = 0$ first and a series of FIDs with the different τ_{CP} values at a constant T_{CP} from second to the 14th FIDs. Once the experiment is initially tested using a standard protein sample, such as ubiquitin, that does not undergo significant conformational exchange in the most of residues, the following spectra should be obtained (Fig. 3): (1) Signal intensity of the 1D Fourier-transformed (FT) slice of the first FID is larger than those of second to the 14th ones. (2) 1D FT spectra of the FIDs from the second to the 14th exhibit almost identical signal intensity to each other. If the signal intensities from the second to the 14th are not uniform for ubiquitin, the vp and vc may not be correctly calculated, i.e., the total T_{CP} is not uniform for the series of τ_{CP} points.

7. NmrPipe processing commands. Note that there are three scripts, "fid.com," "all.com," and "ft.com" below. Among them, "ft.com" is read in the "all.com." Therefore, run only "fid.com" and subsequently "all.com."

```
—————————— fid.com ——————————
#!/bin/csh
bruk2pipe -in ./ser \
  -bad 0.0 -noaswap -DMX -decim 16 -dspfvs 12 -grpdly -1 \
  -xN        1024        -yN        1400               \
  -xT        512         -yT        700                \
  -xMODE     DQD         -yMODE     Complex            \
  -xSW       8389.262    -ySW       1946.661           \
  -xOBS      600.233     -yOBS      60.828             \
  -xCAR      4.658       -yCAR      119.959            \
  -xLAB      1H          -yLAB      15 N               \
  -ndim      2           -aq2D      States             \
  -out ./test.fid -verb -ov

sleep 5

—————————— all.com ——————————
#!/bin/csh

ft.com 1  1 0 0 0 0 0 0
ft.com 2  0 1 0 0 0 0 0
ft.com 3  0 0 1 0 0 0 0
ft.com 4  0 0 0 1 0 0 0
ft.com 5  0 0 0 0 1 0 0
ft.com 6  0 0 0 0 0 1 0
ft.com 7  0 0 0 0 0 0 1

—————————— ft.com ——————————
#!/bin/csh
```

```
echo Processing experiment $argv[1].

nmrPipe  -fn COADD -cList $argv[2-8] -axis Y -time -in test.
fid                                                        \
| nmrPipe -fn SOL                                          \
| nmrPipe -fn SP -off 0.40 -end 0.99 -pow 1 -c 1.0         \
| nmrPipe -fn ZF -size 2048                                \
| nmrPipe -fn FT -verb                                     \
| nmrPipe -fn PS -p0 126.0  -p1 0.0                        \
| nmrPipe -fn EXT -x1 6ppm -xn 11ppm -sw -di               \
| nmrPipe -fn TP                                           \
| nmrPipe -fn LP                                           \
| nmrPipe -fn SP -off 0.40 -end 0.99 -pow 1 -c 0.5         \
| nmrPipe -fn ZF -size 512                                 \
| nmrPipe -fn FT                                           \
| nmrPipe -fn PS -p0 0.0 -p1 0.0 -di                       \
| nmrPipe -fn REV                                          \
| nmrPipe -fn TP                                           \
| nmrPipe -fn POLY -auto -out  B_$argv[1].DAT  -ov
```

8. Typically, instead of peak volumes, peak heights are used to determine the relaxation rate because the heights are proportional to volumes. Although in theory, it may be recommended to take average of several data points around a peak top to obtain accurate peak height for each peak. However, it is also true that taking average of too many points will introduce errors by signal shoulder and overlap. In measuring peak heights, detected peak height can reflect noise error but is better not to contain artifact from signal shoulder. Thus, it is best not to average over many points.

Acknowledgments

The authors would like to thank Stefan Bagby and Dennis A. Torchia for critical reading of the manuscript. This project was supported by University of Pittsburgh.

References

1. Szyperski S, Luginbühl P, Otting G, Güntert P, Wüthrich K (1993) Protein dynamics studied by rotating frame. J Biomol NMR 3:151–164
2. Davis DG, Perlman ME, London RE (1994) Direct measurements of the dissociation-rate constant for inhibitor-enzyme complexes via the T-1-Rho and T-2 (CPMG) methods. J Magn Reson B 104:266–275
3. Orekhov VY, Pervushin KV, Arseniev AS (1994) Backbone dynamics of (1–71)bacterioopsin studied by two-dimensional 1H-15N NMR spectroscopy. Eur J Biochem 219:887–896
4. Akke M, Palmer AG 3rd (1996) Monitoring macromolecular motions on microsecond to millisecond time scales by R1ρ–R1 constant relaxation time NMR spectroscopy. J Am Chem Soc 118:911–912
5. Zinn-Justin S, Berthault P, Guenneugues M, Desvaus H (1997) Off-resonance rf fields in

heteronuclear NMR. Application to the study of slow motions. J Biomol NMR 10:363–372
6. Mulder FA, van Tilborg PJ, Kaptein R, Boelens R (1999) Microsecond time scale dynamics in the RXR DNA-binding domain from a combination of spin-echo and off-resonance rotating frame relaxation measurements. J Biomol NMR 13:275–288
7. Loria JP, Rance M, Palmer AG 3rd (1999) A relaxation-compensated Carr-Purcell-Meiboom-Gill sequence for characterizing chemical exchange by NMR spectroscopy. J Am Chem Soc 121:2331–2332
8. Mulder FA, Skrynnikov NR, Hon B, Dahlquist FW, Kay LE (2001) Measurement of slow (micros-ms) time scale dynamics in protein side chains by (15)N relaxation dispersion NMR spectroscopy: application to Asn and Gln residues in a cavity mutant of T4 lysozyme. J Am Chem Soc 123:967–975
9. Tollinger M, Skrynnikov NR, Mulder FA, Forman-Kay JD, Kay LE (2001) Slow dynamics in folded and unfolded states of an SH3 domain. J Am Chem Soc 123: 11341–11352
10. Eisenmesser EZ, Bosco DA, Akke M, Kern D (2002) Enzyme dynamics during catalysis. Science 295:1520–1523
11. Mulder FA, Hon B, Mittermaier A, Dahlquist FW, Kay LE (2002) Slow internal dynamics in proteins: application of NMR relaxation dispersion spectroscopy to methyl groups in a cavity mutant of T4 lysozyme. J Am Chem Soc 124:1443–1451
12. Bosco DA, Eisenmesser EZ, Pochapsky S, Sundquist WI, Kern D (2002) Catalysis of cis/trans isomerization in native HIV-1 capsid by human cyclophilin A. Proc Natl Acad Sci U S A 99:5247–5252
13. Wang CY, Rance M, Palmer AG 3rd (2003) Mapping chemical exchange in proteins with MW > 50 kD. J Am Chem Soc 125: 8968–8969
14. Korzhnev DM, Salvatella X, Vendruscolo M, Di Nardo AA, Davidson AR, Dobson CM, Kay LE (2004) Low-populated folding intermediates of Fyn SH3 characterized by relaxation. Nature 430:586–590
15. Beach H, Cole R, Gill ML, Loria JP (2005) Conservation of mu s-ms enzyme motions in the apo- and substrate-mimicked state. J Am Chem Soc 127:9167–9176
16. Lipari G, Szabo A (1982) Model-free approach to the interpretation of nuclear magnetic resonance relaxation in macromolecules. 1. Theory and range of validity. J Am Chem Soc 104:4546–4559
17. Lipari G, Szabo A (1982) Model-free approach to the interpretation of nuclear magnetic resonance relaxation in macromolecules. 2. Analysis of experimental results. J Am Chem Soc 104:4559–4570
18. Mandel AM, Akke M, Palmer AG 3rd (1995) Backbone dynamics of Escherichia-coli ribonuclease Hi - correlations with structure and function in an active enzyme. J Mol Biol 246:144–163
19. Tjandra N, Wingfield P, Stahl S, Bax A (1996) Anisotropic rotational diffusion of perdeuterated HIV protease from 15N NMR relaxation measurements at two magnetic fields. J Biomol NMR 8:273–284
20. Freedberg DI, Ishima R, Jacob J, Wang YX, Kustanovich I, Louis JM, Torchia DA (2002) Rapid structural fluctuations of the free HIV protease flaps in solution. Protein Sci 11:221–232
21. Koenig SH, Schillinger WE (1969) Nuclear magnetic relaxation dispersion in protein solutions. J Biol Chem 244:3283–3289
22. Kimmich R (1979) Field cycling in NMR relaxation spectroscopy: applications in biological, chemical and polymer physics. Bull Magn Reson 1:195–218
23. Noack F (1986) NMR field-cycling spectroscopy: principles and applications. Prog NMR Spectrosc 18:171–276
24. Bertini I, Briganti F, Xia ZC, Luchinat C (1993) Nuclear magnetic relaxation dispersion studies of hexaaquo Mn(II) ions in water-glycerol mixtures. J Magn Reson A 101:198–201
25. Hodges MW, Cafiso DS, Polnaszek CF, Lester CC, Bryant RG (1997) Water translational motion at the bilayer interface: an NMR relaxation dispersion measurement. Biophys J 75:2575–2579
26. Koenig SH, Brown RD (1990) Field-cycling relaxometry of protein solutions and tissue: implications for MRI. Prog NMR spect, 22:487057
27. Halle B, Denisov VP (1995) A new view of water dynamics in immobilized proteins. Biophys J 69:242–249
28. Roberts MF, Redfield AG (2004) Phospholipid bilayer surface configuration probed quantitatively by P-31 field-cycling NMR. Proc Natl Acad Sci U S A 101:17066–17071
29. Kimmich R, Anoardo E (2004) Field-cycling NMR relaxometry. Prog NMR Spectrosc 44:257–320
30. Diakova G, Goddard YA, Korb JP, Bryant RG (2010) Water and backbone dynamics in a hydrated protein. Biophys J 98:138–146
31. Ishima R, Torchia DA (2003) Extending the range of amide proton relaxation dispersion

experiments in proteins using a constant-time relaxation-compensated CPMG approach. J Biomol NMR 25:243–248

32. Myint W, Gong Q, Ishima R (2009) Practical aspects of 15N CPMG transverse relaxation experiments for proteins in solution. Concepts Magn Reson 34A:63–75
33. Hansen DF, Vallurupalli P, Kay LE (2008) An improved (15)N relaxation dispersion experiment for the measurement of millisecond time-scale dynamics in proteins. J Phys Chem B 112:5898–5904
34. Ishima R, Torchia DA (2006) Accuracy of optimized chemical-exchange parameters derived by fitting CPMG R2 dispersion profiles when R2 (0a) not = R2(0b). J Biomol NMR 34:209–219
35. McConnell HM (1958) Reaction rates by nuclear magnetic resonance. J Chem Phys 28:430–431
36. Luz Z, Meiboom S (1963) Nuclear magnetic resonance study of the protolysis of trimethylammonium ion in aqueous solution - order of the reaction with respect to solvent. J Chem Phys 39:366–370
37. Carver JP, Richards RE (1972) General 2-site solution for chemical exchange produced dependence of T2 upon Carr-Purcell pulse separation. J Magn Reson 6:89–105
38. Kovrigin EL, Kempf JG, Grey MJ, Loria JP (2006) Faithful estimation of dynamics parameters from CPMG relaxation dispersion measurements. J Magn Reson 180:83–104
39. Bieri M, Gooley PR (2011) Automated NMR relaxation dispersion data analysis using NESSY. BMC Bioinformatics 12:421
40. Hansen DF, Lundström P, Velyvis A, Kay LE (2012) Quantifying millisecond exchange dynamics in proteins by CPMG relaxation dispersion NMR using side-chain 1H probes. J Am Chem Soc 134:3178–3189
41. Kleckner IR, Foster MP (2012) GUARDD: user-friendly MATLAB software for rigorous analysis of CPMG RD NMR data. J Biomol NMR 52:11–22
42. Hoult DI, Lauterbur PC (1979) The sensitivity of the zeugmatographic experiment involving human samples. J Magn Reson 34: 425–433
43. Gadian DG, Robinson FNH (1979) Radiofrequency losses in NMR experiments on electrically conducting samples. J Magn Reson 34:449–455
44. Kelly AE, Ou HD, Withers R, Dotsch V (2002) Low-conductivity buffers for high-sensitivity NMR measurements. J Am Chem Soc 124: 12013–12019
45. Horiuchi T, Takahashi M, Kikuchi J, Yokoyama S, Maeda H (2005) Effect of dielectric properties of solvents on the quality factor for a beyond 900 MHz cryogenic probe model. J Magn Reson 174:34–42
46. Ishima R (2011) Recent developments in (15) N NMR relaxation studies that probe protein backbone dynamics. Top Curr Chem 326: 99–122
47. Delaglio F, Grzesiek S, Vuister GW, Zhu G, Pfeifer J, Bax A (1995) Nmrpipe – a multidimensional spectral processing system based on Unix pipes. J Biomol NMR 6:277–293
48. Wishart DS, Bigam CG, Yao J, Abildgaard F, Dyson HJ, Oldfield E, Markley JL, Sykes BD (1995) 1H, 13C and 15N chemical shift referencing in biomolecular NMR. J Biomol NMR 6:135–140

Chapter 3

Confocal Single-Molecule FRET for Protein Conformational Dynamics

Yan-Wen Tan, Jeffrey A. Hanson, Jhih-Wei Chu, and Haw Yang

Abstract

Single-molecule Förster-type resonance energy transfer (smFRET) is a unique technique capable of following conformational motions of individual protein molecules. The direct observation of individual proteins provides rich information that would be washed away in ensemble measurements, hence opening up new avenues for establishing protein structure-function relationships through dynamics. Retrieving dynamics information of biomolecular motions via smFRET, though, requires careful experiment design and rigorous treatment of single-molecule statistics. Here, we describe the rudimentary steps for an smFRET experiment, including sample preparation for the microscope, building of critical parts for single-molecule FRET detection, and a robust methodology for photon-by-photon data analysis.

Key words Protein immobilization, Single photon counting, Poisson statistics, Model free

1 Introduction

It has long been hypothesized that conformation changes of a protein plays an indispensible role in the protein's function [1, 2]. The manner by which a sequence of conformation-change events may lead to functional output of the protein, however, remains elusive.

Förster-type resonance energy transfer, FRET [3, 4], is a mechanism by which the energy of a chromophore (the energy donor) in its electronically excited state can be transferred to another (the energy acceptor) through dipole-dipole coupling. A spectroscopy-based technique, FRET serves as a molecular ruler that is sensitive to donor-acceptor distances in the range of ~2–8 nm—the size of a protein. When used in single-molecule experiments [5], it enables monitoring the intramolecular distance changes of a single protein in real time, providing uniquely new insights that could bridge the aforementioned knowledge gap.

The work described in the manuscript was conducted at both University of California at Berkeley and Princeton University, as well in part at Fudan University, China.

Dennis R. Livesay (ed.), *Protein Dynamics: Methods and Protocols*, Methods in Molecular Biology, vol. 1084,
DOI 10.1007/978-1-62703-658-0_3, © Springer Science+Business Media New York 2014

There are two distinct ways to do fluorescent single-molecule FRET experiments. They are (a) recording the photon bursts from individual diffusing proteins and (b) following the time-dependent signal from immobilized single proteins. The former burst-detection approach enjoys the advantages of high throughput, simple sample preparation, and no surface interactions; however, it is limited by the length of the trajectory (determined by the time for a protein molecule to diffuse across the laser focal volume), and its signal is complicated by the nonuniform laser illumination at the focus. The latter surface-immobilization approach, on the other hand, does not have the noted limitation and complications when special attentions are paid to making sure that the dynamics of the immobilized proteins are not affected by the surface. With the scientific goal of following protein motions on the functionally relevant timescales (sub-millisecond to minutes), we opt to focus on smFRET experiments on immobilized protein molecules.

In terms of data acquisition mode, the immobilized single molecules can be imaged by a wide-field camera under a total internal reflection fluorescence microscopy (TIRFM) configuration [6], or they can be detected by photon-counting avalanche photodiodes (APD) under a confocal configuration [7]. Between these two methods, the camera-based approach allows the simultaneous imaging of many single molecules at the same time, thus offering high throughput. The time resolution, however, is limited by the camera frame rate, which is usually on the order of ~30 ms (30 frames per second). The camera-based approach is therefore oblivious to protein motions in the sub-millisecond to millisecond region—timescales that are expected to be important for enzyme catalysis and protein functional dynamics. On the other hand, the confocal approach locates individual molecules through faster scanning and acquires data one molecule at a time. While a relatively lower-throughput method, this latter approach is amenable to advanced statistical analysis of the series of detected photon [8, 9], delivering the highest possible time resolution and FRET distance precision.

In this chapter, we outline the steps to set up a confocal microscope for photon-counting single-molecule FRET experiments and, without loss of generality, use the adenylate kinase enzyme as an example to illustrate the experimental procedures [10]. The data is then analyzed using a model-free method that is unbiased and reliable and has been successfully applied on a number of protein systems. Following the outlined steps, protein conformational dynamics can be measured one molecule at a time with < 5-Å distance precision and < 1 ms temporal resolutions. These limits are determined by the florescent intensities of the employed dyes and the photon detection machinery in the experiment.

2 Materials

2.1 Reagents

All chemicals are purchased from Sigma and used as received unless specified. All aqueous solutions are prepared using deionized ultrapure water with a resistivity of 18.2 M-Ω-cm at room temperature.

1. Protein buffer: Prepare a buffer solution containing 100 mM KCl, 2 mM $MgCl_2$, and 100 mM tris(hydroxymethyl)aminomethane hydrochloride (Tris–HCl) and adjust it to appropriate pH (e.g., pH 7.5) (*see* **Note 1**).
2. Fluorescent labels: Thiol-reactive maleimide derivatives of Alexa Fluor 555 and Alexa Fluor 647 (Invitrogen, $R_0 = 51$ Å) are used as the donor and acceptor, respectively, for single-molecule FRET (smFRET) experiments. For the 1-mg packaging from the vendor, 20 μL of dimethyl sulfoxide (DMSO) is used to dissolve it and stored in −20 °C freezer for future use (*see* **Note 2**).
3. SulfoLink resin (Pierce).
4. Superdex 75 gel-filtration column (Amersham Pharmacia).
5. Quartz coverslips (1″ × 1″ × 0.15 mm, Technical Glass Products).
6. Profusion chamber: Store the CoverWell profusion chamber (Grace BioLabs) in 20 % ethanol when not in use.
7. Fluorescent spheres for beam alignment: FluoSpheres carboxylate-modified microspheres, nominal size 0.02 μm, Nile Red Fluorescent (535/575) (Invitrogen, F8784).
8. Silane solution: Dissolute 3 % (v/v) 3-aminopropyltriethoxysilane (APES) (80 mL volume) in acetone as stock; handle with care to avoid water and humidity.
9. Polyethylene glycol (PEG) solution: Dissolve 4 mg biotin-PEG-SCM and 36 mg mPEG-SCM (both from Laysan Bio Inc.; SCM = succinimidyl carboxymethyl ester) in 0.2 mL 0.1 M $NaHCO_3$ (pH 8.3).
10. Streptavidin: Prepare the streptavidin (Jackson ImmunoResearch) stock solution at a concentration of 10 mg/mL. For incubation with the protein sample, aliquot 3 μL of 10 mg/mL stock into 0.5 mL protein buffer.
11. Biotinylated α-His antibody (Rockland): for incubation with protein sample, aliquot 1 μL of 1 μM stock into 0.5 mL buffer.

2.2 Equipment

2.2.1 General

1. Laser light source at 532 nm: Coherent Compass 315M-100 Green DPSS CW-laser system (*see* **Note 3**).
2. Optical table with a surface area for at least 1.5 m × 1.5 m.
3. Inverted fluorescence microscope system (Olympus IX71).

4. Piezoelectric stage with nanometer positioning precision (Physik Instrumente, P-517K021).
5. High numerical aperture oil-immersion microscope objective (Olympus PlanApo 60×, NA = 1.4).
6. Microscope objective immersion oil (Cargille) with index of refraction matching quartz and low autofluorescence.
7. Counter: Agilent 5314A universal counter for avalanche photodiode alignment.

2.2.2 The Laser Excitation Optics

1. Neutral density (ND) filters for tuning the laser power (Thorlabs).
2. A cleanup filter and a Notch filter for 532 nm (Semrock).
3. A linear polarizer and quarter-wave plates (Thorlabs).
4. A dichroic filter (Semrock) for 532-nm excitation.
5. A beam expander consisted of a 50-mm and a 200-mm focal length, Ø1″, plano-convex lenses (Thorlabs).
6. A periscope set (holder from Newport, mirrors from Thorlabs) to raise the beam height matching the microscope back port.

2.2.3 The Detection Unit

1. Two single photon-counting avalanche photodiode (APD) modules for donor and acceptor channels (Excelitas Technologies, SPCM-AQRH-14), as well proper DC power supply for the APDs.
2. Two 1″-travel *XYZ* translational stages to mount and align APD modules (Thorlabs).
3. A 200-mm focal length, Ø1″, plano-convex lens to refocus image (Thorlabs).
4. A dichroic filter 650dcxr from Chroma.
5. Bandpass filters: FF01-580/60-25-D from Semrock; ET705/100m from Chroma.
6. A 30-mm cage cube (Thorlabs C4W) with B4C platform and cage-compatible dichroic filter holder FFM1 to mount the dichroic mirror.
7. SM1 tubes of various lengths and light blocking paper and clothes to block ambient light.

3 Methods

3.1 Protein Sample Preparation

1. To allow for single-molecule FRET experiments, the protein sample should have only two cysteine residues per molecule. The cysteine residues are located at the sites between which the distance is to be measured. Using the AK enzyme from *Escherichia coli,* for example, site-directed mutagenesis is used to

introduce cysteine mutations in the Lid (A127C) and the Core (A194C) domains for site-specific labeling. A further mutation is introduced (C77A) to remove the only native cysteine from the gene. The resulting mutant is then appended with a $(His)_6$ tag for immobilization at the C-terminus. Following expression and purification, the AK sample is stored in protein buffer in 4 °C refrigerator.

2. To conjugate the reporting fluorescent probes, the FRET dyes are dissolved in DMSO and reacted with ≈ 1 mM AK at a five- to tenfold molar excess for 3 h at room temperature. A fivefold excess of tris(2-carboxyethyl)phosphine (TCEP) hydrochloride is also added to the reaction to prevent the formation of intermolecular disulfide bonds.
3. Run the protein sample/dye mixture through a Superdex 75 gel-filtration column to remove unreacted dyes.
4. To further enrich doubly labeled proteins, the sample from the previous step kept in SulfoLink resin for 30 min at room temperature to remove proteins with unconjugated free thiol groups (*see* **Note 4**).

3.2 Coverslip Passivation

1. Place a clean batch (usually 8) of quartz coverslips in a Teflon holder (Invitrogen, C-14784), then place the holder containing coverslips in a 100-mL glass beaker.
2. Rinse three times with dry acetone to remove all traces of water before silanization.
3. Incubate the coverslips in 80 mL acetone containing 3 % 3-aminopropyltriethoxysilane for 30 min.
4. Rinse the coverslips once with acetone.
5. Rinse thoroughly with ultrapure water.
6. Dry coverslips with N_2.
7. Bring PEG-SCM and biotin-PEG-SCM to room temperature in a glove bag (e.g., the AtmosBag from Sigma-Aldrich) store under nitrogen.
8. For 8 coverslips, dissolve 40 mg PEG power in 200 μL, 0.1 M $NaHCO_3$ (pH 8.3) right before next step.
9. Sandwich between two coverslips 50 μL of the PEG/biotin-PEG solution.
10. Incubate 3 h at room temperature in a humid environment (e.g., a closed pipette tip box with water at the bottom).
11. Immerse the coverslip sandwiches to ultrapure water and separate them in water. It is important to note which side has PEG.
12. Rinse thoroughly.
13. The coverslips are now passivated and ready for experiments. They can be stored in sealed ultrapure water for about 3 weeks.

3.3 Single-Protein Immobilization on Quartz Coverslips

1. Rinse the passivated coverslips 8 times, each time with 0.5 mL protein buffer.
2. Incubate streptavidin on the quartz slide for 15 min.
3. Meanwhile, preincubate biotinylated α-His antibody to dye-labeled proteins in the dark for 15 min. The antibody should be at a concentration of ~20 nM, which can be prepared by dissolving 1 μL of 1 μM stock into 0.5 mL buffer. The protein solution should be at 50 pM to 1 nM, prepared via serial dilution immediately before use.
4. At the end of incubation, rinse the coverslip eight times with 0.5 mL of protein buffer.
5. Incubate protein/antibody solution on the coverslip for 10 min (*see* **Note 5**).
6. Rinse the coverslips eight times with 0.5 mL protein buffer (*see* **Note 6**).
7. At the center of the slide, add a 30-μL droplet of protein buffer. Seal the sample with a piece of the CoverWell profusion chamber to prevent evaporation. Now the single-molecule sample is ready for measurements [11] (*see* **Note 7**).

3.4 Laser Beam Collimation and Alignment

1. Install the laser and the 532-nm cleanup filter (*see* **Note 8**; Fig. 1).
2. Install the microscope body and make sure there is at least 1.5 m from the output of the laser to the inlet of the microscope back port.
3. With the help of an optical power meter, install the linear polarizer and set the quarter-wave plate so that the laser light is circularly polarized at the sample.

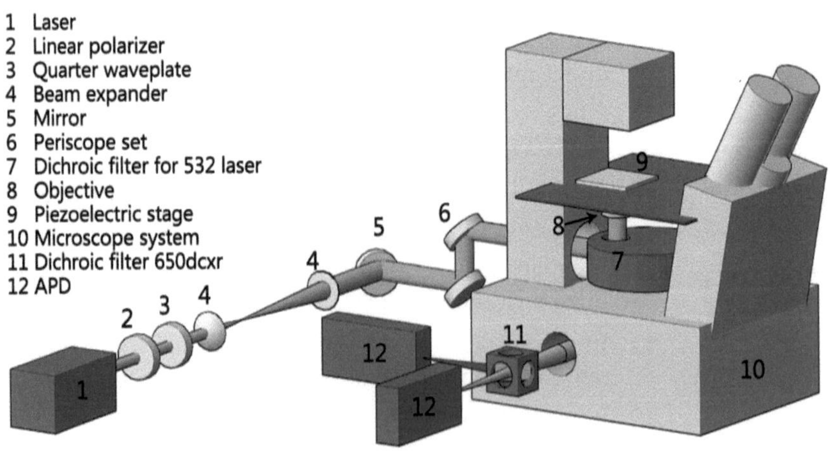

Fig. 1 A schematic drawing for the crucial parts of a microscope setup for single-molecule FRET experiment

4. Install the pair of plano-convex lenses to form a collimated beam with diameter matching the back aperture of the microscope objective.
5. Install the dichroic mirror in the dichroic/filter cube assembly supplied by the microscope.
6. Use the periscope to bring the beam to the level of the microscope back port. Adjust the mirrors so that the beam makes a centered normal incident through the port for the objective.
7. Install the two-dimensional piezoelectric stage.
8. Before every measurement, clean the objective with isopropanol and methanol on lens tissue. Place the objective on the revolving nosepiece of the microscope.

3.5 APD Coarse Alignment and Fine Alignment

1. For coarse alignments, use an SM1 tube mounted with a mirror facing downward in place of the microscope objective (an RMS thread to SM1 thread adapter will be required). Turn on the laser such that the laser beam will be reflected by the mirror and directed towards the various optical ports of the microscope.
2. At the camera port of the microscope, refocus the reflected laser light with a 200-mm focal length, Ø1″, plano-convex lens.
3. Install the cage cube and mount the dichroic mirror so the light will make a 45° incident on the dichroic mirror. The cage should be in a comfortable position to allow room for two APD modules on the optical table.
4. Mount each APD on a separate *XYZ* translational stage.
5. Adjust the *XYZ* axes to the central position before fixing the stage to the optical table to allow for maximal flexibility during the alignment process.
6. Remove the SM1 tube with mirror. Put the objective in place and place a piece of glass coverslip (not quartz) on the piezo stage. (An adaptor sample plate is required to hold the coverslip in place.)
7. Adjust the objective focus on the top of the coverslip at the glass–air interface.
8. Fix the stage so that the sensor area of each APD is roughly at the focus of the beam (*see* **Note 9**).
9. Use SM1 tubes to isolate ambient light from the beam path. Use matte black paper tube to cover the end of SM1 tube connecting to the APD sensor head. Wrap the ends with black cloth.
10. Use cardboards (or dark-colored acrylic) and light blocking cloth to make a case housing of the APD/dichroic modules.
11. Turn off room light. Make sure that the laser beam is dim enough and there is no light leaking into the protective housing.

12. Turn on APDs.
13. Optimize the *Z*-axis (along the beam path) of the *XYZ* translation stage for highest counts from both channels when objective is focused onto the glass–air interface of the glass coverslip (*see* **Note 10**). Tune *X*- and *Y*-axis if necessary.
14. Turn off APDs and block the laser beam.
15. Install the emission filters (FF01-580/60-25-D for donor channel; ET705/100m for acceptor channel).
16. Replace the glass coverslip with a clean quartz coverslip coated with dilute FluoSpheres (1/1,000 dilution of the stock, 3 × 20 μL spin coated to the coverslip).
17. Turn on APDs and unblock the laser beam.
18. Adjust the *X*- and *Y*-axis of the APD translational stage to get highest counts on both channels.

3.6 Single-Molecule FRET Time Trace Acquisition

1. Block the laser beam.
2. Place the prepared sample on the piezoelectric stage.
3. Unblock the beam.
4. Scan a small area (~ 10 × 10 μm^2) with low excitation power (~ 1 μW at the sample). There should be several bright spots in the view with a uniform diffraction-limited spot size. If not, increase the protein sample incubation time and concentration (*see* **Note 11**).
5. Seek spots with acceptor photon counts above the baseline. Move the piezo stage over to the center of those spots.
6. Increase the excitation power (~ 2.5 μW at the sample) to get a trajectory with strong intensity and reasonable length of photobleaching time.
7. Record the arrival time of each detected photon for both the donor and the acceptor channels.
8. Continue recording data until both the donor and the acceptor dyes bleach. Move the piezo stage to the next molecule and record another trajectory.
9. Repeat **steps 5–8** until enough trajectories have been obtained for reasonable statistics.

3.7 Data Analysis

1. To quantify the conformational motions around the millisecond range, one first runs an equal-information binning [12], a necessary step before the maximum entropy deconvolution to remove photon-counting noise [10]. The kinetics of the conformational transitions can be extracted by modeling the conformational distributions with a motional narrowing theory [13, 14] (*see* **Note 12**).

2. Valid trajectories are first binned with equal uncertainty level, denoted by the variance α^2 [15]. For example, the uncertainty α^2 may run from 8 to 15 %. If we define FRET efficiency as E, then the variance α^2 can be calculated from the inverse of Fisher information $J(E)$:

$$J(E) = T\left[I_d^\beta \frac{\left(1 - B_d/I_d^\beta\right)^2}{E(1 - B_d/I_d^\beta) - 1} + I_a^\beta \frac{(1 - B_a/I_a^\beta)^2}{E(1 - B_a/I_a^\beta) + B_a/I_a^\beta}\right]$$
$$= \frac{1}{\alpha^2},$$

where I_d^β (I_d^β) is the detected donor (acceptor) intensity including background contribution for that particular molecule, B_d (B_a) is the donor (acceptor) background level. T can thus be understood as the duration associated with the jth measurement with a relative uncertainty α.

3. Normalize each set of data so that the curve represents the probability density function (pdf). For the jth distance measurement within the mth molecule, the normalized distance is evaluated using

$$\hat{x}_{m,j} = \frac{R_{m,j}}{R_0} = \left[\frac{I_{a,m}^\beta/B_{a,m} I_{d,m}^\beta n_{a,m,j} - I_{a,m}^\beta n_{a,m,j} I_{d,m}^\beta/B_{d,m}}{I_{d,m}^\beta/B_{d,m} I_{a,m}^\beta n_{d,m,j} - I_{d,m}^\beta n_{a,m,j} I_{a,m}^\beta/B_{a,m}}\right]^{1/6},$$

where R is the donor–acceptor distance and R_0 is the Förster radius, a donor–acceptor distance at which the energy transfer efficiency is 50 %. Omitting the m and j indices, again, I_d^β (I_a^β) is the donor (acceptor) intensity including background contribution for that particular molecule, B_d (B_a) is the donor (acceptor) background level, and n_d (n_a) is the number of donor (acceptor) photons in a time interval T that gives a distance estimate with a predetermined variance α^2.

4. Distance data from each uncertainty level are then collected, based on which a probability density function can be generated using the Gaussian kernel estimator. The width of the Gaussian kernel is defined by the variance α^2 of the data set. A set of eight (8–15 %) smooth pdf versus distance profiles will be generated.

5. Run the maximum entropy deconvolution to remove the broadening resulting from the kernel estimator and also from the photon-counting noise (*see* **Note 13**). The noise removed probability-distance distributions that are the model-free distributions for the protein conformation at different timescales. The uncertainties to the conformational distribution can be obtained by running a bootstrap simulation.

6. If there are two states in the higher-uncertainty histograms, the set of eight pdfs are then subjected to a global fitting using the

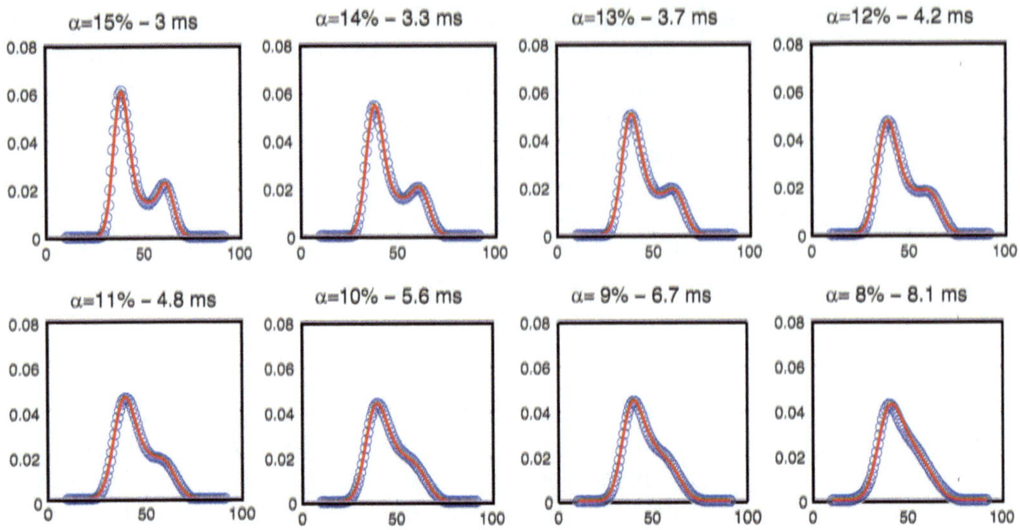

Fig. 2 Motional narrowing fittings from a mutant adenylate kinase. The *curves* linked by *blue circles* are the probability distribution functions generated under different α value setting, the distance measurement uncertainty relative to the Förster radius, from 0.08 to 0.15. The *red solid lines* are global fitting results from a two-state motional narrowing model

motional narrowing theory (Fig. 2). Here, the time resolution is taken as the average time, $T_{\text{avg}} = \sum_j T_j/n$, which has a one-to-one correspondence to the uncertainty level represented by α. The fitting method closely follows that devised by Geva and Skinner [16].

4 Notes

1. This protocol is written according to our experiment on adenylate kinase. Please use functional reaction buffer for your own protein sample.
2. Alexa Fluor 555 and Alexa Fluor 647 is a popular FRET pair with a Förster radius $R_0 \sim 51$ Å under the protein buffer condition. Different FRET pairs can be chosen for different experiments; however, the excitation wavelength, dichroic optics, and emission filters should be modified accordingly.
3. Again, the excitation green light at 532 nm is chosen for Alexa Fluor 555/647 FRET pair. Other wavelengths can be used if different dyes are required.
4. Dye-labeled enzymes were found to retain their activity; however, we cannot currently quantify the effects of dye-labeling on the enzyme's conformational distribution or dynamics on our experimental timescale, if any.

5. It is important to protect labeled protein samples from light as much as possible during the sample preparation steps to prevent premature photobleaching.
6. After proteins are bound to the quartz slide, be careful not to dry the center part of the slide. Always keep a droplet of buffer on top of the slide.
7. The his-antibody-histidine interaction has finite strength and proteins immobilized that way may dissociate from the coverslip surface. For protein–protein interaction studies, the biotin-streptavidin scheme is recommended.
8. This dichroic should be chosen to match the excitation wavelength used in the experimental system.
9. Find the clearest light spot around the focal point of the lens with a piece of white business card. Place the APD head (black in color) there until the bright spot disappear at the sensor head. The bright spot is due to reflection by the metallic casing around the APD-sensing region. The reflection disappears when the laser strikes at the photon-sensing region (about ~180 μm in diameter).
10. You will need a neutral density filter (~O.D. 2) in front of the donor channel to protect the APD and make photon counts on both channels to be at the same order of magnitude.
11. Several factors can be considered for troubleshooting purpose here. The protein sample concentration during incubation and incubation time are the most obvious. Besides that, try to increase biotin-PEG ratio in the passivation step. On the contrary, if too many proteins are bound, try to reduce incubation concentration. Check the effectiveness of surface passivation, too.
12. Various data analysis scheme has been successfully applied to single-molecule FRET dynamics measurements. Here, we apply the method which we consider preserves most information. This method is conservative yet judicious in terms of drawing conclusions.
13. The contribution of photon-counting noise to the broadening of the distance estimate $\hat{x}_{m,j}$ is asymptotically normal (Gaussian) because the distances are calculated using a maximum-likelihood estimator (the Central-Limit Theorem), and that the probability function for the estimator is smooth on the parameter space.

Acknowledgements

We thank Song Song for creating the microscope setup figure used in this chapter. This work was sponsored by Shanghai Pujiang Program grant 11PJ1401000 and National Natural Science

Foundation of China grant 11104039 (to Y.W.T.), as well the National Institutes of Health and the Fudan University Key Laboratory Senior Visiting Scholarship (to H.Y.).

References

1. Koshland DE (1958) Application of a theory of enzyme specificity to protein synthesis. Proc Natl Acad Sci U S A 44:98–104
2. Koshland DE (1995) The Key-lock theory and the induced Fit theory. Angew Chem-Int Edit Engl 33:2375–2378
3. Förster T (1948) Zwischenmolekulare energiewanderung und fluoreszenz. Ann Phys 437:55–75
4. Clegg RM (1992) Fluorescence resonance energy-transfer and nucleic-acids. Methods Enzymol 211:353–388
5. Ha T, Ting AY, Liang J, Caldwell WB, Deniz AA, Chemla DS, Schultz PG, Weiss S (1999) Single-molecule fluorescence spectroscopy of enzyme conformational dynamics and cleavage mechanism. Proc Natl Acad Sci U S A 96:893–898
6. Funatsu T, Harada Y, Tokunaga M, Saito K, Yanagida T (1995) Imaging of single fluorescent molecules and individual ATP turnovers by single myosin molecules in aqueous solution. Nature 374:555–559
7. Nie SM, Chiu DT, Zare RN (1994) Probing individual molecules with confocal fluorescence microscopy. Science 266:1018–1021
8. Barkai E, Brown FLH, Orrit M, Yang H (eds) (2008) Theory and evaluation of single-molecule signals. World Scientific, Singapore
9. Yang H (2011) Change-point localization and wavelet spectral analysis of single-molecule time series. In: Komatsuzaki T, Kawakami M, Takahashi S, Yang H, Silbey RJ (eds) Single-molecule biophysics: experiment and theory, vol 146, Advances in chemical physics. Wiley, New York, pp 129–143
10. Hanson JA, Duderstadt K, Watkins LP, Bhattacharyya S, Brokaw J, Chu J-W, Yang H (2007) Illuminating the mechanistic roles of enzyme conformational dynamics. Proc Natl Acad Sci U S A 104:18055–18060
11. Pal P, Lesoine JF, Lieb MA, Novotny L, Knauf PA (2005) A novel immobilization method for single protein spFRET studies. Biophys J 89: L11–L13
12. Watkins LP, Yang H (2004) Information bounds and optimal analysis of dynamic single molecule measurements. Biophys J 86:4015–4029
13. Kubo R (1954) Note on the stochastic theory of resonance absorption. J Phys Soc Jpn 9:935–944
14. Anderson PW (1954) A mathematical model for the narrowing of spectral lines by exchange or motion. J Phys Soc Jpn 9:316–339
15. Watkins LP, Chang H, Yang H (2006) Quantitative single-molecule conformational distributions: a case study with poly-(l-proline). J Phys Chem A 110:5191–5203
16. Geva E, Skinner JL (1998) Two-state dynamics of single biomolecules in solution. Chem Phys Lett 288:225–229

Chapter 4

Protein Structural Dynamics Revealed by Site-Directed Spin Labeling and Multifrequency EPR

Yuri E. Nesmelov

Abstract

Multifrequency electron paramagnetic resonance (EPR) of spin-labeled protein is a powerful spectroscopic technique to study protein dynamics on the rotational correlation time scale from 100 ps to 100 ns. Nitroxide spin probe, attached to cysteine residue, reports on local topology within the labeling site, dynamics of protein domains reorientation, and protein global tumbling in solution. Due to spin probe's magnetic tensors anisotropy, its mobility is directly reflected by the EPR lineshape. The multifrequency approach significantly decreases ambiguity of EPR spectra interpretation. The approach, described in this chapter, provides a practical guideline that can be followed to carry out the experiments and data analysis.

Key words Electron paramagnetic resonance (EPR), Electron spin resonance (ESR), Multifrequency, Spin, Label, Probe, Nitroxide

1 Introduction

The flexibility of surface loops and motion of protein structural elements is usually required for protein function, such as interaction with ligands and conformational changes. The combination of the site-directed spin labeling and the electron paramagnetic resonance (EPR) is a well-established method to study protein dynamics [1]. The common way is to covalently attach nitroxide spin probe to the cysteine amino acid residue of the single-cysteine protein or protein mutant. The spin probe reports on local topology within the labeling site, on conformational dynamics of protein domains and on global protein tumbling. The exceptional sensitivity to the spin probe orientation in the external magnetic field makes EPR high-resolution technique for determination of the spin probe's rotational motion.

1.1 Stable Nitroxide Radical

In the nitroxide radical, an unpaired electron is localized within the highly anisotropic π orbital of the N–O bond (Fig. 1). In the external magnetic field, the electron spin is polarized and flips

Dennis R. Livesay (ed.), *Protein Dynamics: Methods and Protocols*, Methods in Molecular Biology, vol. 1084,
DOI 10.1007/978-1-62703-658-0_4,

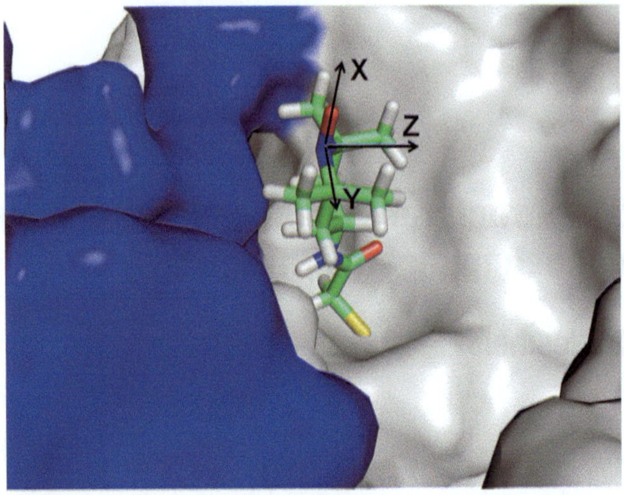

Fig. 1 Spin probe at the myosin labeling site C707, surrounded by the converter domain (*blue*) and the N-terminal domain (*gray*). Axes x, y, and z represent molecular frame of the spin probe. Spin probe's unpaired electron density is localized at the N–O bond, colored *blue* (nitrogen) and *red* (oxygen)

between parallel and antiparallel orientations when resonant microwave field is applied. Those spin flips cause absorption of microwave radiation, and the asymmetry of nitroxide's electron density determines orientational dependence of microwave absorption in the external magnetic field (Eq. 1, first term). The proportionality constant (g-factor tensor) reflects the anisotropy of electron density:

$$H = (h\nu)/(g\beta_e) + m_I A, \tag{1}$$

where H is the resonant magnetic field, ν the resonant frequency, h Planck's constant, β_e Bohr's magneton, $m_I = (-1, 0, +1)$ nitrogen nuclear spin quantum number, and A hyperfine splitting tensor. Equation 1 determines the magnetic field of the resonance; the higher the EPR frequency, the better resolution for g-factor components (first term in Eq. 1). The interaction of the unpaired electron and the nitrogen nucleus (nitrogen nuclear spin $I = 1$) results in splitting of the EPR absorption line into a triplet (Eq. 1, second term). In the result, the EPR absorption spectrum of a nitroxide radical is a triplet, which spectral position and splitting value depend on the projection of g and A tensor values on the direction of the external magnetic field. Then, the EPR spectrum of randomly oriented nitroxide probes or spin-labeled protein molecules (powder spectrum) is the sum of many triplets, which are splitted and positioned according to the spin probes orientational distribution in the external magnetic field [2]. Fast unrestricted motion averages both g and A tensors and results in a three-line EPR spectrum. All intermediate cases (restricted and/or slow motion of spin-labeled protein) result in various lineshapes, depending on nitroxide's orientation in the external

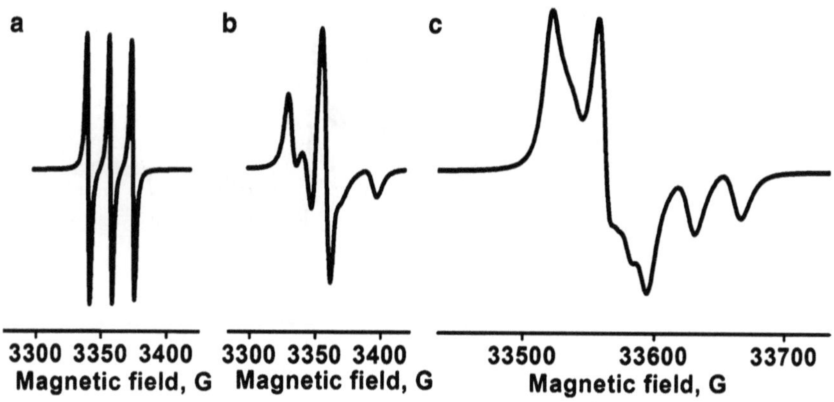

Fig. 2 EPR spectra of a nitroxide radical, effect of isotropic motion. (**a**) X-band, rotation correlation time 50 ps, (**b**) X-band, rotation correlation time 10 ns, (**c**) W-band, rotation correlation time 10 ns

magnetic field and the degree of tensor averaging due to the spin probe motion.

1.2 Effect of the Spin Probe's Motion

Fast isotropic motion averages g and A tensors of the nitroxide spin probe, reducing them to scalar values $g_0 = 1/3(g_x + g_y + g_z)$ and $A_0 = 1/3(A_x + A_y + A_z)$ and producing a three-line EPR spectrum, similar to the EPR spectrum of an oriented nitroxide in the absence of motion (Fig. 2a). The frequency of averaging isotropic motion for A and g tensors is

$$2\pi(A_z - A_x)\beta_e g_0/h \quad \text{and} \quad 2\pi(g_x - g_z)\beta_e H/h \tag{2}$$

accordingly, where A_i and g_i are the components of A and g tensors. Isotropic motion of a spin probe, faster than the averaging frequencies, does not change the EPR spectrum lineshape. Slower motion significantly changes the EPR lineshape, until the powder limit is reached (Figs. 2 and 3). The restrictions of spin probe motion affect the EPR lineshape as well. Another factor affecting the EPR lineshape is the spectral broadening, usually unknown due to the unknown spin probe's relaxation time T_2. The spectral broadening masks sharp spectral features, reducing differences between spectral response of the fast and slow motion of the spin probe.

1.3 Slow and Restricted Motion of a Spin Probe at the Protein

Fit of EPR spectra in the approach of the slow motion [3] (rotation correlation time $\tau_c > 1$ ns) produces two parameters of spin probe motion: the coefficient of rotational diffusion D_R and the equilibrium distribution of orientation probability P_0. In the single-frequency EPR experiment, the separation of these two parameters is ambiguous, mostly because of undetermined spin–spin relaxation time T_2, responsible for spectral broadening. Multifrequency approach allows separation of these parameters, due to different sensitivity of high- and low-frequency EPR to the rate of spin probe motion [3, 4]. Simultaneous fit of high- and low-frequency EPR

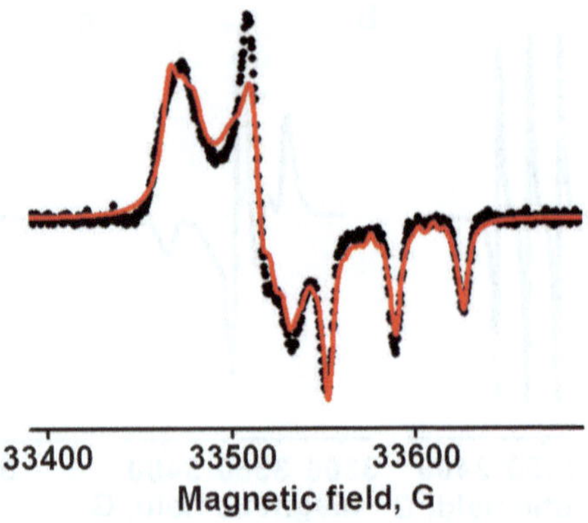

Fig. 3 W-band EPR spectrum of IASL-labeled myosin, acquired at T = 80 K to determine g and A tensor values. Black—experiment. Red—fit as a powder spectrum

spectra produces unambiguous result for $\{D_R, P_0\}$, corresponding to both high- and low-frequency EPR spectra [4].

1.4 Spin Probe Dynamics Determined by EPR Spectra Simulation and Fit

The spin probe's slow motion approach was realized with NLSL software [5], utilizing the MOMD (macroscopic order, microscopic disorder) model [3]. The MOMD model is based on consideration of the time evolution of spin magnetization and subsequent averaging over the spin ensemble [3, 6]. In this model the restricted motion of the spin probe at the protein labeling site is considered, but not the motion of the protein itself, assuming random protein orientations. Slow protein tumbling is considered in the SRLS (slow relaxing local structure) model [7], realized with NLSL-SRLS software. Two motions are considered in the SRLS: restricted local motion of the spin probe and unrestricted tumbling of protein in solution. The restrictions of rotational motion of spin probe are defined in these models by expansion coefficients c_{LK} of spherical harmonics D_K^L (Ω),

$$U(\Omega) = -kT \sum_{L,K} c_{LK} D_K^L(\Omega) \tag{3}$$

where $U(\Omega)$ is the restoring potential and Ω is a set of angles needed to describe the spin probe orientation relative to the applied magnetic field [5]. The equilibrium distribution of orientation probability defined from the restoring potential as [5]

$$P_0(\Omega) = \frac{\exp[-U(\Omega)/kT]}{\int d\Omega \exp[-U(\Omega)/kT]} \tag{4}$$

1.5 Multifrequency Approach

The multifrequency EPR decreases ambiguity of the EPR lineshape analysis because of different sensitivity of high- and low-frequency EPR to the spin probe motion. Indeed, according to Eq. 2, rotational motion with the correlation time $\tau = 1$ ns completely averages g tensor for X-band EPR, but not at W-band. The complete averaging of g tensor for W-band will happen when spin probe rotates with the correlation time $\tau > 0.5$ ns. This example shows different reflection of spin probe motion by X- and W-band EPR spectra, and thus, the simultaneous multifrequency analysis decreases ambiguity of the spectra interpretation. X- and W-band EPR spectra are not redundant and their simultaneous fit is helpful in the determination of motional parameters of the probe, which reflects mobility of the protein structural elements and local protein topology.

1.6 Spin Label with Flexible Linker as a Probe for Local Structure of a Protein

Myosin is a motor enzyme involved in muscle contraction and cell motility through a cycle of actin-activated ATP hydrolysis. Myosin's head domain, subfragment 1 (S1), is responsible for the coupling of ATP hydrolysis and the force generation. During the ATPase cycle, myosin experiences two conformational changes, the power stroke, when the force is generated, and the recovery stroke, when myosin molecule primes to perform the power stroke. Three structural states of myosin were originally identified by intrinsic fluorescence [8, 9], M, the apo state; M*, the post-powerstroke state, trapped with ADP nucleotide; and M**, the pre-powerstroke state, trapped with ADP.AlF_4. These structural states are assumed to be tightly coupled to biochemical states, which are defined by the nucleotide or nucleotide analog, bound to the active site. These conformational states of S1 have also been observed and characterized by EPR through the changes in mobility of a nitroxide spin probe (IASL) at SH1 labeling site (C707) [10–14] in the force-generating region of myosin. The multifrequency EPR approach was employed to probe local structure of the force generation region in skeletal muscle myosin and *D. discoideum* cellular myosin [13, 14]. Both myosins were labeled at the same site: C707 in the skeletal muscle myosin and T688C in the constructed single-cysteine *D. discoideum* myosin mutant. Myosin was labeled with IASL and then trapped in several biochemical states, apo (no bound nucleotide), post-recovery stroke (bound ADP.AlF_4 nucleotide analog of ADP.Pi), and ADP-bound state. Figures 4, 5, and 6 show EPR spectra, fits, and determined probability of the spin label orientations distribution for each biochemical state of IASL-S1. In the apo S1 biochemical state (Fig. 4), multifrequency EPR unambiguously detects a single structural state of myosin in the force-generating region, M. The motion of the spin label is complex; the molecular frame of the spin label is tilted in the diffusion frame ($\beta_d = 32°$), the spin probe moves slowly on the nanosecond time scale ($\tau_\perp = 9.9$ ns, $\tau_{||} = 64$ ns), and the motion is restricted ($c_{20} = 4.36$ kT). These slow restricted motions probably reflect those of the secondary structural elements in close contact with the labeled site. Indeed, our modeling of the structure of *D. discoideum* myosin in the apo S1 state (1FMV, [15]) shows that

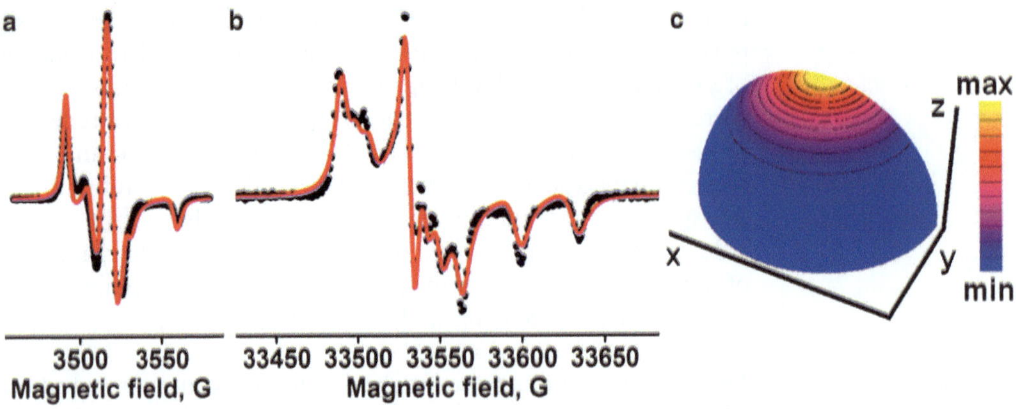

Fig. 4 IASL-labeled myosin in the apo biochemical state (M), no nucleotide bound. Black—experiment. Red—spectral fits for the model of slow restricted motion. (**a**) X-band, (**b**) W-band. (**c**) Orientational distribution of the spin probe, determined from the simultaneous fit of X-band and W-band spectra

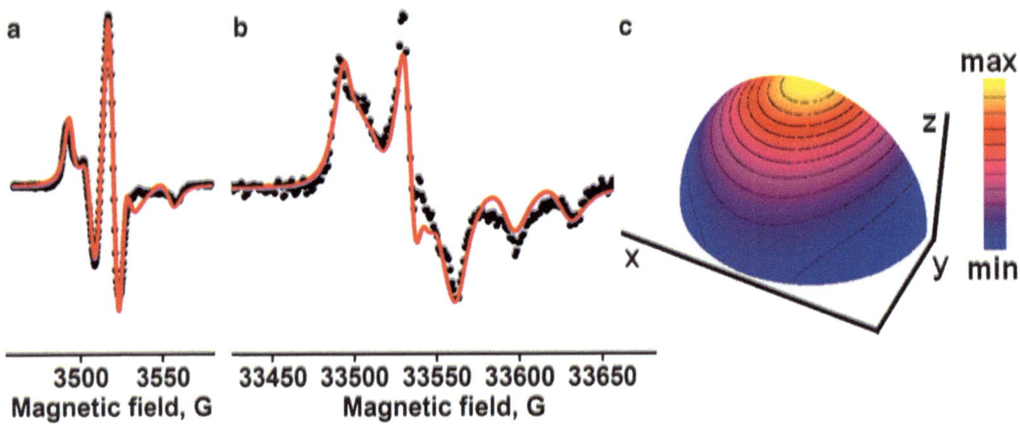

Fig. 5 IASL-labeled myosin in the ADP-bound biochemical state (M*). Black—experiment. Red—spectral fits for the model of slow restricted motion. (**a**) X-band, (**b**) W-band. (**c**) Orientational distribution of the spin probe, determined from the simultaneous fit of X-band and W-band spectra

the spin probe is located between the converter and N-terminal domains of myosin (Fig. 1). The spin label is squeezed between myosin domains, but not buried under the surface, since g and A tensor values do not reflect changes in polarity for different myosin states in the W-band low-temperature experiments. In the S1.ADP state, a single structural state (M*) is determined. In this conformation of the force-generating region, the spin probe moves slightly faster, its motion is less restricted relative to the apo S1, and the tilt of the molecular frame of the spin label in the diffusion frame is less ($\tau_\perp = 6.8$ ns, $\tau_\| = 25.2$ ns, $c_{20} = 1.88$ kT, $c_{22} = -0.36$ kT, $\beta_d = 5°$), Fig. 5. The structure of *D. discoideum* myosin with bound ADP (1MMA, [16]) shows a slightly different position of the converter relative to N-terminal domain, compared to that of

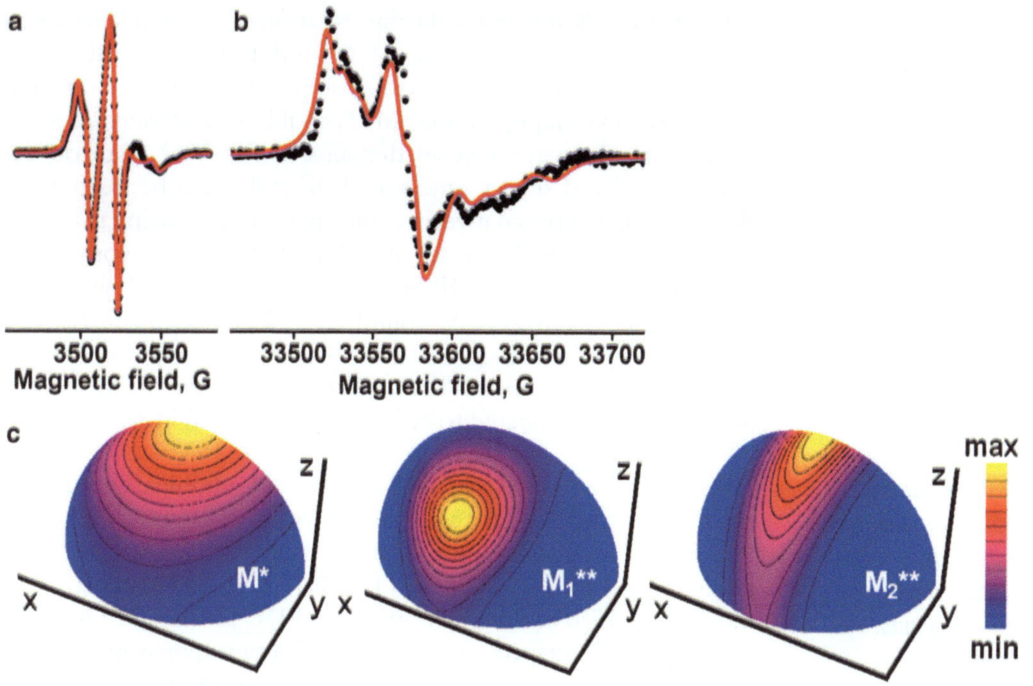

Fig. 6 IASL-labeled myosin in the ADP.AlF_4-bound biochemical state (M**). Black—experiment. Red—spectral fits for the model of slow restricted motion. (**a**) X-band, (**b**) W-band. (**c**) Orientational distribution of the spin probe, determined from the simultaneous fit of X-band and W-band spectra. Note that the fit gives two populations for the M** state, with similar dynamics of the spin probe. M_1** populates 78 % and M_2** populates 6 %. M* populates 16 % of myosin with ADP.AlF_4 bound

apo S1, allowing more space for the spin label to move, in qualitatively agreement with EPR data. A much more complex structural picture emerges in the S1.nucleotide analog complex, which is presumed to correspond to structural state M**. The EPR spectra of both myosins with the nucleotide analog show at least two and probably three structural states, while only one structural state is revealed by corresponding crystal structure [17]. One spectral component (component 1) is similar to that of IASL-S1.ADP, suggesting the presence of M* structural state. The two other spectral components have different dynamics of the spin probe (component 2, $\tau_x = 1.0$ ns, $\tau_y = 1.8$ ns, $\tau_z > 300$ ns, $c_{20} = 4.96$ kT, $c_{22} = -4.91$ kT, $c_{40} = -1.93$ kT; component 3, $\tau_x = 0.2$ ns, $\tau_y = 0.1$ ns, $\tau_z > 300$ ns, $c_{20} = 6.42$ kT, $c_{22} = -6.24$ kT); there is no tilt of the molecular frame relative to the diffusion frame. These spectral components could be assigned as the M** structural state of myosin [11]. According to the crystal structure of the S1.ADP.AlF_4 state, the spin probe is located between the converter and N-terminal domains, while the relay helix is bent in this structural state and is not likely to affect spin label motion. The two observed distributions of spin label orientation in the M** structural state (most prominent in the skeletal S1) could reflect a difference in relative position of the converter and N-terminal domains: rotation of the spin label about the *x* axis is allowed in one state (M_1**) and stopped in another state (M_2**, Fig. 1). This slight difference in relative position of domains within one myosin population (M**) could be interpreted in terms of flexibility between myosin domains in the M** state on the microsecond and slower time scale [18] and could be treated as a slow exchange between myosin conformations, producing two spectral components. EPR spectra of skeletal and cellular myosins were the same for apo and ADP-bound states and showed significant difference in the ADP.AlF_4-bound state. EPR spectra fit showed that both myosins adopt the same conformations, but differently populated. Skeletal myosin showed larger population of M** component (84 %), compared to cellular myosin (28 %) [13].

1.7 Spin Label, Rigidly Coupled to the Protein, as a Probe for the Protein Conformational Dynamics

Phospholamban (PLB) is a 52-amino acid amphipathic integral membrane protein that regulates the active transport of calcium in the heart muscle through the interaction with the cardiac sarcoplasmic reticulum Ca-ATPase [19, 20]. The NMR structure of monomeric PLB in detergent micelles (PDB 1N7L) [21] reveals a predominant L-shaped conformation with two approximately perpendicular helices, cytoplasmic (residues 1–16) and transmembrane (residues 21–52), connected by a flexible hinge. According to solid-state NMR [22, 23] and EPR [24, 25], PLB's predominant conformation in the membrane has the transmembrane domain approximately perpendicular to the membrane. NMR data show the cytoplasmic domain oriented almost perpendicular to the transmembrane domain, lying along the membrane surface and interacting with lipid headgroups. EPR quenching studies confirm interaction of cytoplasmic domain with the surface of membrane [24]. However, cross-linking [26] and functional mutagenesis [27] data suggest that the cytoplasmic domain of PLB extends well above the membrane surface when it interacts with Ca-ATPase, so most models of the PLB–Ca-ATPase complex show PLB straightened almost linearly on the surface of Ca-ATPase, with the entire molecule, including the cytoplasmic domain, nearly perpendicular to the membrane surface [24, 28–31].

To understand how PLB undergoes this dramatic structural transformation in a membrane, PLB labeled with TOAC (paramagnetic amino acid that reports directly the dynamics of the peptide backbone [24]) was synthesized. TOAC-PLB was reconstituted into DOPC/DOPE lipid vesicles (1/200 protein/lipid ratio), and EPR spectra were acquired at 9.4 and 94 GHz at $T = 4$C. The spin probe was strongly immobilized, when TOAC is inserted into the transmembrane domain of PLB, consistent with a well-ordered α-helix. It was determined that the transmembrane domain rotates slowly ($\tau_{||} = 105$ ns) around the helix axis and performs a rapid ($\tau_{\perp} = 2$ ns) but very restricted ($c_{20} = 14$ kT) wobbling motion perpendicular to this axis in the membrane (Fig. 7).

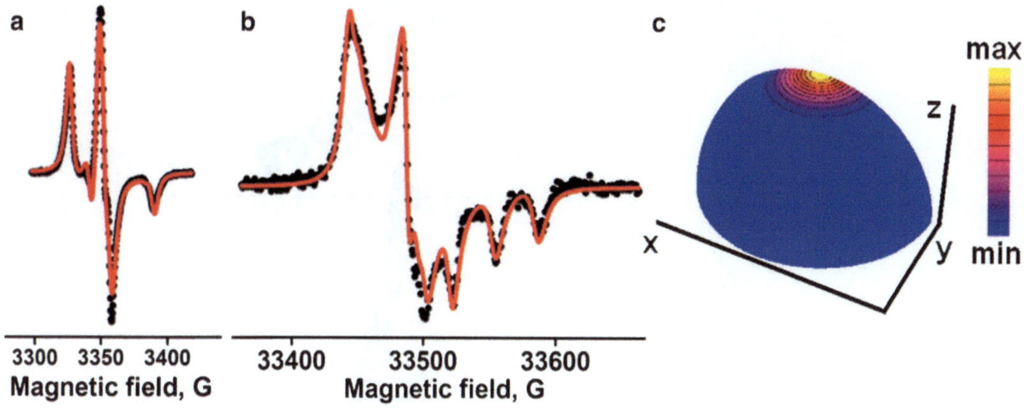

Fig. 7 TOAC-labeled PLB, cytoplasmic domain. Black—experiment. Red—spectral fits for the model of slow restricted motion. A. X-band, B. W-band. C. Orientational distribution of the spin probe, determined from the simultaneous fit of X-band and W-band spectra. EPR spectra of spin probe, rigidly attached to the cytoplasmic domain of PLB, exhibit two populations, T (84 %) and R (16 %), with different dynamics and restrictions for motion

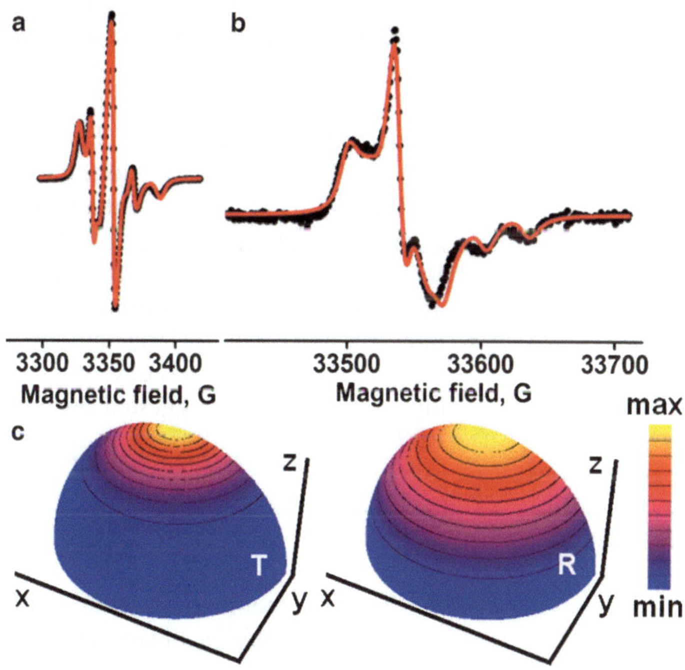

Fig. 8 TOAC-labeled PLB, transmembrane domain. Black—experiment. Red—spectral fits for the model of slow restricted motion. (**a**) X-band, (**b**) W-band. (**c**) Orientational distribution of the spin probe, determined from the simultaneous fit of X-band and W-band spectra. EPR spectra of spin probe, rigidly attached to the transmembrane domain of PLB, exhibit single population, with very restricted motion

TOAC-PLB with the labeled site in the cytoplasmic domain shows clear evidence for two resolved conformations, one a well-ordered helix (the *T* conformation, $\tau_{iso} = 3.5$ ns, $c_{20} = 4.2$ kT, 84 % mole fraction, Figs. 8 and 9) and the other dynamically disordered on the

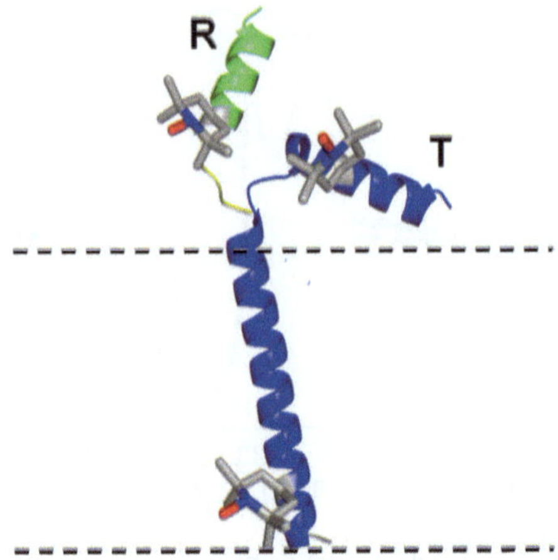

Fig. 9 Two-state model for PLB structural dynamics in a membrane. Dashed lines indicate the membrane surface. The transmembrane domain performs slow uniaxial rotation about the membrane normal and rapid restricted wobbling motion in both PLB conformations. The cytoplasmic domain wobbles rapidly and almost unrestrictedly in the *R* conformation. In the T conformation the motion of the cytoplasmic domain is slower and more restricted

nanosecond time scale (the *R* conformation, $\tau_{iso} = 0.68$ ns, $c_{20} = 0.52$ kT, 16 % mole fraction, Figs. 8 and 9). The disordered *R* conformation binds preferentially to Ca-ATPase [30, 31] and is enhanced by phosphorylation of PLB [31], resulting in Ca-ATPase activation. The *T* conformation interacts with the surface of the membrane, slowing its motion and stabilizing its helical conformation. It is likely that the large rate and amplitude of the cytoplasmic helix rotational motion is due, at least in part to the partial unfolding of this helix, as suggested previously by both NMR [32] and EPR [24] data.

2 Materials

1. Proteins: membrane protein phospholamban was synthesized with the TOAC spin label attached either at position 46 in the transmembrane domain (46-TOAC-AFA-PLB) or at position 11 in the cytoplasmic domain (11-TOAC-AFA-PLB) [24, 33]. Rabbit skeletal myosin and *D. discoideum* myosin single-cysteine myosin construct were prepared as described elsewhere [13, 14].
2. Buffers: phospholamban was reconstituted into lipid vesicles containing DOPC/DOPE (Avanti Polar Lipids, 4:1, 200 lipids per PLB) in 25 mM imidazole at pH 7.0. Myosin labeling buffer: 50 mM MOPS at pH7.0, 50 mM KCl, 1 mM EGTA.

Myosin experimental buffer: 20 mM EPPS at pH 8.0, 50 mM KCl, 5 mM $MgCl_2$, 1 mM EGTA.

3. Chemicals: glycerol, $MgCl_2$, ADP, NaF, $AlCl_3$.
4. Myosin labeling consumables: Amicon spin concentrators (cut-off 30 kDa), Millipore, Billerica MA.
5. Myosin spin label: IASL, 4-(2-iodoacetamido)-TEMPO. Stock solution 200 mM in DMF.
6. EPR spectrometers: Bruker EleXsys E500 (X-band, 9.4 GHz) and E600 (W-band, 94 GHz).
7. Quartz sample tubes (VitroCom, Mountain Lakes, NJ) and tube sealant (Oxford Labware, St. Louis, MO).
8. Analysis software: NLSL, NLSL-SRLS (http://www.acert.cornell.edu).

3 Methods

3.1 Myosin Labeling

1. Mix protein with spin label, molar ratio 1:6, protein to label. Dilute spin label stock solution in labeling buffer to keep final DMF concentration at 1 % or less.
2. Label protein for 12 h at $T = 4$ °C.
3. Clean labeled protein from the excess of label and transfer to experimental buffer with spin concentrators. Final amount of free spin label should be less than 1 %.
4. Clarify protein by ultracentrifugation at 100,000 × g to remove possible aggregates of denatured protein.
5. Measure protein concentration spectrophotometrically at 280 nm $(A_{280}-A_{320})\cdot\varepsilon$, where ε is the extinction coefficient of myosin, $\varepsilon = 0.69\ (\text{mg/ml})^{-1}\ \text{cm}^{-1}$ for *D .discoideum* myosin, and $\varepsilon = 0.74\ (\text{mg/ml})^{-1}\ \text{cm}^{-1}$ for skeletal myosin.
6. Concentrate protein with spin concentrator for EPR experiment, final concentration is 100 μM.
7. Add 10–40 % glycerol (v/v) or 30–60 % sucrose (w/v) as cryoprotectant for low-temperature EPR measurement.
8. Load sample in a quartz tube (0.8 mm OD, 0.6 mm ID for X-band, and 0.25 mm OD, 0.15 mm ID for W-band).

3.2 Formation of Stable Myosin Complexes with ADP and ADP.AlF_4

1. Incubate myosin with 5 mM MgADP + 20 mM NaF for 5 min at 25 °C in the experimental buffer.
2. Add 5 mM $AlCl_3$.
3. Incubate at 25 °C for additional 20 min.

3.3 Electron Paramagnetic Resonance

At X-band, a SHQ cavity with a variable temperature accessory was used. 30 μL sample was contained in a quartz capillary with OD/ID = 0.84/0.6 mm, sealed with Critoseal. The scan width was 120 G, the peak-to-peak modulation amplitude was 1 G, and the modulation frequency was 100 kHz. Spectra were recorded below saturation, at $H_1 = 0.05$ G, at 20 °C (myosin) or 4 °C (phospholamban). At W-band, the standard TE_{011}-mode cylindrical cavity resonator was used with a variable temperature TeraFlex probehead; the 0.2 μL sample was contained in a quartz capillary with OD/ID = 0.25/0.15 mm. The sample was loaded at one end of the capillary, then the other end of the capillary was flame-sealed and the sample was sedimented by low-speed centrifugation (300 × *g*). The scan width was 400 G, the peak-to-peak modulation amplitude was 1 G, and the modulation frequency was 100 kHz. Spectra were recorded below saturation, at $H_1 = 0.045$ G, at 20 °C (myosin) or 4 °C (phospholamban). Spectra were recorded at 80 K as well to obtain components of g and A tensors. X-band EPR experiments were performed immediately after sample preparation, W-band EPR experiments were performed within 24 h after sample preparation, and the sample was stored on ice during that time. For low-temperature EPR experiments, 30 % glycerol (v/v) was added to the sample as a cryoprotectant. Baseline EPR spectra at X-band and W-band were obtained with buffer solution at the same conditions as the protein spectra acquisitions. Baseline spectra were subtracted from the protein spectra before analysis. Before lineshape analysis, the phase of the W-band spectrum was adjusted to correct for a slight admixture of dispersion [34]. The quadrature spectrum was calculated from a Hilbert transformation of the experimental spectrum [35], and the weighted quadrature spectrum was subtracted from the experimental spectrum until the integrated spectrum displayed a flat baseline in the low field part.

Low-temperature W-band spectra were analyzed according to the procedure described below. Ambient temperature X-band and W-band spectra were analyzed with NLSL and NLSL-SRLS software.

3.4 Determination of g and A Tensors Components

1. Simulate the "powder spectrum" for W-band with the generic g and A tensors, $g = \{2.002, 2.006, 2.009\}$ and $A = \{6\text{ G}, 6\text{ G}, 37\text{ G}\}$, according to Eq. 5:

$$\begin{aligned} H_{\rm res}(\theta,\phi) &= h\nu/[g(\theta,\phi)\beta] - m_I A(\theta,\phi),\ (m_{\rm I} = -1, 0, +1) \\ g(\theta,\phi) &= g_{\rm x}\sin^2\theta\cos^2\phi + g_{\rm y}\sin^2\theta\sin^2\phi + g_{\rm z}\cos^2\theta \\ A(\theta,\phi) &= \left(A_{\rm x}^2\sin^2\theta\cos^2\phi + A_{\rm y}^2\sin^2\theta\sin^2\phi + A_{\rm z}^2\cos^2\theta\right)^{1/2} \end{aligned} \quad (5)$$

where $H_{\rm res}$ is the magnetic field at resonance, ν is the applied frequency, β is the Bohr magneton, and angles θ and ϕ define orientations of the nitroxide probe relative to the applied magnetic field, directed parallel to z axis of the laboratory frame.

2. Sum simulated "powder spectra" for 10,000 random orientations of the spin label in the laboratory frame.
3. Convolute obtained spectrum with the Lorentz function of desired width (example: 1 G).
4. Calculate first derivative of the simulated spectrum, convoluted with the Lorentz function.
5. Fit simulated EPR spectrum to the experimental W-band spectrum, acquired at $T = 80$ K.
6. Calculate mean-square deviation of the experimental spectrum and the simulated spectrum.
7. Minimize mean-square deviation by changing the fitting parameters (components of g and A tensors and Lorentz linewidth) according to chosen minimization procedure (e.g., Levenberg–Marquardt method), until reasonable agreement between experimental and simulated spectrum is reached.
8. Alternatively, fit g and A tensor components and Lorentz broadening with EasySpin [36] or NLSL [5] software, freely available (*see* **Note 1**).

3.5 Determination of Spin Probe Motional Parameters

Spin probe's motional parameters are obtained from simultaneous fits of X-band and W-band EPR spectra at ambient temperature, using g and A tensor components determined from low-temperature W-band spectra fits earlier. These g and A tensor components were fixed in all spectral simulations. To describe the motion of the membrane protein (phospholamban), we have used NLSL software [5]. The software simulates EPR spectra according to specified parameters and uses nonlinear least-squares minimization algorithm to fit simulated lineshape to the experimental EPR spectrum. The software provides the best fit parameters, the least-square deviation of simulated and experimental spectra, and correlation coefficients between the order parameter and the rotational correlation time. Only motion of phospholamban with respect to the membrane was analyzed, since the lipid vesicles were large enough that their correlation times are in the μs–ms time scale [37], which is undetectable by conventional EPR at both X-band and W-band (*see* **Note 2**). To fit EPR spectra of phospholamban, labeled within the cytoplasmic domain, we assumed isotropic motion (rbar) of the cytoplasmic domain in the axial restoring potential. We used only the first term of the series (*c*20), making the standard assumption that the restoring potential should be symmetric with respect to the membrane normal (local director). This eliminated non-axial terms *c*22, *c*42, and *c*44. The inclusion of higher-order terms with axial symmetry (e.g., *c*40) did not improve the fit. Under these conditions, $P_0(\Omega)$ is completely defined by the order parameter S [6, 38]. For EPR spectra of phospholamban, labeled within the transmembrane domain, the anisotropic tensor of rotational diffusion (*see* **Note 3**) of the transmembrane helix produced better fit. The axial restoring potential *c*20 was sufficient for the fit. One- and two-component EPR spectra were considered in the fits. The fit of EPR spectra of the cytoplasmic part clearly showed two components (two populations) with distinct motional parameters. EPR spectra of the transmembrane domain showed single spectral component, corresponding to one population.

We have used NLSL-SRLS software to describe the motion of the spin probe attached to myosin S1, accounting for the spin probe motion and the tumbling of S1 in solution. The rate of S1 rotational motion was determined from the Stokes–Einstein equation [14], considering S1 as a prolate ellipsoid, $D_{R\perp} = 1.0{\cdot}10^6$ rad^2 s^{-1}, $\tau_{c\perp} = 167$ ns, and $D_{R||} = 2.45{\cdot}10^6$ rad^2 s^{-1}, $\tau_{c||} = 68$ ns. NLSL-SRLS software operates under the same principles as NLSL. The determined components of g and A magnetic tensors and the calculated diffusion coefficients of myosin tumbling (*see* **Note 4**) were kept constant. We varied the coefficients of the rotational diffusion (*see* **Note 5**), restrictions of spin label motion within myosin (*c*20, *c*22, *c*40), isotropic Lorentzian line broadening, the angle between the magnetic frame and the diffusion frame of the spin label (betad), and the number of spectral components, corresponding to myosin with the different structure of the force-generating region.

4 Notes

1. When fitting powder spectra with NLSL, in the script file specify the number of sites (1, one conformation of the label), center field in Gauss, generic values for g and A tensor components, and number of orientations of the spin probe (we find 40 orientations satisfactory), and set order parameter to zero and the coefficient of rotational diffusion (rbar, $\log_{10}$ of the coefficient in seconds^{-1}) to a low value, to account for the absence of the spin probe motion. Rbar < 6 corresponds to the coefficient of rotational diffusion in the microsecond range, way beyond the sensitivity of W-band EPR measurement to spin probe motion. For better fit one can use tensor values for the line broadening. Complete documentation of NLSL software is available for download (http://www.acert.cornell.edu).
2. In the spectra simulations, we have fixed spin probe's diffusion tilt angle (betad), specifying Euler angle β_d between the diffusion axis of the α-helix and the z axis of the spin probe [39].

The coefficients of rotational diffusion, line broadening (gib0), and the coefficients of the restoring potential were the variable parameters. The number of MOMD orientations (nort) was set to 40.

3. Two coefficients of the rotational diffusion were considered, parallel (rpll) and perpendicular (rprp) to the axis of rotation of the transmembrane helix. This model corresponds to nonrestricted rotation about the axis of the transmembrane helix and the rocking motion of the transmembrane helix in the membrane.
4. Parameters R0prp, R0pll, are analogous to the parameters rprp and rpll considered in the **Note 3**.
5. Parameters rxx, ryy, and rzz. We found that isotropic rotational diffusion (parameter rbar) does not produce satisfactory fit, as well as axial parameters rprp and rpll.

Acknowledgements

This work was supported by NIH grants AR53562, AR59621. NLSL and NLSL-SRLS software was kindly provided by Dr. Z. Liang and Dr. J.H. Freed (Cornell University).

References

1. Berliner LJ (ed) (1976) Spin labeling theory and applications. Academic, New York
2. Van SP, Birrell GB, Griffith OH (1974) Rapid anisotropic motion of spin labels. Models for motion averaging of the ESR parameters. J Magn Res 15:444–459
3. Freed JH (1976) Theory of slow tumbling ESR spectra for nitroxides. In: Berliner LJ (ed) Spin labeling: theory and applications. Academic, New York, pp 53–132
4. Nesmelov YE, Karim C, Song L, Fajer PG, Thomas DD (2007) Rotational dynamics of phospholamban determined by multifrequency electron paramagnetic resonance. Biophys J 93:2805–2812
5. Budil D, Lee S, Saxena S, Freed J (1996) Nonlinear-least-squares analysis of slow-motion EPR spectra in one and two dimensions using a modified Levenberg-Marquardt algorithm. J Magn Reson A120:155–189
6. Schneider DJ, Freed JH (1989) Calculating slow motional magnetic resonance spectra: a user's guide. In: Berliner LJ (ed) Biological magnetic resonance, vol 8. Plenum Publishing Corporation, New York, pp 1–76
7. Polimeno A, Freed JH (1995) Slow motional ESR in complex fluids: the slowly relaxing local structure model of solvent cage effects. J Phys Chem 99:10995–11006
8. Johnson KA, Taylor EW (1978) Intermediate states of subfragment 1 and actosubfragment 1 ATPase: reevaluation of the mechanism. Biochemistry 17(17):3432–3442
9. Bagshaw CR, Trentham DR (1974) The characterization of myosin-product complexes and of product-release steps during the magnesium ion-dependent adenosine triphosphatase reaction. Biochem J 141(2): 331–349
10. Seidel J, Chopek M, Gergely J (1970) Effect of nucleotides and pyrophosphate on spin labels bound to S1 thiol groups of myosin. Biochemistry 9(16):3265–3272
11. Barnett VA, Thomas DD (1987) Resolution of conformational states of spin-labeled myosin during steady-state ATP hydrolysis. Biochemistry 26(1):314–323
12. Ostap EM, White HD, Thomas DD (1993) Transient detection of spin-labeled myosin subfragment 1 conformational states during

ATP hydrolysis. Biochemistry 32(26): 6712–6720

13. Agafonov RV, Nesmelov YE, Titus MA, Thomas DD (2008) Muscle and nonmuscle myosins probed by a spin label at equivalent sites in the force-generating domain. Proc Natl Acad Sci U S A 105(36):13397–13402
14. Nesmelov YE, Agafonov RV, Burr AR, Weber RT, Thomas DD (2008) Structure and dynamics of the force-generating domain of myosin probed by multifrequency electron paramagnetic resonance. Biophys J 95(1):247–256
15. Bauer CB, Holden HM, Thoden JB, Smith R, Rayment I (2000) X-ray structures of the apo and MgATP-bound states of Dictyostelium discoideum myosin motor domain. J Biol Chem 275(49):38494–38499
16. Gulick AM, Bauer CB, Thoden JB, Rayment I (1997) X-ray structures of the MgADP, MgATPgammaS, and MgAMPPNP complexes of the Dictyostelium discoideum myosin motor domain. Biochemistry 36(39): 11619–11628
17. Fisher AJ, Smith CA, Thoden JB, Smith R, Sutoh K, Holden HM, Rayment I (1995) X-ray structures of the myosin motor domain of Dictyostelium discoideum complexed with MgADP.BeFx and MgADP.AlF4. Biochemistry 34(28):8960–8972
18. Houdusse A, Sweeney HL (2001) Myosin motors: missing structures and hidden springs. Curr Opin Struct Biol 11(2):182–194
19. Reddy LG, Jones LR, Thomas DD (1999) Depolymerization of phospholamban in the presence of calcium pump: a fluorescence energy transfer study. Biochemistry 38(13): 3954–3962
20. MacLennan DH, Kranias EG (2003) Phospholamban: a crucial regulator of cardiac contractility. Nat Rev Mol Cell Biol 4(7):566–577
21. Zamoon J, Mascioni A, Thomas DD, Veglia G (2003) NMR solution structure and topological orientation of monomeric phospholamban in dodecylphosphocholine micelles. Biophys J 85(4):2589–2598
22. Mascioni A, Karim C, Zamoon J, Thomas DD, Veglia G (2002) Solid-state NMR and rigid body molecular dynamics to determine domain orientations of monomeric phospholamban. J Am Chem Soc 124(32):9392–9393
23. Traaseth NJ, Buffy JJ, Zamoon J, Veglia G (2006) Structural dynamics and topology of phospholamban in oriented lipid bilayers using multidimensional solid-state NMR. Biochemistry 45(46):13827–13834
24. Karim CB, Kirby TL, Zhang Z, Nesmelov Y, Thomas DD (2004) Phospholamban structural dynamics in lipid bilayers probed by a spin label rigidly coupled to the peptide backbone. Proc Natl Acad Sci U S A 101(40): 14437–14442
25. Kirby TL, Karim CB, Thomas DD (2004) Electron paramagnetic resonance reveals a large-scale conformational change in the cytoplasmic domain of phospholamban upon binding to the sarcoplasmic reticulum Ca-ATPase. Biochemistry 43(19):5842–5852
26. James P, Inui M, Tada M, Chiesi M, Carafoli E (1989) Nature and site of phospholamban regulation of the Ca2+ pump of sarcoplasmic reticulum. Nature 342(6245): 90–92
27. Toyofuku T, Kurzydlowski K, Tada M, MacLennan DH (1994) Amino acids Lys-Asp-Asp-Lys-Pro-Val402 in the Ca(2+)-ATPase of cardiac sarcoplasmic reticulum are critical for functional association with phospholamban. J Biol Chem 269(37):22929–22932
28. Toyoshima C, Asahi M, Sugita Y, Khanna R, Tsuda T, MacLennan DH (2003) Modeling of the inhibitory interaction of phospholamban with the Ca2+ ATPase. Proc Natl Acad Sci U S A 100(2):467–472
29. Hutter MC, Krebs J, Meiler J, Griesinger C, Carafoli E, Helms V (2002) A structural model of the complex formed by phospholamban and the calcium pump of sarcoplasmic reticulum obtained by molecular mechanics. ChemBioChem 3(12):1200–1208
30. Zamoon J, Nitu F, Karim C, Thomas DD, Veglia G (2005) Mapping the interaction surface of a membrane protein: unveiling the conformational switch of phospholamban in calcium pump regulation. Proc Natl Acad Sci U S A 102(13):4747–4752
31. Karim CB, Zhang Z, Howard EC, Torgersen KD, Thomas DD (2006) Phosphorylation-dependent conformational switch in spin-labeled phospholamban bound to SERCA. J Mol Biol 358(4):1032–1040
32. Metcalfe EE, Zamoon J, Thomas DD, Veglia G (2004) (1)H/(15)N heteronuclear NMR spectroscopy shows four dynamic domains for phospholamban reconstituted in dodecylphosphocholine micelles. Biophys J 87(2): 1205–1214
33. Karim CB, Zhang Z, Thomas DD (2007) Synthesis of TOAC-spin-labeled proteins and reconstitution in lipid membranes. Nat Protoc 2:42–49
34. Earle KA, Budil DE, Freed JH (1993) 250-GHz EPR of nitroxides in the slow-motional regime: models of rotational diffusion. J Phys Chem 97:13289–13297

35. Ernst RR, B G, Wokaun A (1987) Principles of nuclear magnetic resonance in one and two dimensions. Oxford University Press, Oxford
36. Stoll S, Schweiger A (2006) EasySpin, a comprehensive software package for spectral simulation and analysis in EPR. J Magn Reson 178 (1):42–55. doi:S1090-7807(05)00289-2 [pii] 10.1016/j.jmr.2005.08.013
37. Birmachu W, Thomas DD (1990) Rotational dynamics of the Ca-ATPase in sarcoplasmic reticulum studied by time-resolved phosphorescence anisotropy. Biochemistry 29(16):3904–3914
38. Lipari G, Szabo A (1982) Model-free approach to the interpretation of nuclear magnetic resonance relaxation in macromolecules. 1. Theory and range of validity. J Am Chem Soc 104: 4546–4559
39. Hanson P, Anderson DJ, Martinez G, Millhauser G, Formaggio F, Crisma M, Toniolo C, Vita C (1998) Electron spin resonance and structural analysis of water soluble, alanine-rich peptides incorporating TOAC. Mol Phys 95(5):957–966

Chapter 5

Probing Backbone Dynamics with Hydrogen/Deuterium Exchange Mass Spectrometry

Harsimran Singh and Laura S. Busenlehner

Abstract

Protein dynamics can be probed by the solution technique amide hydrogen/deuterium exchange. The exchange rate of hydrogen for deuterium along a peptide backbone is dependent on the extent of hydrogen bonding from secondary structure, accessibility by D_2O, and protein motions. Both global and local conformational changes that alter bonding or structure will lead to changes in the amount of deuterium incorporated. The deuterium can be localized via pepsin digestion of the protein and quantified by electrospray ionization mass spectrometry through the mass shifts of the resulting peptides. The technique is emerging as an essential tool to study protein structure in solution due to the exceptional capability of examining both dynamic and structural changes related to protein function.

Key words Hydrogen/deuterium exchange, Mass spectrometry, Tandem MS/MS sequencing, HDX-MS, Protein dynamics, Backbone amide

1 Introduction

1.1 Hydrogen/Deuterium Exchange Mass Spectrometry

The discovery in the late 1970s that protein structures are dynamic was a welcome addition to the static representation of protein structures that were appearing from recent advances in X-ray diffraction. In fact, it has been widely accepted that there is an intimate relationship between protein dynamics and cellular function. The biophysical toolkit with which to study protein dynamics has traditionally centered on NMR [1–4], fluorescence [5, 6], and IR [7, 8] spectroscopies along with crystallographic structure analysis. However, not all proteins are amenable to these techniques. Improvements in the field of electrospray ionization mass spectrometry have given rise to another experimental tool with which to characterize these essential molecular motions, hydrogen/deuterium exchange mass spectrometry (HDX-MS) [9].

Proteins contain a large number of labile hydrogens that can exchange with protons from surrounding water molecules. In particular, the amide hydrogen is an excellent indicator of protein

Dennis R. Livesay (ed.), *Protein Dynamics: Methods and Protocols*, Methods in Molecular Biology, vol. 1084, DOI 10.1007/978-1-62703-658-0_5,

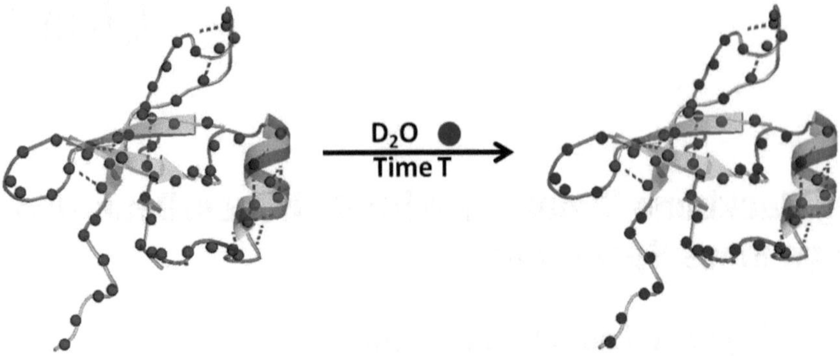

Fig. 1 Hydrogen/deuterium exchange. The amide hydrogens of a protein exchange with deuterons from the D_2O solvent depending on the extent of hydrogen bonding within the secondary and tertiary structural elements. Hydrogen-bonded amide protons exchange for deuterons much slower than solvent-accessible amide protons, and it is the dynamic motions that allow these bonds to break and reform

structure and dynamics since its solvent exchange rate is dependent on hydrogen bonding within secondary structural elements, as well as dynamic motions (Fig. 1) [10, 11]. Hydrogen-bonded backbone amides may exchange with solvent as a result of random, thermal fluctuations that break and reform these bonds in the native state [12]. This rate is most often monitored by incorporation of deuterium upon replacing the protic solvent with deuterium oxide (D_2O). In order for a hydrogen-bonded amide proton to exchange for a deuteron (HDX), the native bond must become exposed as part of the local or global folding/unfolding equilibrium

$$\mathrm{N-H_{closed}} \overset{k_o}{\leftrightarrow} \mathrm{N-H_{open}} \overset{k_i}{\leftrightarrow} \mathrm{N-D_{open}} \overset{k_c}{\leftrightarrow} \mathrm{N-D_{closed}}$$

where k_o, k_c, and k_i are the opening, closing, and intrinsic exchange rate constants, respectively [10, 13, 14]. The observed rate of deuterium exchange (k_{ex}) is demonstrated by Eq. 1:

$$k_{ex} = \frac{k_o k_i}{k_o + k_i + k_c}$$

In native proteins, the rate of folding is much faster than the intrinsic exchange rate of an amide ($k_c >> k_i$); thus, the observed rate of exchange can be simplified to

$$k_{ex} = K_{eq} k_i$$

where K_{eq} is the equilibrium constant between the open and the closed state (k_o/k_c) [9, 12].

A variety of factors including pH, temperature, and adjacent amino acid side chains affect backbone amide HDX kinetics [15, 16]. The intrinsic rate of exchange (k_i) has its minimum at pH 2.3, and it increases by an order of magnitude for every 1 pH unit [12, 17]. Most HDX-MS assays are performed at neutral pD where deuterium exchange is catalyzed by base abstraction (OD^-)

of the amide proton [13]. Since HDX-MS is a discontinuous assay, it is necessary to halt deuterium incorporation after the appropriate incubation time. Deuterium exchange can be slowed 5 orders of magnitude by simply reducing the pH to 2.3 and temperature from 25 to 0 °C [9, 10, 14]. This essentially traps the deuterium label at the amide nitrogen (half-life ~10 min), where it can be detected via an increase in mass of the protein. Further localization of the deuterium label can be obtained through digestion with an acid protease like pepsin, which is stable at pH 2.3, to create small peptides whose mass will increase by 1 amu for each deuteron incorporated. The extent of HDX is measured via electrospray mass spectrometry by the change in mass of each pepsin-generated peptide as a function of incubation time in D_2O (pH/D 7, 25 °C) [9]. The corresponding progress curves can be fit for HDX rates to reveal the localized dynamics.

The rate of HDX is indicative of the extent of hydrogen bonding and solvent accessibility of amide protons in the native protein [9]. HDX rates (k_{ex}) of greater than 4 min^{-1} are more reflective of solvent accessibility and millisecond timescale dynamics [18]. These rapid motions are too fast to be measured by manual HDX, but can be determined with rapid quench instrumentation, which is now at the forefront of the technique [19]. HDX rates slower than 10^{-4} min^{-1} usually constitute the stable or solvent-inaccessible protein core. HDX rates between these two extremes (4 $min^{-1} > k_{ex} > 10^{-3}$ min^{-1}) are the result of localized folding/unfolding of the protein backbone to expose amide hydrogens to deuterated solvent. This dynamic behavior is heavily influenced by hydrogen bond breakage and formation. Thus, one can gain valuable information on the areas with conformational flexibility, especially when dissecting structural dynamic changes for two different states of a protein.

2 Materials

All the solvents should be made in ultrapure water (sensitivity 18 MΩ at 25 °C). All reagents used should be HPLC grade or analytical grade. Unless otherwise indicated, all the materials are to be prepared and stored at room temperature.

2.1 Major Equipment and Supplies

1. Ion trap mass spectrometer with an electrospray ion source capable of collision-activated dissociation or electron capture/transfer dissociation (*see* **Note 1**).
2. HPLC with 1/16″ internal diameter tubing, but this may vary depending on the HPLC.
3. Microbore C18 reversed-phase column with internal diameter between 0.5 and 2.0 mm. A guard column or inline filter is recommended to protect the C18 column (*see* **Note 2**).

4. External or internal HPLC injector (*see* **Note 3**) with a 20 μL stainless steel sample loop.
5. Cooling system or ice bath.
6. Gas tight syringe.
7. Computer with instrument-specific MS data analysis software.
8. Software: tandem MS/MS analysis software, manual centroiding software or automated HDX analysis programs, and spreadsheet program (*see* **Note 4**).

2.2 Chemicals and Stock Solutions

Prepare all solutions using ultrapure water (sensitivity of 18 MΩ cm at 25 °C) and analytical grade reagents:

1. 99.9 atom % D deuterium oxide (D_2O). Aliquot and store in desiccator.
2. *Phosphate buffer*: 0.01 M potassium phosphate, pH 7. Dissolve 0.234 g KH_2PO_4 and 0.818 g K_2HPO_4 in 500 mL water.
3. *Pepsin* (2,000–4,000 units/mg). Dissolve 1 mg pepsin in 0.01 M potassium phosphate, pH 7, to a final concentration of 1–5 mg/mL (*see* **Note 5**). Prepare fresh and store on ice.
4. *Quench solution*: 0.1 M potassium phosphate, pH 2.3. Add about 80 mL of water to a small glass beaker. Add 1.34 g KH_2PO_4 and mix. Adjust pH to 2.3 with phosphoric acid. Make up to 100 mL with water (*see* **Note 6**).
5. *HPLC solvent A*: 98 % water, 2 % acetonitrile, and 0.4 % formic acid. Combine 976 mL HPLC grade water with 20 mL HPLC grade acetonitrile and 4 mL formic acid. Degas the solution.
6. *HPLC solvent B*: 98 % acetonitrile, 2 % water, and 0.4 % formic acid. Combine 976 mL HPLC grade acetonitrile with 20 mL HPLC grade water and 4 mL formic acid. Degas the solution (*see* **Note 7**).
7. *Protein stock*: >10 μM in suitable buffer (*see* **Note 8**).

3 Methods

The HDX-MS experiment can be divided into three parts: (Subheading 3.1) MS/MS peptide sequencing, (Subheading 3.2) discontinuous HDX time course, and (Subheadings 3.3–3.5) data analysis and interpretation.

3.1 MS/MS Sequencing to Build a Peptide Map

The amino acid sequences of each pepsin-generated peptide must be determined before HDX experiments can be undertaken [20]. This method is designed for ion trap mass spectrometers with electrospray ionization sources capable of data-dependent collision-activated dissociation (CAD) and/or electron transfer dissociation (ETD). The same principles can be applied to other mass spectrometers. This process is called tandem MS/MS or product-ion scanning. Because a cycle composed of a survey scan and a product-ion scan is fast (~10 ms) compared with the chromatographic elution time of a particular peptide (~30 s), and because many precursor ions are typically detected in a MS scan, one MS scan can be followed by several MS/MS fragmentation scans. Temperature control is not required for MS/MS sequencing, so the HPLC solvents, injector, column, and syringe can be at room temperature:

1. *Instrument setup*: Connect HPLC leads to solvents A and B and then purge each at 1–2 mL/min for 5 min to remove air (*see* **Note 9**). Connect the injector fitted with a 20 μL sample loop to the HPLC (*see* **Note 10**). Connect the C18 column with a guard column or inline filter to the injector. Connect the column to the electrospray ionization source.
2. *Column equilibration*: Wash the column with a 50:50 % mixture of solvents A and B for 15 min at 0.1 mL/min. Switch the HPLC to 100 % solvent A for 15 min to equilibrate the column (*see* **Note 11**).
3. *Mass spectrometry settings*: Set the ion trap scanning range to 100–2,000 *m/z* in positive-ion mode with a maximum accumulation time of ~100 msec and averages of at least 5 spectra. Make sure the nebulizer pressure, drying gas flow rate, and capillary temperature are set appropriately for the MS used (*see* **Note 12**). Select data-dependent MS/MS scanning (i.e., product-ion scanning or Auto-MS^n). The number of peptide molecular ions to fragment per scan is dependent on the MS molecular ion scan rate, but typically the 5–10 most intense precursor ions will be sequenced per molecular ion scan.
4. *HPLC gradient*: Set up an HPLC elution method (*see* Table 1A). Typically, this requires a linear gradient of 0–65 % of solvent B over 8 min at 0.1 mL/min (*see* **Note 13**).
5. Prepare fresh pepsin solution (1–5 mg/mL) in 0.01 M potassium phosphate, pH 7, as described in Subheading 2.2.3. Keep this stock on ice.
6. If required, dilute the protein with HPLC grade water so that the final concentration is 0.1–1 μM with a 50 μL volume (*see* **Note 14**).
7. Add 50 μL of quench solution preincubated at 0 °C (1:1 *v/v*) and mix with the pipette.

Table 1
Example HPLC gradients

Time (min)	% A	% B	Notes
(A) MS/MS:			0.1 mL/min
0	100	0	Equilibration (hold at 100 % A)
2.0	100	0	Divert 2 min to waste to flush salt
32.0	35	65	Linear gradient 0–65 % B
33.0	0	100	Ramp up to 100 % B
43.0	0	100	Wash with 100 % B
44.0	100	0	Switch to 100 % A
55.0	100	0	Re-equilibrate with 100 % A
(B) HDX-MS:			0.1 mL/min
0	100	0	Equilibration (hold at 100 % A)
2.0	100	0	Divert 2 min to waste to flush salt
12.0	35	65	Linear gradient 0–65 % B
13.0	0	100	Ramp up to 100 % B
23.0	0	100	Wash with 100 % B
24.0	100	0	Switch to 100 % A
35.0	100	0	Re-equilibrate with 100 % A

Typical gradients for (A) tandem MS/MS sequencing of the peptides generated from a pepsin digest (*see* Subheading 3.1) and (B) separation of peptides after HDX (*see* Subheading 3.2) are shown. Solvent A: 98 % H_2O, 2 % ACN, and 0.4 % formic acid. Solvent B: 2 % H_2O, 98 % ACN, and 0.4 % formic acid

8. Add pepsin at a final concentration of 0.1–1 mg/mL. Digest on ice for 5 min (*see* **Note 15**).
9. Using the syringe, load the entire 100 μL sample to the injector in "load" position. Start the HPLC gradient and immediately switch the injector to the "inject" position to flush the peptides onto the C18 column. Begin MS/MS data acquisition. Collect data from 3 to 35 min to account for HPLC dead time. To avoid introduction of salt into the ion source, it is recommended that the first several minutes of elution be diverted to waste manually or by use of a divert valve.
10. Upload the tandem MS/MS sequencing file to PEAKS Online or other analysis program to create a pepsin peptide map of the protein (*see* **Note 16**).

3.2 HDX Time Course Experiment

The general design of the HDX experiment is outlined in Fig. 2. For the HDX time course, the solvents, injector, syringe, protein stock, and quench buffer must be kept on ice or in a cooling chamber

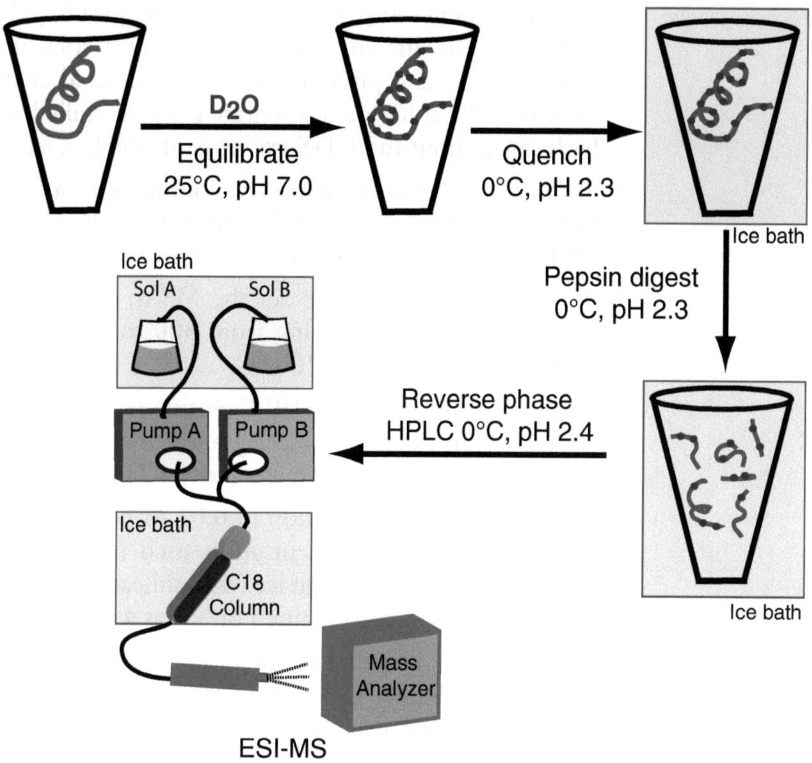

Fig. 2 General design of HDX-MS experiment. The protein is incubated with 10–20-fold excess of D_2O for a specific amount of time. The reaction is then stopped by reducing the temperature to 0 °C and pH to 2.3 with an acidic quenching solution. The reduction in pH and temperature traps the deuterium label at exchanged amides. The quenched reaction is digested on ice with pepsin, which yields peptides between 3 and 15 amino acids in length. The peptides are loaded onto a reversed-phase C18 HPLC column (0 °C) inline to an electrospray mass spectrometer. The peptides are eluted using an increasing concentration of acetonitrile, ionized in the gas phase, and introduced into the mass spectrometer where the mass-to-charge ratio of the peptides is recorded. The isotope distribution pattern for each peptide is used to calculate the number of deuterons incorporated per peptide per incubation time point

maintained at 0 °C to minimize loss (back-exchange) of the deuterium label. Store the HPLC solvents A and B at 4 °C. D_2O should be kept at 25 °C, and protein samples should be pre-warmed to 25 °C before the addition of D_2O to start the HDX time course. Maintain all HDX reactions at 25 °C. It is important that all HDX time point samples and the control reactions are all run on the same day, if possible. Be consistent with timing of **steps 7–9**. Two control experiments corresponding to no deuteration (m_0) and full deuteration (m_{100}) are also required for every full HDX experiment. It is imperative that each HDX sample is set up so that by the time the HDX sample has digested for 5 min, the previous HPLC separation is complete. It is recommended that the timing of an HDX experiment is planned using MSTools [21]:

1. The instrument setup is as described in Subheading 3.1, **item 1**, with the following modifications. Connect solvents A and B to the HPLC and place in an ice bath or cold chamber. The injector, C18 column, and syringe must be chilled using an ice bath throughout the HDX experiment. Replace ice as needed.
2. *Column equilibration*: Wash the column with a 50:50 % mixture of solvents A and B for 15 min at 0.1 mL/min. Switch the HPLC to 100 % solvent A for 15 min to equilibrate the column.
3. *Mass spectrometry settings*: Set the MS to average 5 spectra in positive ion mode, scanning from 300 to 2,000 *m/z*. Set the capillary temperature to 200 °C.
4. *HPLC settings*: Set up an HPLC elution method (*see* Table 1B). Typically, this requires a linear gradient of 0–65 % of solvent B over 8 min at 0.1 mL/min (*see* **Note 17**).
5. Prepare fresh pepsin solution in 0.01 M potassium phosphate, pH 7, at the same concentration used for tandem MS/MS (1–5 mg/mL) and keep on ice (*see* Subheading 2.2.3). Quench solution should be maintained on ice as well.
6. *Sample*: Dilute protein sample if necessary so that the stock for HDX is 10–30 μM. The protein can be maintained at 4 °C and then equilibrated at 25 °C before the addition of deuterium to start the HDX reaction.
7. *Individual HDX time course samples*: The incubation time with D_2O is varied for each run from 15 s to 6 h (*see* **Note 18**). Dilute the protein with D_2O at 25 °C so that the deuterium level is 90 % (e.g., 5 μL of 10–30 μM protein + 45 μL D_2O). Mix by gently pipetting and start a timer to count down from the desired incubation time. Place the reaction at 25 °C.
8. Following the incubation, quench the exchange reaction with 50 μL (1:1 *v/v*) of cold quench solution, mix, and place on ice. The quench step should be ~30 s. Proceed immediately to the pepsin digestion (*see* **Note 19**).
9. Add pepsin at the same concentration used for sequencing (~0.1–1 mg/mL), mix by pipetting, and digest on ice for exactly 5 min (*see* **Note 20**).
10. Using the cold syringe, load the entire 100 μL sample into the injector in "load" position. Start the HPLC gradient and immediately switch the injector to the "inject" position. Begin MS data acquisition (*see* **Note 21**).
11. *Control reactions*: Two control experiments corresponding to no deuteration (m_0) and full deuteration (m_{100}) are also run (*see* **Note 22**). The m_0 control is made as in **step 7**, but the protein is diluted in H_2O instead of D_2O. Follow **steps 8–10** as above. The m_{100} control is made as in **step 7**,

except that it is incubated at 50–65 °C for >5 h. Proceed with **steps 8–10** as above.

12. After the samples are complete, proceed to data analysis either by manually centroiding the isotopic envelopes of each peptide at every HDX time point (*see* Subheading 3.3) or by processing the data with an automated HDX analysis program (*see* Subheading 3.4).

3.3 Manual HDX Data Analysis

1. Create a spreadsheet with the following columns: (A) the mass of the peptide ion that was sequenced; (B) the charge state of the ion (+1, +2, +3, etc.); (C) its corresponding $[M + H]^+$ monoisotopic mass; (D) the amino acid residue numbers; (E) the number of exchangeable amide hydrogens, N_{ex}; (F) the retention time; (G–H) the experimentally determined monoisotopic mass for the m_0 and m_{100} controls; and (I–n) remaining columns for the monoisotopic mass determined for each HDX time point (*see* Table 2). Fill in columns A–E for the identified peptides from the MS/MS data (*see* **Note 23**).
2. Open the MS data files using the software provided with the mass spectrometer. Display the data as a total ion chromatogram (Fig. 3a). Search the TIC for the mass of each peptide, usually the m/z that was sequenced during tandem MS/MS (Fig. 3b). This generates an extracted ion chromatogram (i.e., retention profile) for each peptide. Enter the retention time in the spreadsheet (*see* **Note 24**).
3. Average the mass spectra within the ion chromatographic peak (Fig. 3c). Import that combined spectrum into an analysis program to determine the centroid for the isotope envelope using the area under the peaks (*see* **Note 25**).
4. Calculate the number of deuterons (D_t) incorporated at each HDX incubation time (t) for each peptide using the equation

$$D_t = N_{ex} \times \left[\frac{(m_t - m_0)}{(m_{100} - m_0)}\right]$$

where N_{ex} is the number of exchangeable amides (column E in the spreadsheet), m_0 (column G), m_{100} (column H), and m_t (columns I–n). The number of deuterons incorporated can also be converted to the percent exchanged: $\% \ D_t = \frac{D_t}{N_{ex}} \times 100$
5. For each peptide, plot the number or percentage of deuterium incorporated against the log (t) in min (*see* **Note 26**) and fit to the first-order rate term [9] using

$$D_i = N_i - \sum_{i=1}^{N} \exp(-k_i t)$$

Table 2
Example spreadsheet for manual HDX-MS data analysis

m/*z* (H_2O)	Charge	$[M + H]^+$	Residues	Amide H (N_{ex})	Retention time (min)	0 % centroid	100 % centroid	15 s D_2O	30 s D_2O	etc.
843.3	1	843.3	LGLNNIAE	7	17.7	843.86	846.89	845.7	845.86	
1,005	2	2,009	LAITHVSNVL GTENPLAEM	17	19.6	2010.14	2017.72	2012.12	2012.56	
666.3	2	1331.7	SREVNGLPLAYL	10	19.8	1332.4	1336.92	1333.66	1333.9	
439.6	3	1316.6	AVMHHPVDVQAL	10	16.8	1317.46	1325.11	1318.45	1318.87	

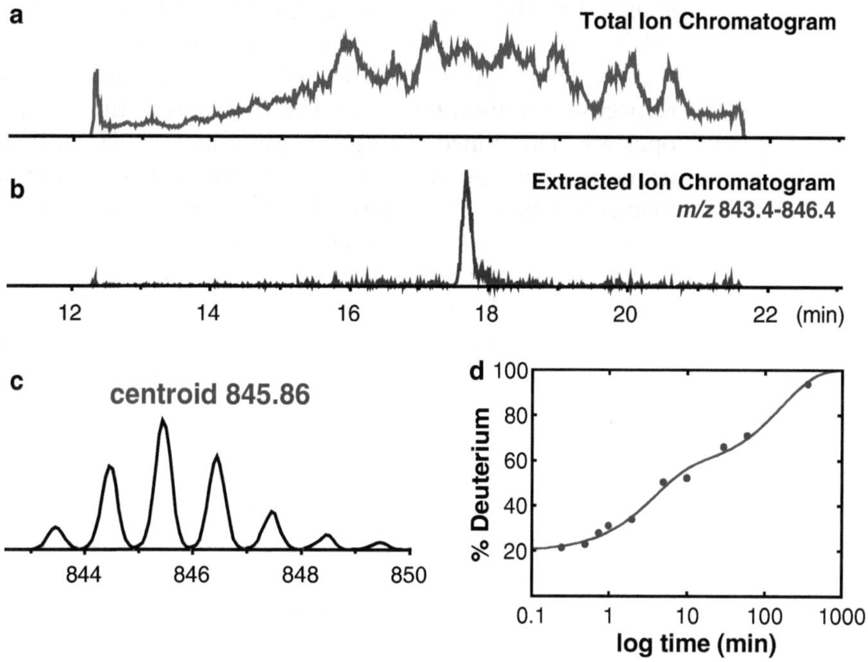

Fig. 3 Manual HDX-MS data analysis. The total ion chromatogram (TIC) is searched for the *m/z* of each peptide ion sequenced during the MS/MS and positively identified. **(a)** TIC for a sample HPLC separation of HDX peptides (Singh, Busenlehner, unpublished data). **(b)** The TIC is searched for mass range 843.4–844.4 *m/z*, which corresponds to the mass of a sample peptide with additional mass added to account for deuterium incorporation, to generate the extracted Ion chromatogram (EIC) for the peptide. **(c)** The MS spectra that are contained with the EIC peak are averaged to yield the isotope envelope for the peptide. This isotope envelope is then exported to get the centroid value and entered into Table 2. **(d)** The percentage of deuterons incorporated is plotted as a function of incubation time and fit to a double-exponential equation

where N_i is the number of amides exchanging at a rate k_i for the isotopic exchange time t (Fig. 3d). Usually the deuterium exchange progress curves can be satisfactorily fit with 1–2 exponential terms. The kinetic curves generated for each identified peptide can be analyzed for regions with slow, medium, and fast exchange (*see* Subheading 3.5).

3.4 Automated HDX Data Analysis

A number of programs are available for analysis of HDX-MS data. Some of the common programs include HDExaminer, HD Analyzer [22], HD Express [23], and HD Desktop [24] (*see* **Note 4**). Each program is unique, but there are basic requirements required by most: (a) The MS data files from the software provided by the instrument may not always be in the correct format for HDX programs. They should be converted to a format accepted by the analysis software, usually *.mzxml* format. (b) The HDX raw data files are uploaded to the software, along with the protein sequence, the list of peptide fragments, and their respective retention times. (c) The centroids of the isotope envelopes are determined for each peptide at each HDX time point using program-specific algorithms. Some programs allow users to manually optimize this step. (d) After analysis, the program will display the results in plots (deuterium vs. time) or in a deuteration level heat map using the provided protein sequence. More rigorous fitting of the HDX kinetic profiles to obtain rates requires the data to be exported into graphing programs. An example of the automated HDX analysis is demonstrated in Fig. 4 using HDExaminer.

3.5 Data Interpretation

The fits to the HDX kinetic profiles provide several pieces of information about the conformational dynamics of a protein and how that might change by altering some condition, such as substrate binding, metal binding, and posttranslational modification. The three basic steps to data analysis include (1) determining solvent accessibility related to hydrogen bonding, (2) identifying peptides whose dynamics are more reflective of local unfolding/unfolding, and (3) identification of the hydrophobic core:

1. Create a solvent accessibility map of the protein to observe areas with fast dynamic motions by plotting the percent deuterium (% D) incorporated at 10 or 15 s for each peptide region. Usually it is best to find a set of contiguous peptides with minimal overlap. The data can be plotted as a bar chart or as a heat map of either the primary sequence (*see* Fig. 5) or the three-dimensional structure, if available (*see* **Note 27**).
2. Identify peptides whose progress curves yield amide exchange rates between (4 $\text{min}^{-1} > k_{ex} > 10^{-3}$ min^{-1}). Amide exchange in this time regime is due to slower dynamic behavior of proteins, the local folding/unfolding processes. Mapping these peptides to the three-dimensional structure of the protein (if available) may provide additional insight (*see* **Note 28**).
3. Peptides that still have amide hydrogens not exchanged for deuterium after 6 or more hours can safely be considered to reside in the protein core [10, 13]. These amides do not exchange because they are not accessible to D_2O and/or they are important for the thermodynamic stability of the protein. For proteins without three-dimensional structures, this is useful information as these regions should cluster and provide information on the protein fold.

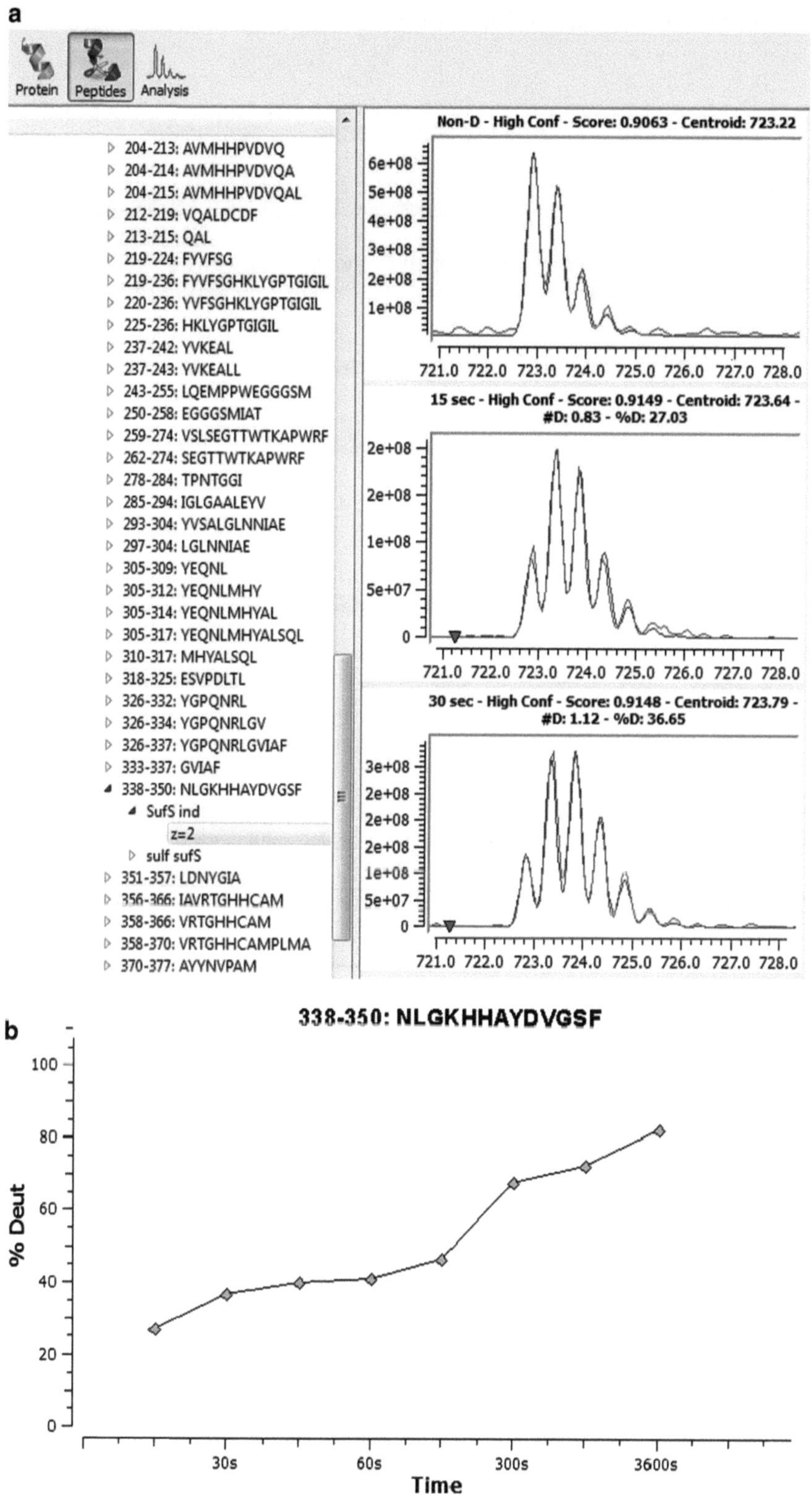

Fig. 4 Example of automated data analysis by HDExaminer. **(a)** The HDX isotope envelopes for a sample peptide are shown (Singh and Busenlehner, unpublished data). The calculated fits for the envelopes used to obtain the centroid are overlayed. The software gives the fitting score, the centroid value, and the percent deuterium incorporation for each time point. Note the increasing centroid value with longer incubations in D_2O. **(b)** HDExaminer created plot of percent deuterium incorporated vs. time in seconds of all HDX time points for this peptide

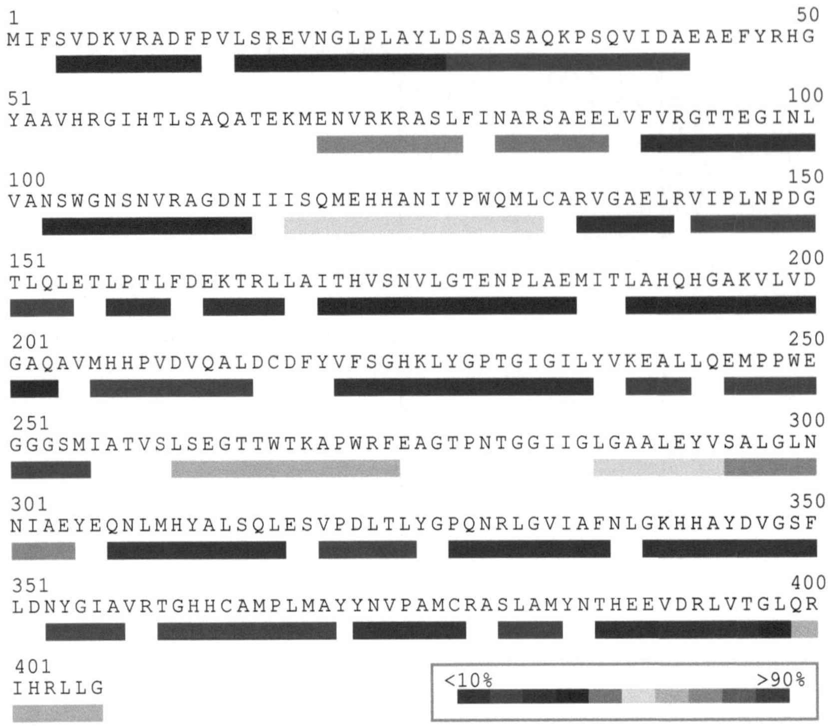

Fig. 5 Heat map of HDX showing dynamic regions. Shown is a sample heat map created using a 15 s incubation in D_2O. The length of each bar denotes the residues included in each peptide. The color of the bar represents the percentage of deuterium incorporated according to the legend. The higher the percent incorporation, the more dynamic the backbone amide is. This figure was created with HDExaminer

4 Notes

1. HDX-MS can also be performed using MALDI ionization, but electrospray is preferred for its interface with an HPLC. The type of mass spectrometer (ion trap, quadrupole, ToF) is not as important as the ionization source.
2. Microbore (4 μm particle size) C18 reversed-phase columns with internal diameter of <2.0 mm are preferred. The standard column is 1 × 50 mm, equipped with a filter or guard column (usually C8) to trap undigested protein and, thus, extend the lifetime of the C18 column.
3. External injectors are preferred so that they can be submerged in an ice bath. It is recommended that the bottom portion be wrapped in parafilm to protect internal components. Internal injectors can be used if in a cooling chamber.
4. The MS/MS analysis software used for this protocol is PEAKS Online from Bioinformatics Solutions Inc. (Waterloo, ON, Canada) (http://www.bioinfor.com/peaks/products/public-server.html). Other online analysis programs include MASCOT

MS/MS Ions Search (http://www.matrixscience.com/search_form_select.html) from Matrix Science and SEQUEST (http://proteomicsresource.washington.edu/sequest.php) hosted at multiple sites. Manual analysis of MS/MS data can be done using MS-Product (http://prospector.ucsf.edu/prospector/cgi-bin/msform.cgi?form=msproduct). Choices of centroiding software for manual analysis of HDX-MS data include MagTran available from Amgen Inc. (http://magtran.software.informer.com) and HX-Express (http://www.hxms.com/HXExpress) developed by Weis et al. [23]. Automated HDX-MS analysis includes HDExaminer (http://www.massspec.com/HDExaminer.html) by Sierra Analytics and HDX Workbench (http://deuterator.florida.scripps.edu/hdx_workbench/Home.html), among others. Graphing programs must be able to fit exponential curves.

5. For some proteins, pepsin may not give optimal digestion. Instead try other acid proteases such as protease XIII and protease XVIII. These enzymes are not as efficient, so the enzyme-to-protein ratio must be determined experimentally [25].
6. Formic acid can also be used to quench. A stock of 0.15 % formic acid is required.
7. Do not use tubes or filters that leach polymers on extended contact with acetonitrile. These contaminating ions will be present in the MS spectra.
8. The protein stock should be in the concentration range of 10–30 μM and as pure as possible. The most suitable buffers are ammonium acetate, HEPES, and ammonium bicarbonate. The presence of additives such as detergents and glycerol should be avoided. If detergents are required, good choices include nonionic or zwitterionic detergents such as *n*-dodecyl-β-D-maltoside (DDM) and CHAPS, respectively [26]. The concentration of monovalent salt in the sample should be as low as possible, but not so low as to lose protein stability. Generally <50 mM NaCl or KCl is preferred. The sample should also be stable at 25 °C for the length of time required for HDX-MS.
9. Additional purging may be required if the solvents are not degassed prior to use.
10. For a typical 6-port injector, the tubing from the HPLC is connected at position 2, the loop at positions 1 and 4, and the line to the column at position 3. Make all tubing past the injector as short as reasonably possible.
11. Usually flow rates for microbore C18 columns are 0.1 mL/min. Do not use a flow rate beyond the maximum flow rate specified for the C18 column. Much lower flow rates are required for capillary LCs and capillary columns.

12. Please refer to the operation manual for the mass spectrometer used. Typical values for electrospray ion traps are a nebulizer pressure of 20–40 psi, drying gas flow rate of 5–8 L/min, and a capillary temperature of 350 °C for 0.1 mL/min LC flow rate.
13. The elution gradient profile can be varied both in terms of gradient percentage and elution time to get good spatial resolution of peptides. Generally a 20–30 min linear gradient of solvent B is sufficient for sequencing proteins under 50 kDa. Particularly hydrophobic proteins may require an elution gradient to 100 % B.
14. Instead of water, buffer can also be used for dilution if the protein is unstable.
15. Pepsin digestion conditions must be optimized [27]. The times and amounts listed here are typical. Increasing the ratio of pepsin to protein may be required if the total ion intensity is low ($<10^5$) or if the spatial coverage of peptides is low. The length of digestion should be kept at 5 min or less.
16. See the product manual for full instructions. Create a new account at http://www.bioinfor.com:9999. Save each MS/MS run in *.pkl*, *.dta*, *.mgf*, or *.mzxml* formats and upload to PEAKS. Set the enzyme to "none" since pepsin cleaves unpredictably. Choose up to 100 missed cleavage sites. Define any fixed or possible modifications (PTMs) the protein sample may have. Input protein sequence in FASTA format or choose to search Swiss-Prot or other protein database. In PEAKS, a matched peptide will have a confidence score of 10lgD of >15. The same settings are also applicable to other programs.
17. Generally an 8–10 min linear gradient of solvent B is sufficient to elute proteins under 50 kDa. Particularly hydrophobic proteins may require an elution gradient to 100 % solvent B. It is imperative that the gradient be as short as possible to minimize deuterium loss during HPLC analysis under aqueous conditions.
18. Generally the D_2O incubation time points for the experiment are 15 s, 30 s, 45 s, 1 min, 2 min, 5 min, 10 min, 15 min, 30 min, 1 h, 2 h, and 6 h. This provides enough points for curve fitting to get groups of HDX rates within a peptide.
19. During the quench on ice, HDX is only slowed down; therefore, one must immediately proceed with the pepsin digestion and injection. The timing of every sample must be the same for consistent HDX. If a quenching step or digest proceeds longer than the others, that sample must be remade.
20. The change in time of digestion may cause a change in the digestion pattern and inconsistent deuterium incorporation.

21. Acquire MS data from 3 to 15 min to account for the HPLC dead time. To avoid introduction of salt to the ion source, it is recommended that the first several minutes of elution be diverted to waste manually or by use of a divert valve.
22. The control reactions, non-deuterated and fully-deuterated, should be run with each time course experiment. These controls account for the day-to-day differences during HDX analysis and correct for the intrinsic isotope abundance (m_0) and back-exchange of deuterium for protium during HPLC (m_{100}) [9].
23. The ion that was sequenced could be the monoisotopic ion $[M + H]^+$ or the higher charge state ions ($[M + 2H]^{2+}$, $[M + 3H]^{3+}$). See reference [28] for a description of multiple charging in electrospray MS. N_{ex} = total number of amide hydrogens minus the number of prolines and minus the N-terminal amine.
24. If there are two or more ions of the same mass, use the retention time and the charge state of that ion to aid in the correct assignment. It is best accomplished using the non-deuterated (m_0) sample described in Subheading 3.2.11. Also, as deuterium is incorporated into the peptides as a function of time, their *m/z* values will increase. Thus, to get the extracted ion chromatogram for peptides at longer incubation times, one must increase the mass range used to search the total ion chromatogram.
25. Click and drag the cursor from the beginning to the end of the extracted ion chromatographic peak to average the spectra. Depending on the centroid analysis program used, either copy the averaged MS spectrum into MagTran or save as a mass list and import. Manually determine the centroid for the given isotope envelope for the peptide by selecting the left edge of the isotopic peak to the right edge of the highest observable isotope peak. Convert to the monoisotopic mass if the ion has a charge state greater than +1. Enter the centroid mass into the spreadsheet. Repeat for all time points and peptides. HX-Express is semiautomated and can accept spectra in (x,y) format to calculate centroids if given the retention times. See the program manual for more detailed instructions.
26. The data can also be plotted against t (min) if the span of HDX time points is small. If the deuterium incubations span several orders of magnitude (e.g., sec–hr), then log (t) is preferred.
27. Peptides of low solvent accessibility will not show significant deuterium uptake (0–10 %). Solvent accessibilities of 10–50 % usually indicate that these areas are more likely to have hydrogen bonding, but also experience conformational dynamics. Areas of high solvent accessibility (>50 %) are usually the least hydrogen bonded and more conformationally flexible.

Comparisons of D_2O accessibility between two protein states are often the most informative.

28. Changes in the rates of deuterium exchange of a given peptide when comparing two or more states of a protein directly report on changes in conformational dynamics. This is the most powerful application of HDX-MS [9, 29–32].

Acknowledgement

This work was supported by NSF grant #0845273 to L.S.B.

References

1. Boehr DD, Dyson HJ, Wright PE (2006) An NMR perspective on enzyme dynamics. Chem Rev 106(8):3055–3079
2. Volkman BF, Lipson D, Wemmer DE et al (2001) Two-state allosteric behavior in a single-domain signaling protein. Science 291(5512):2429–2433
3. Tzeng SR, Kalodimos CG (2011) Protein dynamics and allostery: an NMR view. Curr Opin Struct Biol 21(1):62–67
4. Rozovsky S, Jogl G, Tong L et al (2001) Solution-state NMR investigations of triosephosphate isomerase active site loop motion: ligand release in relation to active site loop dynamics. J Mol Biol 310(1):271–280
5. Yang H, Luo G, Karnchanaphanurach P et al (2003) Protein conformational dynamics probed by single-molecule electron transfer. Science 302(5643):262–266
6. Weiss S (2000) Measuring conformational dynamics of biomolecules by single molecule fluorescence spectroscopy. Nat Struct Biol 7(9):724–729
7. Ganim Z, Chung HS, Smith AW et al (2008) Amide I two-dimensional infrared spectroscopy of proteins. Acc Chem Res 41(3):432–441
8. Arrondo JL, Goni FM (1999) Structure and dynamics of membrane proteins as studied by infrared spectroscopy. Prog Biophys Mol Biol 72(4):367–405
9. Busenlehner LS, Armstrong RN (2005) Insights into enzyme structure and dynamics elucidated by amide H/D exchange mass spectrometry. Arch Biochem Biophys 433(1):34–46
10. Zhang Z, Smith DL (1993) Determination of amide hydrogen exchange by mass spectrometry: a new tool for protein structure elucidation. Protein Sci 2(4):522–531
11. Englander SW, Mayne L, Rumbley JN (2002) Submolecular cooperativity produces multistate protein unfolding and refolding. Biophys Chem 101–102:57–65
12. Konermann L, Pan J, Liu YH (2011) Hydrogen exchange mass spectrometry for studying protein structure and dynamics. Chem Soc Rev 40(3):1224–1234
13. Maity H, Lim WK, Rumbley JN et al (2003) Protein hydrogen exchange mechanism: local fluctuations. Protein Sci 12(1):153–160
14. Englander SW, Kallenbach NR (1983) Hydrogen exchange and structural dynamics of proteins and nucleic acids. Q Rev Biophys 16(4):521–655
15. Molday RS, Englander SW, Kallen RG (1972) Primary structure effects on peptide group hydrogen exchange. Biochemistry 11(2):150–158
16. Bai Y, Milne JS, Mayne L et al (1993) Primary structure effects on peptide group hydrogen exchange. Proteins 17(1):75–86
17. Smith DL, Deng Y, Zhang Z (1997) Probing the non-covalent structure of proteins by amide hydrogen exchange and mass spectrometry. J Mass Spectrom 32(2):135–146
18. Suchanova B, Tuma R (2008) Folding and assembly of large macromolecular complexes monitored by hydrogen-deuterium exchange and mass spectrometry. Microb Cell Fact 7:12
19. Morgan CR, Engen JR (2009) Investigating solution-phase protein structure and dynamics by hydrogen exchange mass spectrometry. Curr Protoc Protein Sci 16:11–17, Chapter 17, Unit 17
20. Cottrell JS (2011) Protein identification using MS/MS data. J Proteomics 74(10):1842–1851

21. Kavan D, Man P (2011) MSTools: Web based application for visualization and presentation of HXMS data. Int J Mass Spectrom 302:53–58
22. Liu S, Liu L, Uzuner U et al (2011) HDX-analyzer: a novel package for statistical analysis of protein structure dynamics. BMC Bioinformatics 12(Suppl 1):S43
23. Weis DD, Engen JR, Kass IJ (2006) Semi-automated data processing of hydrogen exchange mass spectra using HX-Express. J Am Soc Mass Spectrom 17(12):1700–1703
24. Pascal BD, Chalmers MJ, Busby SA et al (2007) The Deuterator: software for the determination of backbone amide deuterium levels from H/D exchange MS data. BMC Bioinformatics 8:156
25. Cravello L, Lascoux D, Forest E (2003) Use of different proteases working in acidic conditions to improve sequence coverage and resolution in hydrogen/deuterium exchange of large proteins. Rapid Commun Mass Spectrom 17 (21):2387–2393
26. Loo RR, Dales N, Andrews PC (1994) Surfactant effects on protein structure examined by electrospray ionization mass spectrometry. Protein Sci 3(11):1975–1983
27. Wang L, Pan H, Smith DL (2002) Hydrogen exchange-mass spectrometry: optimization of digestion conditions. Mol Cell Proteomics 1 (2):132–138
28. Strupat K (2005) Molecular weight determination of peptides and proteins by ESI and MALDI. Methods Enzymol 405:1–36
29. Asuru AP, Busenlehner LS (2011) Analysis of human ferrochelatase iron binding via amide hydrogen/deuterium exchange mass spectrometry. Intl J Mass Spectrom 302:76–84
30. Asuru AP, An M, Busenlehner LS (2012) Dissection of porphyrin-induced conformational dynamics in the heme biosynthesis enzyme ferrochelatase. Biochemistry 51(36): 7116–7127
31. Busenlehner LS, Salomonsson L, Brzezinski P et al (2006) Mapping protein dynamics in catalytic intermediates of the redox-driven proton pump cytochrome c oxidase. Proc Natl Acad Sci U S A 103(42):15398–15403
32. Busenlehner LS, Codreanu SG, Holm PJ et al (2004) Stress sensor triggers conformational response of the integral membrane protein microsomal glutathione transferase 1. Biochemistry 43(35):11145–11152

Chapter 6

Carbon–Deuterium Bonds as Non-perturbative Infrared Probes of Protein Dynamics, Electrostatics, Heterogeneity, and Folding

Jörg Zimmermann and Floyd E. Romesberg

Abstract

Vibrational spectroscopy is uniquely able to characterize protein dynamics and microenvironmental heterogeneity because it possesses an inherently high temporal resolution and employs probes of ultimately high structural resolution—the bonds themselves. The use of carbon–deuterium (C–D) bonds as vibrational labels circumvents the spectral congestion that otherwise precludes the use of vibrational spectroscopy to proteins and makes the observation of single vibrations within a protein possible while being wholly non-perturbative. Thus, C–D probes can be used to site-specifically characterize conformational heterogeneity and thermodynamic stability. C–D probes are also uniquely useful in characterizing the electrostatic microenvironment experienced by a specific residue side chain or backbone due to its effect on the C–D absorption frequency. In this chapter we describe the experimental procedures required to use C–D bonds and FT IR spectroscopy to characterize protein dynamics, structural and electrostatic heterogeneity, ligand binding, and folding.

Key words Carbon–deuterium, FT IR, Vibrational spectroscopy, Protein dynamics, Electrostatics, Protein folding

1 Introduction

There is a growing appreciation of the central role played by protein dynamics and microenvironmental heterogeneity in biological function [1–4], although their exact contributions have remained ambiguous and controversial, mostly due to the difficulties associated with their direct experimental characterization. In comparison to NMR spectroscopy, which has high structural but only moderate temporal resolution, and UV/vis spectroscopy, which must rely on large and perturbative probes, vibrational spectroscopy possesses an inherently high temporal resolution and employs probes of ultimately high structural resolution—the bonds themselves. However, the absorption frequencies of most single-bond vibrations are not unique within a protein, with a few notable

Dennis R. Livesay (ed.), *Protein Dynamics: Methods and Protocols*, Methods in Molecular Biology, vol. 1084,
DOI 10.1007/978-1-62703-658-0_6,

exceptions such as the S–H stretching absorptions of sulfhydryl groups [5] or in some cases the O–H stretch of hydrogen-bonded water [6], and the resulting spectral congestion precludes a straightforward application of IR or Raman spectroscopy to the site-specific characterization of proteins. To circumvent the spectral complexity problem, nonnative, environmentally sensitive probes can be used that absorb IR light between 1,800 and 2,600 cm^{-1}, a region free of obscuring protein absorptions. Probes absorbing light in this spectral region may be characterized directly by subtracting only a small and typically smooth background signal. Towards this goal, cyano (CN) groups, which absorb light around 2,230 cm^{-1}, were first proposed as probes of protein electrostatics [7]. The first IR probes actually incorporated into a protein were carbon–deuterium (C–D) bonds, which were used to replace non-exchangeable C–H bonds in cytochrome *c* [8, 9]. While C–D substitution is reminiscent of other isotopic substitutions (e.g., ^{13}C, ^{15}N, or ^{18}O) [10], it produces a much larger spectral shift, resulting in C–D stretching vibrations that absorb around 2,200 cm^{-1}, as opposed to ~3,100 cm^{-1} for the corresponding C–H stretches. More recently, thiocyano (SCN) and azide (N_3) groups, whose stretching vibrations absorb around 2,160 cm^{-1} and 2,130 cm^{-1}, respectively, have also been suggested as IR probes [11, 12]. Recent efforts have explored the incorporation of C–D bonds into additional proteins [13–15]; the incorporation of CN groups into amino acids [16], model peptides [17–19], or proteins [20, 21]; and the incorporation of N_3 groups into amino acids and model peptides [12] or proteins [22–24].

Characterizing single absorptions within a protein is by definition challenging, and this is made easier by the relatively larger transition dipole moment of CN, SCN, and N_3 stretching absorptions (~0.1–1 D), compared to C–D stretching absorptions (~0.01–0.1 D). However, CN, SCN, and N_3 moieties are exogenous probes, and their addition to an amino acid raises the potential for perturbation, which is exacerbated by the size of SCN or N_3 groups, or the relatively strong dipole moment and metal- or H-bonding propensity of CN and SCN groups [19, 21]. Moreover, when CN or SCN groups bind metals or engage in H bonds, these nonnative interactions may significantly alter the IR absorption spectra [25] and thereby may obscure the interactions intended to be studied. In contrast, C–D bonds will not engage in nonnative interactions, and at least within the Born–Oppenheimer approximation, there is no risk of perturbation; thus, their spectra are more straightforward to interpret. Moreover, C–D labels may be used to visualize any non-exchangeable C–H bonds at any position in a protein, side chain or backbone, while potential labeling sites for CN, SCN, or N_3 probes are much more limited. Thus, C–D bonds are the most versatile and least perturbative among the vibrational probes that absorb in the 1,800–2,600 cm^{-1} spectral window.

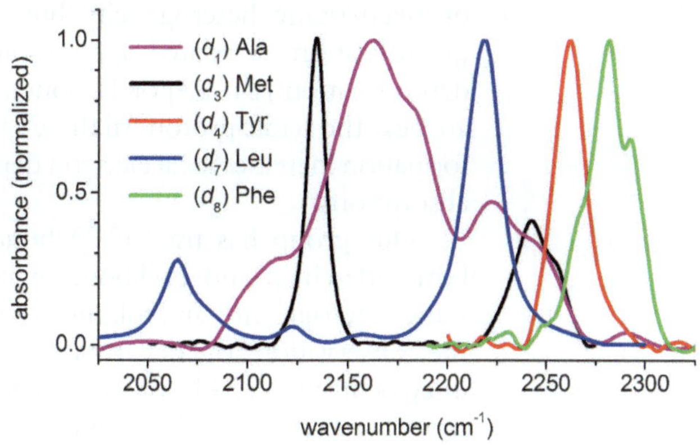

Fig. 1 FT IR difference spectra of some commonly used deuterated amino acids

Figure 1 shows the absorption spectra of some commonly used C–D-labeled residues. In general, residues labeled with CD_3 groups (e.g., Ala, Met, Leu, Val) or CD_2 groups (e.g., Lys) show a lower-energy (2,050–2,150 cm^{-1}) symmetric and a higher-energy (2,200–2,300 cm^{-1}) asymmetric C–D stretch absorption band, originating from a single symmetric stretch and two nearly degenerate asymmetric stretches per CD_3 group, or a single symmetric and a single antisymmetric stretch per CD_2 group, respectively [26, 27]. The asymmetric band is typically more intense ($\varepsilon \sim$ 10–80 M^{-1} cm^{-1}) than the symmetric band, with the notable exception of the CD_3 group of methionine, where the symmetric stretch tends to be more intense [28]. Residues with aromatic side chains bearing aryl C–D labels (e.g., Tyr, His) tend to show broader and thus weaker absorptions. Similarly, absorptions of C_α–D-labeled residues tend to be broadened, likely due to conformational heterogeneity of the peptide backbone; however, this makes them excellent probes of local secondary structure [13, 29, 30].

In general, C–D probes are useful for detecting conformational heterogeneity, which may affect the number of observed absorptions and their line width [13, 26, 28, 31]. The ability of C–D probes to characterize the number of conformations experienced at a given site follows from their inherent time resolution which is sufficient to resolve states that interconvert even on the fastest timescales. In addition to characterizing the conformational heterogeneity of a protein's native state, this sensitivity to conformation also makes C–D bonds excellent probes of protein folding, as described later in this chapter. Moreover, the sensitivity of the C–D absorption frequency to the local electric field makes them useful for the characterization of electrostatic microenvironments within a protein. In general, a more polar local environment results in a greater blue shift of the absorption [14]. When combined with calculations that establish the direction of local electrostatic fields, the observable frequencies should allow not only for the detection of electrostatic heterogeneity but also for its perturbation-free quantification. A related application of C–D bonds is to detect deprotonation [31–33] or H bonding [15] when the C–D bonds are near the acidic proton. In these cases, deprotonation or H-bond formation increases local electron density and results in blue-shifted absorptions.

Our group has used C–D bonds extensively to characterize both side-chain and backbone dynamics, structural and electrostatic heterogeneity, and folding of several proteins. Results include the observation of denaturant-specific unfolding mechanisms in cytochrome *c* [34], substrate-binding-induced changes in dihydrofolate reductase [15], and structural heterogeneity in a Src homology 3 domain [13]. These and other observations were based on observed changes in the frequency, line width, or number of unique absorptions for C–D bonds incorporated in each protein at different positions and would not have been apparent using techniques with a lower temporal and structural resolution.

In this chapter we describe the experimental procedures required to use C–D bonds and FT IR spectroscopy to characterize protein dynamics, structural and electrostatic heterogeneity, ligand binding, and folding. We briefly discuss methods to site-selectively label proteins with C–D bonds, and references are provided to publications with more detailed descriptions. Sample preparation, data acquisition, and data analysis for steady-state difference FT IR spectroscopy are discussed in more detail, as are the challenges and potential pitfalls of the method and their potential solutions. While the focus of this chapter is the use and interpretation of C–D bonds, much of the discussion is equally applicable to CN, SCN, and N_3 probes as well.

2 Materials

1. FT IR spectrometer. A research-grade FT IR spectrometer equipped with a liquid-nitrogen-cooled MCT detector (e.g., Bruker Equinox or Vertex series). The instrument (both optics and sample chambers) should be purged constantly with dry and CO_2-depleted nitrogen (e.g., by running nitrogen gas through an Ascarite filter (Thomas Sci.) for CO_2 removal and then a Drierite filter (Hammond, Inc.) for water vapor removal).
2. Temperature-controlled demountable liquid FT IR sample cell (e.g., TFC-M13, Harrick Scientific Products, Inc.).
3. FT IR cell and Teflon spacers (e.g., Pike Technologies, Inc.) and CaF_2 windows (½ or 1 in. diameter, 2 mm thickness, e.g., ISP Optics Corp.).

4. Deuterated amino acids, either Boc or Fmoc protected for solid-phase peptide synthesis, or bearing a photolabile protecting group for nonsense suppression (e.g., C/D/N Isotopes, Inc.).

3 Methods

3.1 Protein Labeling

To date, two techniques have been used to site-specifically introduce C–D bonds or other IR probes into proteins, solid-phase peptide synthesis and recombinant expression using the nonsense suppression methodology. Both have advantages and limitations, and the approach employed depends on the nature of the protein to be deuterated. A detailed description of these techniques is beyond the scope of this text, but each is briefly discussed below with references to the literature where more details may be found.

3.1.1 Solid-Phase Peptide Synthesis

Solid-phase peptide synthesis utilizing Boc or Fmoc chemistry may be used to access proteins of small to moderate size (up to ~60 amino acids) that are site-selectively deuterated at any backbone or side-chain position. Generally, α-helix-rich proteins are more straightforward to synthesize than proteins rich in β-sheet secondary structure, which are prone to aggregate and/or precipitate. Larger proteins can be accessed from synthesized fragments using a variety of ligation techniques, including native chemical ligation [35] and expressed protein ligation [36, 37].

3.1.2 Nonsense Suppression

This route to site-selective deuteration relies on amber suppressor transfer RNA (e.g., $tRNA^{CUA}$–tRNA synthetase pairs), where the $tRNA^{CUA}$ is selectively charged with an unnatural amino acid, which it then incorporates into a protein during ribosomal synthesis in response to an amber codon (TAG) introduced at the desired position in the gene of interest by site-directed mutagenesis. When the "unnatural" amino acid is a deuterated natural amino acid made unnatural by the presence of a photolabile protecting group, deprotection results in the site-selectively deuterated but otherwise fully natural protein [15]. $tRNA^{CUA}$–tRNA synthetase pairs have been reported for photoprotected Tyr, Lys, Ser, and Cys, thus permitting site-specific labeling at positions containing one of these amino acids [38]. For example, the studies of DHFR reported by our group were based on the incorporation of photoprotected, deuterated Tyr ((d_4) *o*-nitrobenzyl O-tyrosine). After purification, photodeprotection ($\lambda = 360$ nm) afforded ~25 mg/L of the site-selectively deuterated protein, as confirmed by SDS-PAGE and ESI mass spectrometry [15]. This method, which allows site-selective deuteration of essentially any protein that may be recombinantly expressed and purified, has been reviewed elsewhere [38, 39].

3.2 Sample Preparation

1. Collect sample holder, rubber O-rings, CaF_2 windows, Teflon spacer (*see* **Note 1**), and micropipetter equipped with 20 μL pipette tips (Fig. 2a).
2. If the protein has been lyophilized, add 10 μL of the chosen buffer to an aliquot of the lyophilized protein, let equilibrate for 5 min, and then centrifuge >5,000 × *g* for 5 min. Otherwise, use 10 μL of the respective protein sample (*see* **Note 2**).
3. Thoroughly clean CaF_2 windows. Rinse extensively with ddH_2O, blot dry with Kimwipes (or other lint-free tissue). Wipe windows with a folded Kimwipes tissue soaked in ethanol and then with a folded Kimwipes tissue soaked in acetone. Visually inspect the windows to ensure that all residuals and smears are removed. Blow with compressed air to remove all lint and other particles from window surface.
4. Clean sample cell holder, Teflon spacer, and rubber O-rings using ddH_2O and ethanol. Extensively blow with compressed air to dry and remove all lint and other particles.

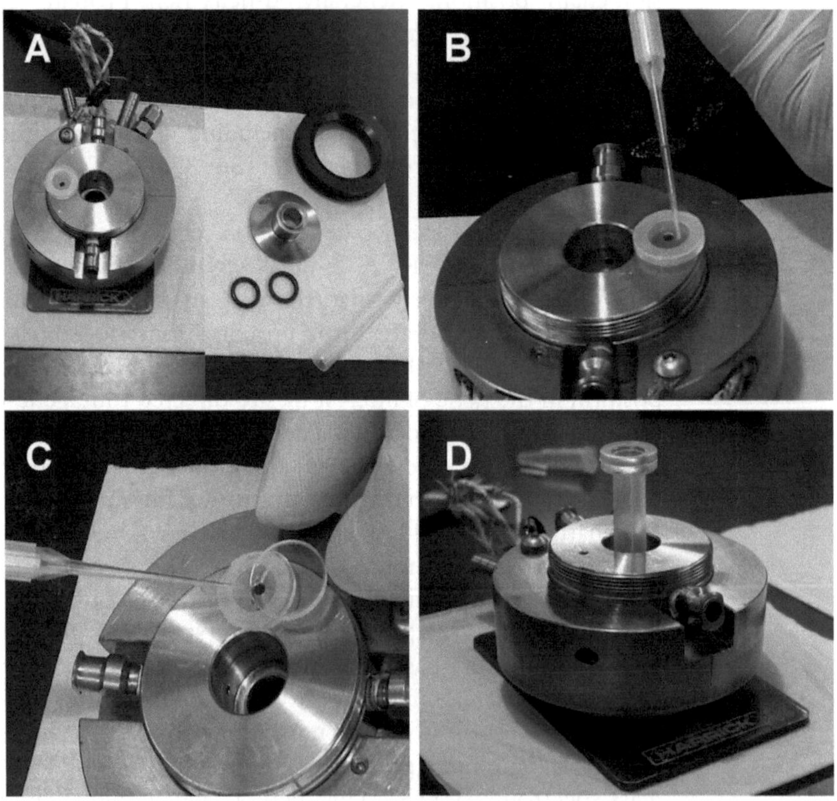

Fig. 2 Sample cell assembly. (**a**) Temperature-controllable sample holder with CaF_2 windows, rubber O-rings, and guide tube. (**b**) The Teflon spacer is adhered to the bottom window with a few drops of sample. (**c**) The sample is filled into the cavity between bottom and top windows. (**d**) The sandwiched sample cell is ready to be lowered into the sample cell holder

5. Insert first rubber O-ring into sample holder, put guide tube in place, and place Teflon spacer on bottom CaF_2 window.
6. Using 10 μL of sample and a micropipetter, adhere the Teflon spacer to bottom window by adding a few drops of sample (1–2 μL) to the inner boundary of Teflon spacer (Fig. 2b). The sample liquid will be absorbed between the Teflon spacer and window, thus "gluing" the Teflon spacer to the window.
7. Slide the top window partially over the bottom window and fill the cavity between the windows with the remaining sample in the pipetter while avoiding the formation of air bubbles (Fig. 2c). Continue sliding the top window over the bottom window and fill the additional space with sample until the entire cavity between the windows is filled. Slide the top window over the bottom window such that both are perfectly aligned with each other. The sample is now ready to be transferred to the sample cell (*see* **Note 3**).
8. Carefully pick up the sandwiched windows while avoiding to pull the windows apart (which would produce air bubbles) and place them on the guide tube (Fig. 2d). Lift the sample holder so that the sandwiched windows slide into position, place the second rubber O-ring on top of the windows, and screw the screw cap on tightly. To improve reproducibility, mark sample cell and screw cap (e.g., with a permanent marker pen) so that the cap will always return to the same marked position (*see* **Note 4**).
9. Blow compressed air across the window surfaces to remove any lint or other particles.
10. Determine protein concentration in the FT IR cell by measuring the OD_{280} for future reference. (This can be accomplished with essentially any UV/vis spectrometer by placing the FT IR cell in front of the sample holder at the correct height.)

3.3 Acquisition of the FT IR Difference Spectrum for a Single Sample

3.3.1 Data Acquisition

1. With no sample inside, purge the instrument for at least 20–30 min with dry, CO_2-depleted nitrogen gas to reduce water vapor and CO_2 concentration to a stable baseline (*see* **Note 5**).
2. Set aperture size to ~90% of maximal signal and verify all other instrument settings (*see* **Note 6**). A typical set of data acquisition parameters are 2 cm^{-1} resolution, 8,000 scans with double-sided forward/backward acquisition (for both background and sample), Blackman–Harris 3-term apodization with Mertz phase correction, 32 cm^{-1} phase resolution, and zero-filling factor of 4.
3. Acquire background spectrum with empty sample chamber (*see* **Note 7**).

4. Place proteo (i.e., unlabeled; *see* **Note 8**) sample into instrument. Purge for at least 20–30 min. Acquire sample spectrum (*see* **Note 9**).
5. Remove sample cell from instrument, clean sample cell, and fill with deutero sample as described above (*see* **Note 10**).
6. Place deutero (i.e., labeled; *see* **Note 8**) sample into instrument. Purge for 20–30 min. Acquire sample spectrum.
7. Subtract the spectrum of the proteo sample from the spectrum of the deutero sample (using the auto-subtract function if available), restricting the spectral range to the area of interest (typically 1,900–2,400 cm^{-1}). This is the FT IR difference spectrum.
8. Visually inspect the sample and determine protein concentration in the FT IR cell by measuring the OD_{280} (*see* Subheading 3.2, **step 10**) to confirm the absence of sample degradation (e.g., due to aggregation, precipitation, or formation of gas bubbles).

3.3.2 Background Correction and Spectral Deconvolution

Further analysis of the FT IR difference spectrum can be performed with a custom-made MATLAB program (MathWorks, Inc.) which is available upon request from the authors. Following is a description of the fit algorithm:

1. Cut the absorption spectrum to the region around the absorption peak of interest (*see* **Notes 11** and **12**), including ~50 cm^{-1} of background on each side of the absorption peak (Fig. 3a, b).
2. Fit an *n*th order polynomial to the spectrum excluding the data points within ~2–3 times the full width at half maximum of the center frequency of the peak of interest (Fig. 3b). The required order of the polynomial should be determined by F-test at the 99% confidence level (*see* **Note 13**).
3. Fit the spectrum including the absorption peak of interest to the sum of the *n*th order polynomial determined in **step 2** and the required number of Gaussians (*see* Subheading 3.4) to the absorption peak of interest (Fig. 3c). The parameters for the *n*th order polynomial are fixed to the values obtained in **step 2**.
4. Repeat **step 3** using the parameters for the *n*th order polynomial obtained in **step 2** and the parameters for the Gaussians obtained in **step 3** as starting parameters, but allow all parameters to float (Fig. 3d).
5. Subtract the *n*th order polynomial from **step 4** from the FT IR difference spectrum. This is the background-corrected FT IR difference spectrum.

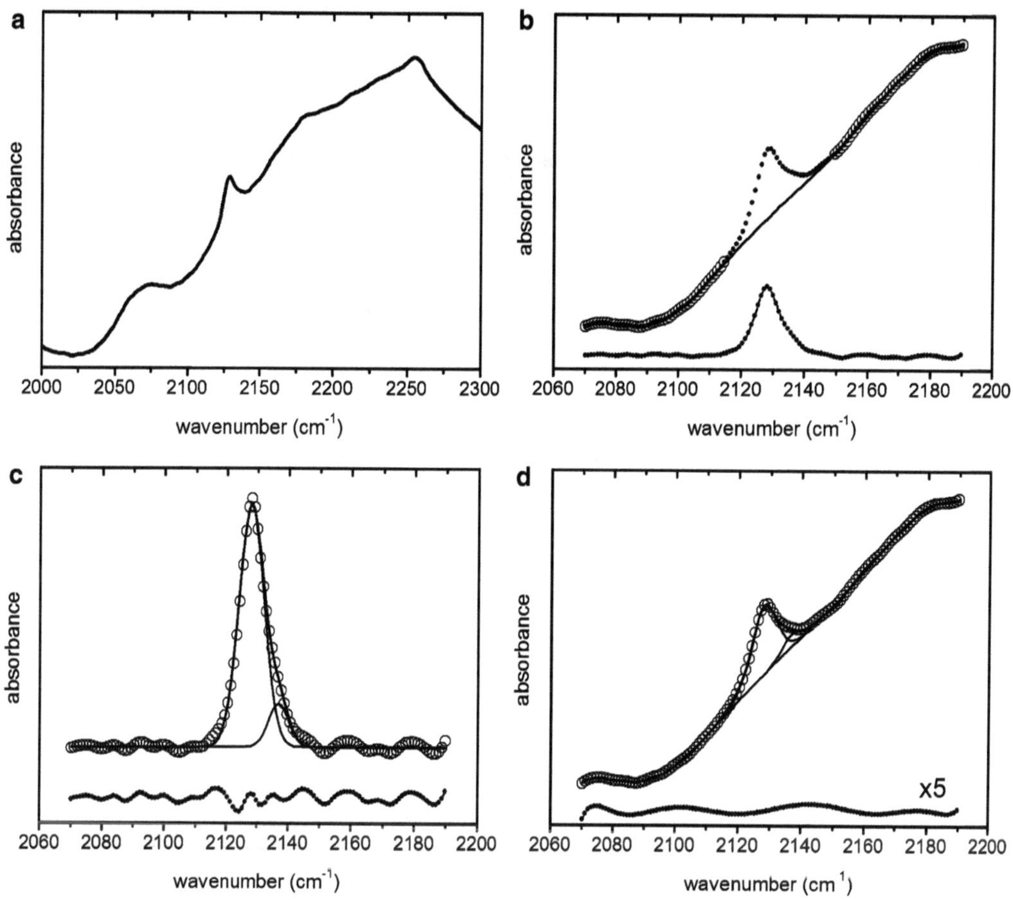

Fig. 3 Background correction and deconvolution. (**a**) FT IR difference spectrum. (**b**) A polynomial (*black line*) is fit to the truncated spectrum excluding the peak of interest (*open circles*). The fit residuals are shown on the *bottom* of the plot. (**c**) Two Gaussians are fit to the residuals in panel (**c**). (**d**) Final fit using the polynomial from panel (**b**) and the Gaussians from panel (**c**) as start parameters. Fit residuals are shown at the bottom of plots (**b–d**)

3.4 Deconvolute FT IR Difference Spectrum into Minimum Number of Gaussian Absorptions

1. Acquire and background-correct the FT IR difference spectrum as described in Subheading 3.3.
2. Fit a single Gaussian to the background-corrected FT IR difference spectrum (*see* **Note 14**).
3. Determine the sum of squared residuals, $SSR = \Sigma_{j=1..N} (a_j - A_j)^2$, of the fit, where n denotes the number of Gaussians used to fit the background-corrected spectrum and a_j and A_j are the experimental and fit values for absorptions at wave number ν_j, with j running over all N data points (Fig. 4).
4. Fit two Gaussians to the background-corrected FT IR spectrum. Determine the SSR of the two-Gaussian fit as described in **step 3**.

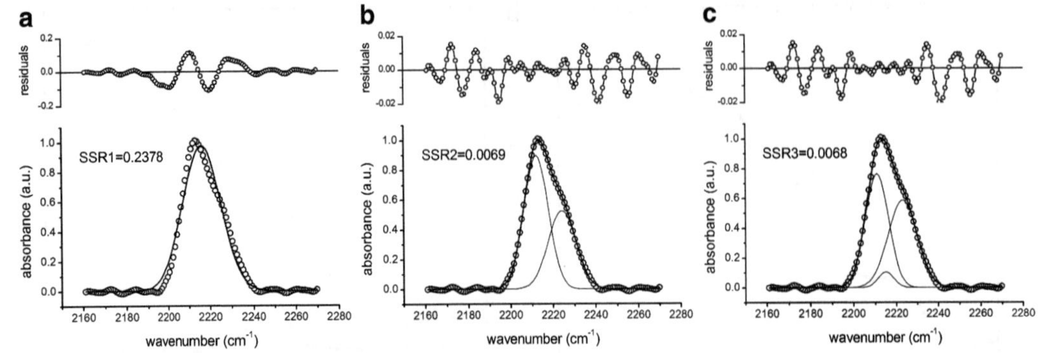

Fig. 4 F-test for a spectrum with $N = 114$ data points acquired with a zero-filling factor $z = 4$. Panels **a–c** show deconvolution of the background-corrected FT IR spectrum into an increasing number of Gaussians, resulting in F numbers $F_1 = 251$ and $F_2 = 0.08$. The critical F-number is $F_c = 4.94$ in both cases. Thus, the deconvolution into two Gaussians is the best statistically justified fit for this spectrum

5. Determine the F-number comparing the statistical determination of the two models, given by

$$F_n = 1/3 \times (N/z - 3 \times (n+1)) \times (\mathrm{SSR}_{n+1}/\mathrm{SSR}_n - 1)$$

where z is the zero-filling factor used in the Fourier transform and SSR_1 and SSR_2 are the SSRs of the one- and two-Gaussian fits, respectively, and $n = 1$.

6. If F_1 is larger than the critical F number, F_c, determined from Table 1, then the fit with two Gaussians is statistically justified at the 99% confidence level; if not, the fit with two Gaussians would result in "overfitting" the spectrum.

7. If the two-Gaussian fit is statistically justified, perform the F-test comparing a three-Gaussian fit to the two-Gaussian fit, i.e., determine SSR_3 and then F_2.

8. Repeat **step** 7, incrementally increasing the number of Gaussians until the F-test fails. The last fit not failing the F-test gives the minimum number of Gaussian absorptions statistically justified at the 99% confidence level to fit the spectrum.

3.5 Site-Specific Characterization of Equilibrium Protein Folding

3.5.1 Urea- or Guanidine Hydrochloride-Induced Unfolding

1. Divide desalted protein samples (proteo and deutero) into aliquots of ~0.3–1.0 nmol protein, lyophilize, and store at −20 °C until use.

2. Prepare solutions of varying denaturant concentrations in the buffer of choice (e.g., 100 mM sodium acetate). Verify that all solutions have the same pH and adjust pH with NaOH/HCl if required.

3. Determine the precise concentration of each denaturant solution by measuring its index of diffraction, using the relations

Table 1
Critical values for *F* statistics at the 99% confidence level

DF[a]	1	2	3	4	5	7	10	15	20	30	60	120	500	1,000
F_c	5,403	99.2	29.5	16.7	12.1	8.45	6.55	5.42	4.94	4.51	4.13	3.95	3.82	3.80

[a]Degrees of freedom, $DF = N/z - 3(n + 1)$

$$c_{\text{urea}} = 117.66 \times \Delta n + 29.753 \times \Delta n^2 + 185.56 \times \Delta n^3$$

$$c_{\text{GdnHCl}} = 57.147 \times \Delta n + 36.68 \times \Delta n^2 - 91.60 \times \Delta n^3$$

where Δn is the difference between the refractive indices of denaturant solution and buffer [40].

4. For each denaturant concentration to be analyzed, add 10 μL of solution to one aliquot of deutero and one aliquot of proteo protein sample, and acquire the FT IR difference spectrum as described in Subheading 3.3.1.
5. Analyze the denaturant-dependent FT IR difference spectra as described in Subheading 3.6.

3.5.2 pH-Induced Unfolding

1. Divide desalted protein samples (proteo and deutero) into aliquots containing ~0.6–2.0 nmol protein, lyophilize, and store at −20 °C until use.
2. Prepare buffered solutions of varying pH (e.g., 50 mM sodium phosphate, pH 6, 100 mM NaCl, and varying amounts of NaOH for alkaline titrations).
3. For each pH solution prepared in **step 2**, add 20 μL of solution to one aliquot of deutero and one aliquot of proteo protein sample and let equilibrate for 10 min.
4. Determine the precise pH of each solution using a micro pH probe (e.g., Mettler Toledo InLab).
5. Acquire the pH-dependent FT IR difference spectra for each pH as described in Subheading 3.3.1.
6. Analyze the pH-dependent FT IR difference spectra as described in Subheading 3.6.

3.5.3 Temperature-Induced Unfolding

1. Divide desalted protein samples (proteo and deutero) into aliquots containing ~0.3–1.0 nmol protein, lyophilize, and store at −20 °C until use.
2. Prepare a proteo sample as described in Subheading 3.2, using a temperature-controlled sample cell.
3. With no sample inside, purge instrument for at least 20–30 min with dry, CO_2-depleted nitrogen gas (*see* **Note 5**).
4. Set aperture size to ~90% of maximal signal and verify all other instrument settings.
5. Acquire background spectrum with empty sample chamber.
6. Place proteo sample into instrument. Purge for at least 20–30 min.
7. Acquire spectra of the proteo sample from 25 to 90 °C with a 3–5 °C step size (*see* **Note 15**). After each increase in temperature, allow sample to equilibrate for 10 min.

8. Cool sample to 25 °C, let sample equilibrate for 20 min, and acquire spectrum.
9. Remove sample cell from instrument, clean, and fill with deutero sample.
10. Repeat **steps 6–8** with the deuterated sample.
11. For each temperature, subtract the spectrum of the proteo sample from the spectrum of the deutero sample (using the auto-subtract function if available) while restricting the spectral range to the area of interest (typically 1,900–2,400 cm^{-1}). This yields the FT IR difference spectra as a function of temperature.
12. Compare the FT IR difference spectra at 25 °C taken before and after heating the sample to gauge reversibility of the heat denaturation (*see* **Note 16**).
13. Analyze the temperature-dependent FT IR difference spectra as described in Subheading 3.6.

3.6 Fitting of Series of Spectra to Determine Transition Midpoints and Relative Stabilities

To analyze a series of related spectra (e.g., acquired at different concentrations of denaturant or at different temperatures), first generate a series of denaturant concentration/temperature-dependent FT IR difference spectra as described in Subheading 3.5 (*see* **Note 16**). All data analysis described here can be performed with a suite of custom-made MATLAB programs (MathWorks, Inc.), which are available upon request from the authors:

1. Deconvolute the limiting spectra at lowest and highest denaturant concentration/temperature into the minimum number of Gaussians as described in Subheading 3.4 (*see* **Note 17**).
2. Fit all spectra at intermediate denaturant concentrations/temperatures to a superposition of the limiting spectra as approximated by the Gaussian deconvolutions from **step 1**. To do so, fix the frequencies, line widths, and relative amplitudes of the Gaussians used to approximate the limiting spectra, and only allow the amplitude of the limiting spectra to float (*see* **Note 18**). This yields a fractional concentration of native and denatured state weighted by the respective extinction coefficient as a function of denaturant concentration/temperature (denoted as d), i.e., $\varepsilon_N \times c_N(d)$ and $\varepsilon_D \times c_D(d)$, respectively.
3. Inspect residuals at intermediate denaturant concentrations/temperatures for evidence of possible folding intermediates (i.e., spectral regions that are not well fit by a superposition of the limiting spectra). In cases where one or more intermediates are observed, fit the spectra at intermediate denaturant concentrations/temperatures as a superposition of the limiting spectra and additional Gaussian(s) representing the intermediate signal(s). Determine the average center frequency and line width of the additional Gaussian(s). Fit all spectra to

Table 2
Denaturant dependence of $\Delta G°$

Urea- or GdnHCl-induced unfolding[a]	$\Delta G°(c_d) = m \times (c_d - c_m)$
pH-induced unfolding[b]	$\Delta G°(\text{pH}) = -2.3\ nRT \times (\text{pH} - \text{pH}_m)$
Temperature-induced unfolding[c]	$\Delta G°(T) = \Delta H^0 \times (1 - T/T_m)$

Note that the transition entropy is given by $\Delta S° = \Delta H°/T_m$
[a] c_d denaturant concentration, m system-dependent proportionality constant, c_m midpoint concentration
[b] pH sample pH, n number of titratable groups involved in the transition, R gas constant, T temperature, pH_m midpoint pH
[c] T temperature, $\Delta H°$ transition enthalpy, T_m midpoint temperature

a superposition of the limiting spectra and the Gaussian(s) representing the intermediate signal(s) while fixing the frequencies, line widths, and for the limiting spectra the relative amplitudes of all Gaussians (*see* above and **Note 18**). This yields denaturant-dependent fractional concentrations for each species weighted by the respective extinction coefficient, i.e., $\varepsilon_i \times c_i(d)$ with $i = N, D, I_1, I_2, \ldots$

4. Fit fractional concentrations to a Boltzmann distribution using the number of experimentally observed states, N (*see* **Note 19**):

$$c_i(d) = \varepsilon_i \frac{\exp(\Delta G°_{N \to Si}(d)/RT)}{\sum_j^N \exp(\Delta G°_{N \to Sj}(d)/RT)}$$

where c_i and ε_i are the fractional concentration and extinction coefficient of state i, respectively, $\Delta G^{\circ}_{N \to S_i}(d)$ is the denaturant concentration/temperature-dependent free energy difference between the native state and state i, R is the gas constant, and T temperature, and the summation runs over all states, $j = 1, \ldots, N$. Table 2 shows various models for the denaturant concentration/temperature dependence of the free energy.

4 Notes

1. It is important to maximize the signal-to-noise (S/N) ratio by using the optimal optical path length. Typical absorbance signals from C–D stretching vibrations under the described experimental conditions are on the order of 10^{-4} to 10^{-3} and are superimposed on a water bending mode that is ~100-fold more intense. The water bending mode thus limits the path length of the cell. Figure 5 shows the predicted S/N ratio as a function of sample absorbance assuming Poisson noise or Poisson noise plus 0.1% dark noise. The S/N ratio is optimal for sample

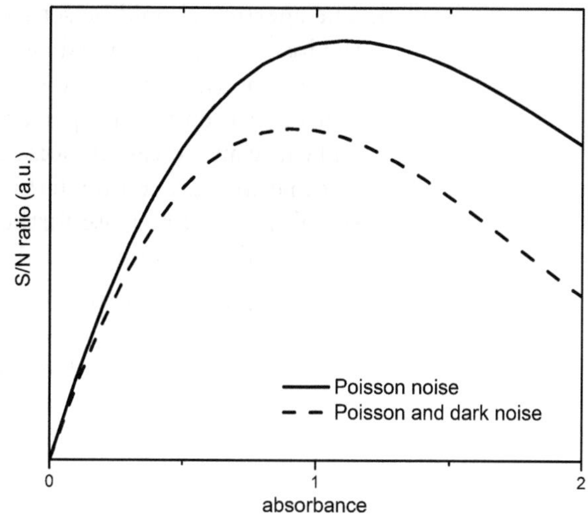

Fig. 5 S/N ratio as a function of sample absorbance assuming Poisson noise ($N = \sqrt{I}$, *solid line*) or Poisson noise and dark noise ($N = \sqrt{I + 10^{-3} I_0}$, *dashed line*) ($N$ noise, I light intensity, I_0 light intensity without sample)

absorbance around 1 in both cases, which corresponds to an optical path length of ~75–100 μm around 2,200 cm^{-1}.

2. Sample volumes in Subheading 3.2 are given for a sample cell with ½″ diameter and a 75 μm spacer and hence need to be adjusted if other cell diameters or spacers are used.
3. We strongly recommend practicing with a non-deuterated but otherwise identical sample until the cell may be reproducibly loaded with sample without the introduction of air bubbles.
4. Alternatively, place the bottom window inside the sample cell, place the Teflon spacer on the bottom window, place a sufficient sample volume in the center of the bottom window, and let the top window drop on the bottom window and then quickly screw the screw cap on. This method works well for larger windows (e.g., 1″ diameter).
5. It is essential to actively purge the instrument, both the sample and optics chambers, with a constant stream of dry nitrogen gas. It is recommended to determine the purge time necessary to return to equilibrium after opening the sample chamber for your instrument. To do so, take a series of spectra after opening the sample chamber. The CO_2 doublet at ~2,350 cm^{-1} will be clearly visible in the spectrum directly after opening/closing of the sample chamber. Create a plot of the intensity of the CO_2 doublet as a function of time; it should decrease exponentially. Determine the time constant τ_{purge} by fitting a single-exponential function to the signal decay. The wait time between closing the sample chamber and recording of the first spectrum of the sample should be 5–10 times τ_{purge}.

6. The aperture should be set to maximize the dynamic range of the instrument (i.e., just below detector saturation). However, if detector saturation cannot be reached due to sample absorbance, ensure that the aperture is not opened too wide, as this will cause an increase in noise due to stray light. To test for the optimal aperture setting in this case, open the aperture until the signal does not increase further and then close the aperture to ~90% of the maximum signal.
7. In principle, it is better to record the background spectrum with the proteo sample in the instrument. This makes it possible to increase the incident light intensity without saturating the detector, resulting in an improved S/N ratio for the same number of scans. However, this approach may be problematic when dealing with small signals; the resulting difference spectrum may be "wavy" and thus not interpretable.
8. We refer to the C–D-labeled samples as deutero and to the unlabeled samples as proteo.
9. Ideally, a proteo background spectrum should be taken directly before after each scan of a C–D-labeled sample. This eliminates artifacts due to instrument drifts. For certain experiments (e.g., temperature titrations), this approach might not be feasible. In this case, the FT IR difference spectra should be calculated using different combinations of deutero and proteo spectra to avoid artifacts due to instrument drift and sample impurities.
10. To minimize errors, it is recommended to use the identical sample cell, CaF_2 windows, and Teflon spacer for both the proteo and deutero sample. CaF_2 windows should be marked (e.g., with an arrow to identify the surface in contact with the sample, and labels to distinguish the windows, writing with a pencil on the side face of the windows) in order to assemble the cell identically each time. In addition, the screw cap of the sample cell and the sample cell holder should be marked such that the screw cap is screwed on and tightened identically each time; this improves the reproducibility of the optical path length.
11. Identification of C–D absorptions. Since C–D absorption bands are weak, it is sometimes difficult at first to distinguish them from background noise or other small absorptions (e.g., water vapor bands). To assign an absorption band in the FT IR difference spectrum to a C–D absorption, the following two experimental assessments are helpful. First, dilutions are an essential tool to support band assignments; the intensity of the C–D absorption should scale with protein concentration (also *see* **Note 12**). Second, the absorption should be sensitive to protein denaturation and will typically assume the frequency and line width of the corresponding free amino acid under denaturing conditions.

12. Impurities in the sample are sometimes difficult to distinguish from C–D absorptions, since they will also scale with protein concentration. An example are traces of acetonitrile in samples that were HPLC purified, which results in the observation of acetonitrile stretch absorptions at 2,259 and 2,296 cm^{-1}. To minimize the possibility that impurities are wrongly assigned to C–D absorptions, it is important to use proteo samples that have been prepared identically to the deutero sample. For example, if the C–D label was introduced via solid-state peptide synthesis, the proteo sample should also be prepared via solid-state peptide synthesis. In this case, if an impurity is present, it should be present in both the deutero and proteo samples, and thus, the corresponding signal in the difference spectrum should not scale linearly with concentration (see above).
13. We recommend to visually inspect the background polynomial to ensure that it reasonably interpolates the background in the area of the C–D absorption; if this is not the case, a change of the order of the polynomial or an increase/decrease of the spectral range used in the background correction may rectify the problem.
14. Vibrational line shapes of C–D stretches tend to be Gaussian; however, other stretch absorptions (e.g., CN stretches) can have Lorentzian character. A more general approach is therefore to deconvolute the spectra into pseudo-Voigt functions:

$$I(\bar{\nu}) = a \times \left[\begin{array}{l} m_L \times \dfrac{2}{\pi} \dfrac{\text{fwhm}}{4(\bar{\nu} - \bar{\nu}_0)^2 + \text{fwhm}^2} + (1 - m_L) \dfrac{2\sqrt{\ln 2}}{\sqrt{\pi}\text{fwhm}} \\ \exp\left(-\dfrac{4 \ln 2 (\bar{\nu} - \bar{\nu}_0)^2}{\text{fwhm}^2} \right) \end{array} \right]$$

where a is the amplitude, fwhm is the full width at half maximum of the observed band, $\bar{\nu}_0$ is the center frequency, and m_L is the fraction of Lorentzian character of the band.
15. It is recommended to decrease the temperature step size around the transition midpoint to more accurately capture the transition instead of maintaining the step size constant over the whole temperature range. However, it is important that the data for all samples under comparison be collected at identical temperatures to avoid fitting biases.
16. The fits to equilibrium distributions discussed in Subheading 3.6 are only valid if the process is reversible, i.e., the total concentration is conserved and no sample is lost due to irreversible conversion of the sample to other species. For this reason, an effort should be made to ensure reversibility.
17. It is best to average spectra from several independent experiments before deconvolution to improve the S/N ratio and to reduce the possibility of spectral artifacts.

18. Occasionally fit results are improved by allowing small deviations in center frequency and line width (on the order of 1 cm^{-1}) due to the fact that the parameters obtained from Gaussian deconvolutions are carrying errors themselves. In addition, absorption features may change systematically with denaturant conditions. For instance, the absorption frequencies are often dependent on the bulk dielectric constant, which needs to be taken into account for the unfolded state in urea- or GdnHCl-induced denaturing. The best strategy is to record absorption spectra of the corresponding deuterated free amino acid as a function of denaturant concentration/temperature to check for denaturant-dependent effects on absorption frequency or line width in the unfolded state.
19. Be aware that site-specific thermodynamic parameters obtained from two-state fits might be meaningless if they are part of several strongly overlapping transitions. In this case, the site-specific fractional concentrations need to be fit globally (for an example, *see* ref. 31).

Acknowledgment

This work was supported by the National Science Foundation under Grant No. 0346967.

References

1. Frauenfelder H, Sligar SG, Wolynes PG (1991) The energy landscape and motions of proteins. Science 254:1598–1603
2. Boehr DD, McElheny D, Dyson HJ et al (2006) The dynamic energy landscape of dihydrofolate reductase catalysis. Science 313:1638–1642
3. Warshel A, Sharma PK, Kato M et al (2006) Electrostatic basis for enzyme catalysis. Chem Rev 106:3210–3235
4. Zimmermann J, Oakman EL, Thorpe IF et al (2006) Antibody evolution constrains conformational heterogeneity by tailoring protein dynamics. Proc Natl Acad Sci U S A 103:13722–13727
5. Barth A (2000) The infrared absorption of amino acid side chains. Prog Biophys Mol Biol 74:141–173
6. Garczarek F, Gerwert K (2006) Functional waters in intraprotein proton transfer monitored by ftir difference spectroscopy. Nature 439:109–112
7. Park ES, Andrews SS, Hu RB, Boxer SG (1999) Vibrational stark spectroscopy in proteins: a probe and calibration for electrostatic fields. J Phys Chem B 103:9813–9817
8. Chin JK, Jimenez R, Romesberg FE (2001) Direct observation of protein vibrations by selective incorporation of spectroscopically observable carbon–deuterium bonds in cytochrome *C*. J Am Chem Soc 123:2426–2427
9. Chin JK, Jimenez R, Romesberg F (2002) Protein dynamics and cytochrome *c*: correlations between ligand vibrations and redox activity. J Am Chem Soc 124:1846–1847
10. Decatur SM (2006) Elucidation of residue-level structure and dynamics of polypeptides via isotope-edited infrared spectroscopy. Acc Chem Res 39:169–175
11. Fafarman AT, Webb LJ, Chuang JI et al (2006) Site-specific conversion of cysteine thiols into thiocyanate creates an IR Probe for electric fields in proteins. J Am Chem Soc 128: 13356–13357
12. Oh KI, Lee JH, Joo C et al (2008) *B*-azidoalanine as an IR Probe: application to amyloid Aβ (16-22) aggregation. J Phys Chem B 112: 10352–10357

13. Cremeens ME, Zimmermann J, Yu W et al (2009) Direct observation of structural heterogeneity in a beta-sheet. J Am Chem Soc 131:5726–5727
14. Thielges MC, Case DA, Romesberg FE (2008) Carbon–deuterium bonds as probes of dihydrofolate reductase. J Am Chem Soc 130: 6597–6603
15. Thielges MC, Groff D, Cellitti S et al (2009) Efforts toward the direct experimental characterization of enzyme microenvironments: tyrosine100 in dihydrofolate reductase. Angew Chem Int Ed 48:3478–3481
16. Getahun Z, Huang CY, Wang T et al (2003) Using nitrile-derivatized amino acids as infrared probes of local environment. J Am Chem Soc 125:405–411
17. Tucker MJ, Getahun Z, Nanda V et al (2004) A new method for determining the local environment and orientation of individual side chains of membrane-binding peptides. J Am Chem Soc 126:5078–5079
18. Mukherjee S, Chowdhury P, DeGrado WF et al (2007) Site-specific hydration status of an amphipathic peptide in AOT reverse micelles. Langmuir 23:11174–11179
19. Fafarman AT, Boxer SG (2010) Nitrile bonds as infrared probes of electrostatics in ribonuclease S. J Phys Chem B 114:13536–13544
20. Schultz KC, Supekova L, Ryu Y et al (2006) A genetically encoded infrared probe. J Am Chem Soc 128:13984–13985
21. Zimmermann J, Thielges MC, Seo YJ et al (2011) Cyano groups as probes of protein microenvironments and dynamics. Angew Chem Int Ed 50:8333–8337
22. Ohno S, Matsui M, Yokogawa T et al (2007) Site-selective post-translational modification of proteins using an unnatural amino acid, 3-azidotyrosine. J Biochem 141:335–343
23. Ye S, Huber T, Vogel P et al (2009) FTIR analysis of GPCR activation using azido probes. Nat Chem Biol 6:397–399
24. Taskent-Sezgin H, Chung J, Banerjee PS et al (2010) Azidohomoalanine: a conformationally sensitive IR Probe of protein folding, protein structure, and electrostatics. Angew Chem Int Ed 49:7473–7475
25. Fafarman AT, Sigala PA, Herschlag D et al (2010) Decomposition of vibrational shifts of nitriles into electrostatic and hydrogen-bonding effects. J Am Chem Soc 132:12811–12813
26. Sagle LB, Zimmermann J, Matsuda S et al (2006) Redox-coupled dynamics and folding in cytochrome *c*. J Am Chem Soc 128:7909–7915
27. Zimmermann J, Gundogdu K, Cremeens ME et al (2009) Efforts toward developing probes of protein dynamics: vibrational dephasing and relaxation of carbon–deuterium stretching modes in deuterated leucine. J Phys Chem B 113:7991–7994
28. Sagle LB, Zimmermann J, Dawson PE et al (2004) A high-resolution probe of protein folding. J Am Chem Soc 126:3384–3385
29. Mirkin NG, Krimm S (2007) Conformation dependence of the **Cαdα** stretch mode in peptides. 1. Isolated alanine peptide structures. J Phys Chem A 111:5300–5303
30. Kinnaman CS, Cremeens ME, Romesberg FE et al (2006) Infrared line shape of an alpha-carbon deuterium-labeled amino acid. J Am Chem Soc 128:13334–13335
31. Weinkam P, Zimmermann J, Sagle LB et al (2008) Characterization of alkaline transitions in ferricytochrome c using carbon–deuterium infrared probes. Biochemistry 47: 13470–13480
32. Terranova ZL, Corcelli SA (2012) Monitoring intramolecular proton transfer with two-dimensional infrared spectroscopy: a computational prediction. J Phys Chem Lett 3: 1842–1846
33. Miller CS, Corcelli SA (2010) Carbon–deuterium vibrational probes of the protonation state of histidine in the gas-phase and in aqueous solution. J Phys Chem B 114: 8565–8573
34. Sagle LB, Zimmermann J, Dawson PE et al (2006) Direct and high resolution characterization of cytochrome *c* equilibrium folding. J Am Chem Soc 128:14232–14233
35. Dawson PE, Kent SB (2000) Synthesis of native proteins by chemical ligation. Annu Rev Biochem 69:923–960
36. Muralidharan V, Muir TW (2006) Protein ligation: an enabling technology for the biophysical analysis of proteins. Nat Methods 3: 429–438
37. Flavell RR, Muir TW (2009) Expressed protein ligation (Epl) in the study of signal transduction, ion conduction, and chromatin biology. Acc Chem Res 42:107–116
38. Liu CC, Schultz PG (2010) Adding new chemistries to the genetic code. Annu Rev Biochem 79:413–444
39. Xie JM, Schultz PG (2005) An expanding genetic code. Methods 36:227–238
40. Pace CN (1986) Determination and analysis of urea and guanidine hydrochloride denaturation curves. Methods Enzymol 131:266–280

Part II

Computational Methods for Characterizing Protein Dynamics

Chapter 7

Balancing Bond, Nonbond, and Gō-Like Terms in Coarse Grain Simulations of Conformational Dynamics

Ronald D. Hills Jr.

Abstract

Characterization of the protein conformational landscape remains a challenging problem, whether it concerns elucidating folding mechanisms, predicting native structures or modeling functional transitions. Coarse-grained molecular dynamics simulation methods enable exhaustive sampling of the energetic landscape at resolutions of biological interest. The general utility of structure-based models is reviewed along with their differing levels of approximation. Simple Gō models incorporate attractive native interactions and repulsive nonnative contacts, resulting in an ideal smooth landscape. Non-Gō coarse-grained models reduce the parameter set as needed but do not include bias to any desired native structure. While non-Gō models have achieved limited success in protein coarse-graining, they can be combined with native structured-based potentials to create a balanced and powerful force field. Recent applications of such Gō-like models have yielded insight into complex folding mechanisms and conformational transitions in large macromolecules. The accuracy and usefulness of reduced representations are also revealed to be a function of the mathematical treatment of the intrinsic bonded topology.

Key words Gō model, Protein folding, Energy landscape, Conformational transition, Coarse-grained molecular dynamics

1 Introduction

Levinthal first commented on the vastness of protein conformational space in 1969. Shortly thereafter, Gō constructed a model for protein folding using only a native contact potential [1]. Since the modern development of energy landscape theory, Gō-like models have become popular for reducing the complex atomistic folding landscape into a small finite set of parameters [2–4]. The widespread success of Gō models has been attributed to the robustness of native over nonnative interactions in dominating the folding mechanisms of proteins on a funneled energy landscape [5]. Many alternative mappings of the atomistic degrees of freedom onto the reduced, or coarse-grained (CG), set are possible [6], and so Gō models have been deployed in a variety of levels of detail and functional forms.

Dennis R. Livesay (ed.), *Protein Dynamics: Methods and Protocols*, Methods in Molecular Biology, vol. 1084, DOI 10.1007/978-1-62703-658-0_7, © Springer Science+Business Media New York 2014

The general utility and application of the Gō model derives from the fact that the simple energy functions used allow standard molecular dynamics (MD) simulations to be performed. MD simulations yield dynamical information that is easily interpretable, can be compared to experiment, and used for hypothesis testing. MD applications with folding-inspired Gō-like approaches have more recently demonstrated their growing use in studies of functionally important conformational transitions [7–15].

1.1 Early CG Models

The importance of native interactions can be seen from primitive CG models used to study folding. Early representations relied on two or three "flavors" to encode interaction potentials [16–21]. Amino acid residues were divided into physicochemical classes, such as hydrophobic, polar, and, optionally, neutral. The polypeptide was represented as a series of interaction centers at the Cα positions connected by stiff harmonic bond and angle potential terms. Computer algorithms were then employed to globally optimize the conformational energy as a function of the backbone topology, defined by the fairly independent Cα pseudodihedral angles between residues [22]. Central to the energy optimization problem is the treatment of nonbonded interactions, in which residue contacts are scored based on physical parameters, or energy weights, for possible chemical pair interactions. Using attractive hydrophobic–hydrophobic and repulsive hydrophobic–polar terms, two- and three-color models have been able to depict nonspecific hydrophobic polymer collapse [23], a crucial event in folding nucleation. Three-color models fail, however, to predict a global energy minimum for the native state, instead predicting numerous compact globule states similar in energy.

Oakley et al. used the basin-hopping algorithm to thoroughly sample the conformational space of a model 69-residue β-barrel protein [24]. Using a three-color model, five local minima were found corresponding to different, effectively degenerate, energy structures. The level of frustration in the landscape is evident from the disconnectivity graph connecting 11,343 metastable intermediate states accessible within eight energy units of the global minimum (Fig. 1a), where the energy unit, ε, is defined as the strength of a hydrophobic-hydrophobic contact. Two preferred low energy topologies were characterized differing in their register shift and flexible loop conformations with a mean Cα RMSD of 2.6 Å (Fig. 1b).

1.2 Simple Gō Models and Native Interactions

Deployment of a simple nonbonded Gō model for the same β-barrel was found to remove frustration in the folding landscape, resulting in a single stable global energy minimum (Fig. 1c). Such Gō models smooth the landscape by only including attractive interactions for residue pairs that form contacts in the native structure. Native interactions in Cα Gō models are commonly encoded via a Lennard-Jones-like nonbond potential:

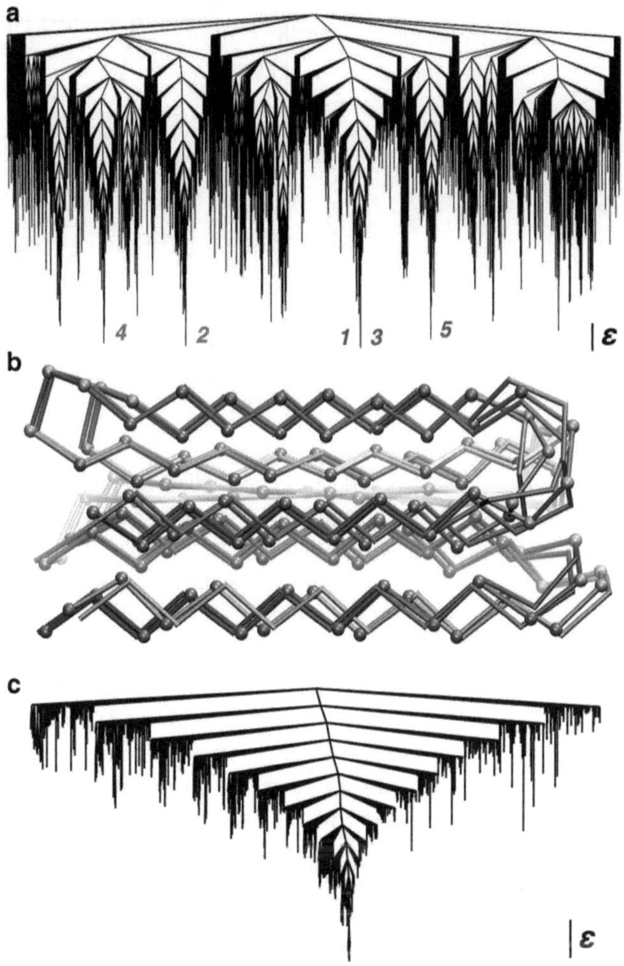

Fig. 1 Disconnectivity graphs comparing the energy landscape (in units of ε) as a function of conformation for a six-stranded antiparallel β-barrel simulated with a non-Gō three-color model (**a**) and a simple Gō model (**c**). (**b**) The five lowest energy minima for the non-Gō model are aligned structurally. Two preferred chain topologies are drawn in *cyan* and rainbow (*red:white:blue*), showing hydrophobic (*gray*), hydrophilic (*pink*), and flexible neutral turn (*green*) residues as *spheres*. Adapted with permission from [24]. © 2011 American Chemical Society (Color figure online)

$$V_{10-12}(r_{ij}) = \varepsilon\left[5\left(\frac{r_{0ij}}{r_{ij}}\right)^{12} - 6\left(\frac{r_{0ij}}{r_{ij}}\right)^{10}\right], \quad (1)$$

where r_{ij} is the instantaneous distance between residues i and j, r_0 is their separation in the reference native structure, and ε is the depth of the potential well at r_0, which defines the temperature and energy scale of the model [22, 25]. The 10–12 well potential is employed in CG models because of its smaller width and dependence on

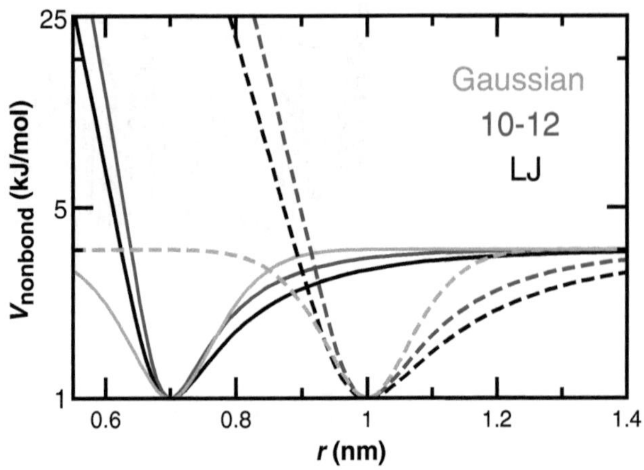

Fig. 2 Nonbond interaction functions. The 6–12 Lennard-Jones potential (*black*) is compared for two contact distances with the 10–12 potential (Eq. 1, *red*) and Gaussian contact potential (Eq. 6, *green*). The potential well of depth $\varepsilon = 2.5$ kJ/mol is shown in log scale (Color figure online)

r_0 compared to the standard 6–12 Lennard-Jones potential (Fig. 2): $V_{\rm LJ} = 4\varepsilon(\sigma^{12}r^{-12} - \sigma^6 r^{-6})$, where σ is the distance at which their contact becomes energetically unfavorable [25–27]. Unlike typical atomistic Lennard-Jones interactions, Cα native contact distances can be up to 12 Å in separation [28]. Simple Gō models assign a uniform ε for all native contacts and are said to lack energetic heterogeneity and frustration [29], leaving room only for topological frustration, or folding traps arising from chain connectivity [30, 31].

Oakley et al. removed folding frustration by neglecting nonspecific hydrophobic interactions for residue pairs separated by 1.167σ in the global energy minimum structure [24]. In other words, nonnative contacting residue pairs experience a repulsive potential of $V_{\rm nonnative} = \varepsilon\sigma^{12}r^{-12}$ to capture the effects of chain volume exclusion. For real protein coordinates from the Protein Data Bank, native contacts can be defined from a list of nonhydrogen sidechain contacts (within 4.5 Å) and backbone hydrogen bonds [22] or by using software to analyze the buried surface area of interatomic contacts in secondary structural units [32]. In addition to stabilizing the global minimum, Gō models containing residue interactions derived from native contact maps have been successful in reproducing the observed order of structure formation in folding mechanisms of diverse proteins [33–36]. Their widespread success is taken as evidence of the dominating influence of native topology [37] on the energy landscape [2, 3] relative to nonspecific, nonnative interactions [38–41]. The role of native topology in dictating folding dynamics is akin to the observation that shape determines function, which has been gleaned from elastic network models of conformational transitions where native contacts are connected by unbreakable harmonic springs [42–44]. The role of competing local and nonlocal interactions [45] has analogously been observed in Gō-like models of RNA secondary structure formation and tertiary folding [46–48].

2 Methodological Considerations

2.1 Bonded Potentials: Dihedrals

Nonbonded native interactions are only defined for residues separated in sequence by at least some exclusion number. Gō models will typically include at a minimum either $j = i + 3$ (1,4) [22] or $j = i + 4$ (1,5) [25] interactions for (i,j) residue contact pairs. The choice of nonbond exclusions depends on the treatment of the Cα pseudodihedral angles between every four successive residues. Analysis of dihedral distributions from the Protein Data Bank shows that Cα virtual dihedrals generally have two conformational minima at approximately $\phi = -135°$ and $\phi = +45°$, corresponding to local β-strand and α-helix geometry [22]. The two minima and the small barrier between them can be modeled as a fourth order cosine series that depends only on the identity of the middle two residues. The representative fourth order dihedral potential for X-Ala-Ala-X is plotted in Fig. 3a, where the energy has been scaled to reproduce distributions from atomistic simulations of polyalanine at 300 K [49]:

$$V_{\mathrm{PDB}}(\phi_{\mathrm{xAAx}})(\mathrm{kJ/mol}) = 1.26\cos(\phi - 287°) + 3.11\cos(2\phi - 272°) + 0.18\cos(3\phi - 180°) + 0.82\cos(4\phi - 108°). \tag{2}$$

The Gō model of Karanicolas and Brooks employs such potentials for 20^2 possible amino acid pairings of the central two residues. As a dihedral is free to sample α and β geometry, the model relies on 1,4 and 1,5 native interactions with a Lennard-Jones-like term to stabilize local α-helical segments. Locally driven helix formation with uniform native interactions, ε, was shown to result in slightly overstabilized α-helixes relative to β-strands, which rely on long-range contacts [22].

Another commonly used Gō model, developed by Clementi et al., neglects 1,4 native interactions but instead includes a stabilizing dihedral potential term:

$$V_{\mathrm{native}}(\phi_{ijkl}) = \varepsilon\left[1.5 - \cos(\phi_{ijkl} - \phi_{0ijkl}) - 0.5\cos 3(\phi_{ijkl} - \phi_{0ijkl})\right], \tag{3}$$

where ε is uniform contact energy and ϕ_0 is the reference Cα dihedral angle formed by residues i-j-k-l in the native structure [25, 26]. It can be difficult to compare the energy scales of different models, but Fig. 3a compares potentials corresponding to native α-helix and β-strand geometries when ε is chosen to be 1 kT.

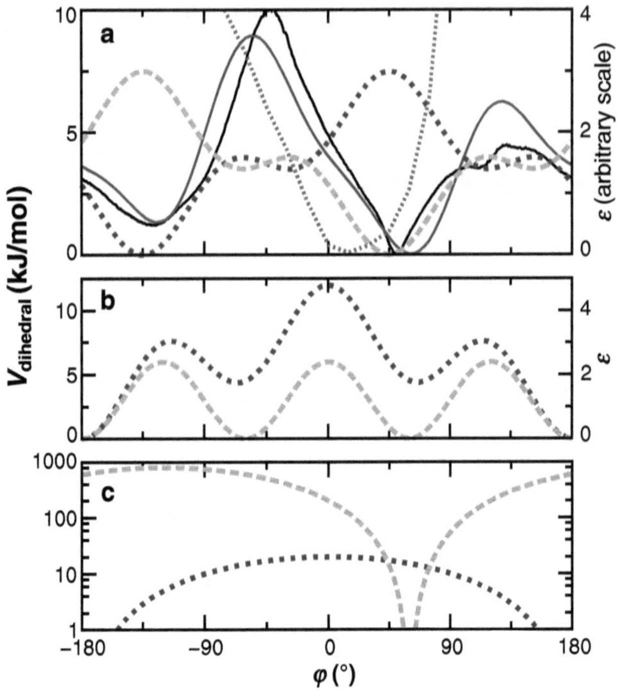

Fig. 3 $C_\alpha C_\alpha C_\alpha C_\alpha$ pseudodihedral potentials of different CG models. (**a**) Stabilizing terms favor native dihedrals (Eq. 3), as shown for $\phi_0 = -135°$ (*blue*) and $\phi_0 = +45°$ (*green*) assuming an energy scale of $\varepsilon = kT_{300K} = 2.5$ kJ/mol [25]. A statistical X-Ala-Ala-X cosine function (Eq. 2) can be scaled (*red*) to fit the Boltzmann-inverted probability distribution ($V = -kT \ln p$) of an atomistic polyalanine simulation (*black*) at 300 K [22, 49]. Compare to the distribution of the Ala-*trans*Ala-*cis*Ala-*trans*Ala peptide configuration in atomistic simulation at 498 K (*purple*). (**b**) Oakley et al. employed flexible potentials for turn residues (*green*) but not β-strands (*blue*) [24]. (**c**) The MARTINI model relies on stiff restraints to maintain α-helix (*green*) and β-sheet (*blue*) secondary structure [55] (Color figure online)

The native dihedral potential is seen to provide a modest stabilizing energy while not precluding other transitions. The physical accuracy of stabilizing torsion potentials is expected to be greater for α-helical structures with locally driven folding, compared to β-sheets that require a slower, nonlocal conformational search [37, 50–52].

2.2 Backbone Bond Angles

To complete the forcefield energy function, some form of Eqs. 1 and 3 are commonly coupled with stiff harmonic Cα bond and angle terms. For *all-trans* peptide bond configurations a uniform bond stretching potential can be applied: $V_{\text{bond}} = 100\varepsilon(r - 3.8\,\text{Å})^2 = 0.5 \times 78{,}000\,(r - 0.38\ \text{nm})^2$ kJ/mol. *Cis* peptide configurations have an equilibrium Cα virtual bond length of 3 Å [53]. The simple harmonic virtual bond connecting adjacent CG beads derives from Boltzmann

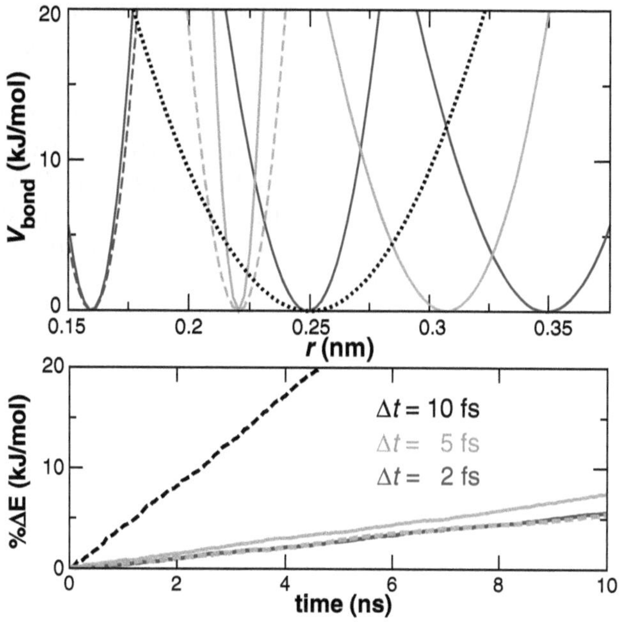

Fig. 4 Harmonic bond potentials. *Upper*: Boltzmann-inverted distributions of representative CG sidechain virtual bonds generated from atomistic MD (*solid colors*) [49]. A large time step can be used if a soft potential (*color dash*) is substituted for spring constants above: 100,000 kJ/mol/nm^2. The stiffest bond used in the original MARTINI model is shown for comparison (*black*) [55]. *Lower*: The relative percent energy drift increases with time step Δt in three simulations of 3B5X.pdb with the Hills et al. sidechain model. A simulation with stiff bonds is also shown for $\Delta t = 5$ fs (*solid green*). Integration of stiff bonds with $\Delta t = 10$ fs resulted in fatal termination of constant NVE simulations (Color figure online)

inversion ($V_{\text{bond}} = -kT\ln p/r^2$) of the narrow Gaussian distribution of site distances defined by the Cα geometric centers [49, 54], which do not depend on rotatable chemical bonds. In stark contrast, defining bonded CG interaction sites by residue center of mass results in broad anharmonic distributions that depend on secondary structure [55, 56]. The reference distributions can be obtained from either the Protein Data Bank or standard atomistic MD simulations, provided the test set is sufficiently general. As bond vibrations are usually the fastest degree of freedom, CG force fields may employ soft potentials in order to achieve computational speedup via a large integration time step (Fig. 4) [57].

In common Gō implementations [22, 25], the reference values for virtual Cα bending angles vary with the residue number in the sequence:

$$V_{\text{angles}} = 20\varepsilon \sum_{i=1}^{N-2} (\theta_{ijk} - \theta_{0ijk})^2, \tag{4}$$

for N residues with $N-2$ three-body reference Cα-Cα-Cα angles adopted in the native structure. Even though the harmonic biasing term is widely used in Gō models, Boltzmann-inverted ($V_{\text{angle}} = -kT \ln p/\sin\theta$) angle probability distributions from atomistic peptide simulations reveal two preferred conformational minima at $\theta = 95°$ and $\theta = 137°$, with a negligible barrier height connecting the α-helix and β-strand geometries [49]. Hills et al. performed unbiased CG simulations with the Gromacs simulation package [58] using a single fourth order polynomial, or shallow quartic angle potential, for all backbone angles:

$$V_{\text{angles}} = \sum_{i=1}^{N-2} \Big[127(\text{kJ/mol–rad}^4)(\theta_i - 116°)^4 - 33 \times (\text{kJ/mol-rad}^2)(\theta_i - 116°)^2 \Big]. \quad (5)$$

Unfolding simulations with the generic CG potential resulted in close correspondence between the observed atomistic and CG angle distributions. Nonetheless, the precise influence on folding mechanism of sampling on such a flexible potential compared to conventional quadratic terms remains largely unexplored [27, 59, 60]. Flexible secondary structure is particularly important for modeling conformational transitions [8]. Functional studies by Okazaki et al. employed the double-well scheme [13] for connecting two harmonic minima in order to capture surface loop flexibility and its effect on actomyosin binding affinity [61]. Detailed studies of stabilizing bonded terms uncover the role of select local interactions in funneling the energy landscape toward the global minimum [62, 63].

2.3 Illustration: Cis–Trans Isomerization

The role of dihedral angle terms on folding simulations can be seen by considering peptide prolyl *cis–trans* isomerization. CheY has been the subject of numerous folding experiments and is a member of the common flavodoxin fold containing five (βα)-repeat segments, two *trans* proline residues, and one *cis* proline [25, 64]. Fluorescence and NMR spectroscopy have identified the trans-cis isomerization of Pro110 as a rate-limiting step in the formation of native structure [65], proceeding from an unfolded state that prefers the trans isomer by a ratio of 90:10 trans/cis. Pro110 resides in the β5-α5 loop that allows the C-terminal helix to dock onto the rest of the α/β/α sandwich.

The off-the-shelf Gō model of Karanicolas and Brooks was used to explore the role of individual βα-repeats in the CheY folding mechanism [31]. The free energy was computed as a function of folding progress by collecting conformational snapshots with a given fraction of native contacts formed, Q [66, 67]. To characterize the high energy transition state structural ensemble for the folding reaction, MD simulations were performed at the

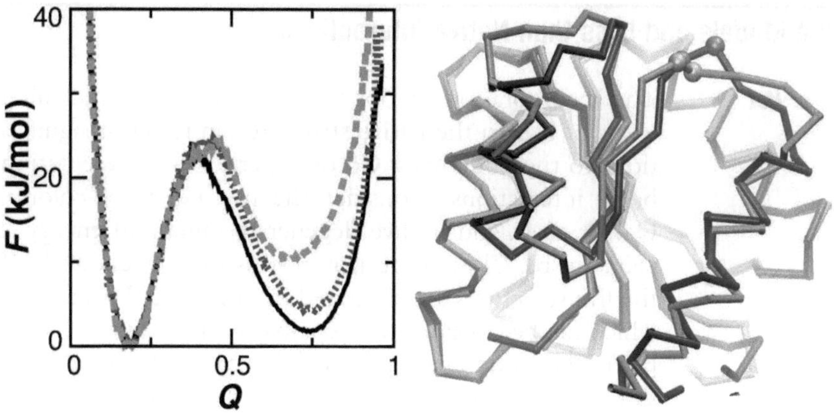

Fig. 5 *Left*: Free energy versus fraction of native contacts formed, *Q*, in Gō model simulations [31]. Destabilization in the native basin relative to unfolded is compared for CheY simulations with Pro110 harmonically restrained in Cα torsions representing native *cis* (*purple*) and nonnative *trans* (*green*) configurations of the peptide bond. Each simulation was performed at the same temperature, the T_f of unrestrained CheY (*black*). *Right*: Final structure of the *trans*-restrained simulation at 0.88 T_f (*green*), aligned with the crystal structure (3CHY.pdb) containing cis Pro110 (*purple*). *Spheres* denote the K109-P110 virtual bond (Color figure online)

folding transition temperature, T_f. In principle the ε interaction energies can be scaled to forecast a particular T_f, but consistency across systems of different size is difficult [68]. Karanicolas and Brooks adjusted by native contact energy per residue. While small proteins exhibit folding temperatures of ~350 K [22, 35], the 128-residue CheY underwent a cooperative folding transition at 301 K (Fig. 5), as determined from the temperature dependence of the heat capacity.

To examine the role of *cis–trans* isomerization, two additional simulations were performed with the K109-P110 torsion harmonically restrained to either the native value corresponding to the cis isomer or rotated by 180° (trans). Reference values of $\phi_{0,\text{cis}} = +57°$ and $\phi_{0,\text{trans}} = \phi_{\text{crystal}} - 180° = -123°$ for the Cα pseudodihedral angle were maintained using a force constant of 20 kcal/mol rad^2 (compare to Fig. 3a). Simulation with trans Pro110 revealed that CheY is still able to access the native basin (N) from the unfolded state (U), albeit with a destabilization relative to simulations with flexible Pro110 of $\Delta\Delta G_{\text{N-U}} = 8.8$ kJ/mol. As a control, minimal perturbation of the folding free energy was observed when Pro110 was restrained in the gauche configuration. The anti K109-P110 torsion causes local Cα backbone strain in the β5-α5 loop (Fig. 5). Larger destabilization may arise from atomistic interactions, as natively structured CheY has not been observed to exhibit the trans configuration at Pro110 [65]. A precedent for *cis–trans* isomerization within the native state can be found in the CheY-like response regulator Spo0A, in which isomerization of its homologous proline results in formation of an α5 helix-swapped dimer [69].

3 Gō-Like Models and Less than Native Interactions

3.1 To Gō or Not to Gō?

The incorporation of bonded potential terms including varying levels of bias to the native structure can result in significant variations to the underlying Gō or general CG representation of non-bond interactions. Consider the non-Gō three-color model of Oakley et al. and its five degenerate minimum energy structures (Fig. 1a, b). Close inspection reveals that in residue RMSD space the five structures are separated by at most 2.7 Å pairwise RMSD, which can be considered acceptable accuracy in CG fold prediction [70]. The modest structural stability can be attributed to the inclusion of stabilizing β-sheet dihedral terms (Fig. 3b). By contrast, two-color lattice models have predicted thousands of degenerate energy minima [23]. Analogously, the non-Gō model of Hills et al. incorporating rotatable Cα dihedrals and a basic description of residue polarity predicted misfolded degenerate energy minima in replica exchange simulations [49]. Some of the more promising non-Gō models for folding simulations limit misfolding by implementing sophisticated potentials to enforce angular constraints arising from regular backbone hydrogen bonding [71–77].

Cα-based models for protein dynamics simulation that lack Gō-like nonbond interactions require some form of structural restraints to stabilize the native state [56]. Consider the sidechain centric model of Marrink and colleagues, dubbed MARTINI [55]. Each of the 20 amino acids is represented by one to four interaction centers depending on size and polarity, and connected by backbone beads located at successive alpha carbons. Nonbond interactions consist of Lennard-Jones and screened Coulomb potentials. MARTINI was parameterized to reproduce amino acid partitioning in lipid bilayers [78, 79]. Other approaches use atomistic MD to compute the potential of mean force in aqueous solution (Fig. 6) [49, 77]. The utility of Cα + Cβ models is that sidechain sterics and polarity enable simulation of the insertion and interaction of peptides and proteins in CG bilayers [80, 81]. A limitation in studying transitions between distinct conformational states, however, is that the generic Cα backbone requires restraints to maintain a particular secondary or tertiary structure (Fig. 3c).

Non-Gō peptide models employ a variety of bonded potential terms such as described above [82]. In models including sidechain interaction sites an additional term is needed to prevent unphysical chirality (*D*) arising from the virtual beta carbon position [83, 84]. For example, Cα chirality can be added to the sidechain model of Hills et al. [49] by simply defining an improper torsion between every three residues: $V_{\text{chiral}} = 96.5$ kJ/mol rad^2 $(\phi - 15°)^2$, for each $C\alpha^{(i)}$-$C\alpha^{(i-1)}$-$C\alpha^{(i+1)}$-$C\beta^{(i)}$. This generic harmonic term is suitable for the overlapping probability distributions of α and β secondary structures (Fig. 7).

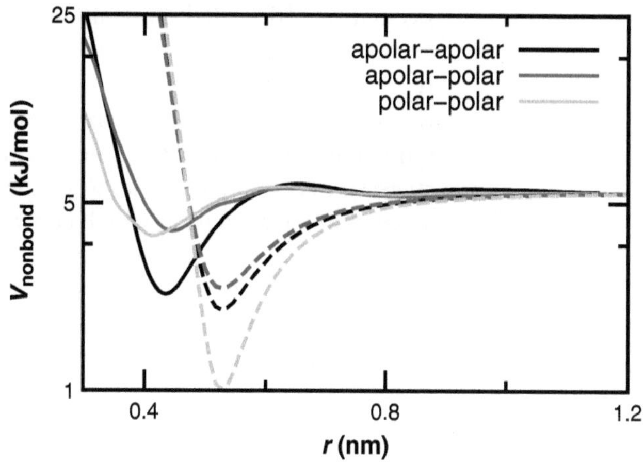

Fig. 6 Non-Gō sidechain pair interaction potentials. Tabulated potentials of mean force developed for aqueous peptides (*solid*) [49] are compared with Cβ Lennard-Jones terms from the membrane protein MARTINI model (*dash*) in log scale. Apolar parameters are taken from Val/Pro amino acids. Polar beads correspond to Ser/Thr

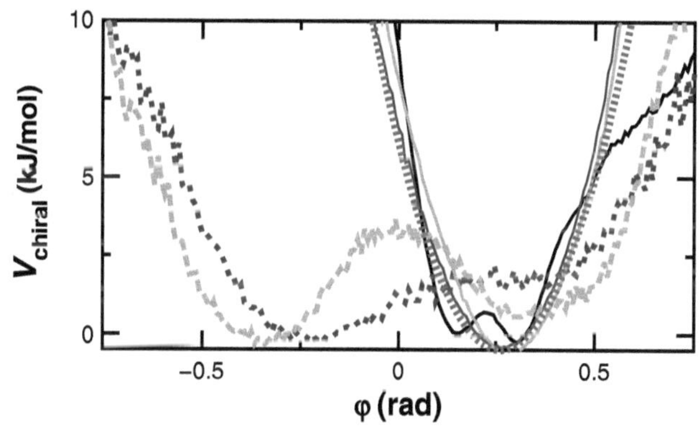

Fig. 7 Boltzmann-inverted $C\alpha^{(i)}$-$C\alpha^{(i-1)}$-$C\alpha^{(i+1)}$-$C\beta^{(i)}$ improper torsion distributions denoting amino acid chirality. Native simulations of Trpzip (*blue*) and Trp-cage (*green*) with the original model of Hills et al. (*dash*) deviate from atomistic MD (*black*) of Trpzip and polyleucine helix [49]. Application of the harmonic potential V_{chiral} (*purple*) is observed to stabilize the *L*-enantiomer (*solid colors*) (Color figure online)

Sequence effects arising from sidechains have also been incorporated into single bead Cα models. Karanicolas and Brooks [22] added energetic heterogeneity to their Gō model by weighting native interactions, ε_{ij}, by the relative abundance of their nonbonded amino acid pair combination in the Protein Data Bank [85]. The so-called Miyazawa-Jernigan contact energies have enjoyed further success in generic non-Gō models for CG simulation [83, 86] and in adding nonnative interactions to existing Gō models [60]. Alternatively, charged sidechains can be represented explicitly using Debye-Hückel electrostatic interactions [61, 87–89], which are important for nucleic acid studies. Chu et al. investigated the role of salt concentration in electrostatic steering between a disordered chaperone and histone [90]. Lastly, atomistic Gō models, in which contacting heavy atoms experience a 6–12 Lennard-Jones attraction, are a growing method of choice for including an explicit detailed sidechain representation [4, 26, 91]. All these approaches add ruggedness to the underlying ideal, smooth and funneled Gō landscape.

3.2 Multiple-Basin Transitions

A powerful outgrowth of Gō-like approaches is the ability to model functional dynamics by including bias to multiple reference structures representing endpoints of a conformational transition. The double-well potential was constructed with a valence bond approach by interpolating two elastic networks comprised of long range harmonic potentials [92, 93], resulting in a smooth transition barrier of adjustable height. The minimum energy path connecting the two conformers could be computed using an optimization algorithm. For general MD simulation of all accessible transition pathways, however, short range Lennard-Jones interactions are employed for dual Gō models corresponding to each structure [13]. For contacts shared between the two structures, clashes with the repulsive branches can be alleviated using exponential weighting [8, 9]. Recent methods simplify the interpolation by conjoining the two contact maps [94, 95]. Different equilibrium distances in contacts shared by the two reference basins can be incorporated safely with the use of an attractive double-basin Gaussian potential (Fig. 2) and a separate repulsive term [4, 12, 41]. One particular application involved manually selecting three disjoint, non-conflicting [96] contact maps and ligand-mediated [97] interdomain contacts to model the open, apo closed, and holo closed states of maltose binding protein [15].

The switching Gō model offers a simple illustration of functional transitions. By sequentially driving subunits in alternate reference conformations, Koga and Takada were able to reproduce the mechanistic rotary motion of F1-ATPase [98]. Consider the generic peptide model of Hills et al., which does not encode any structural information concerning the native state, as recently applied to the hinge bending motion of transmembrane helices in MsbA (Fig. 8a) [99]. Simulations with a modestly stabilizing Cα elastic network [56] of the open conformer exhibited hinge closing but could not predict domain orientation in the closed state (10.5 Å RMSD) [81]. A 50 ns switching simulation was performed starting from the open state with a Gaussian network of closed

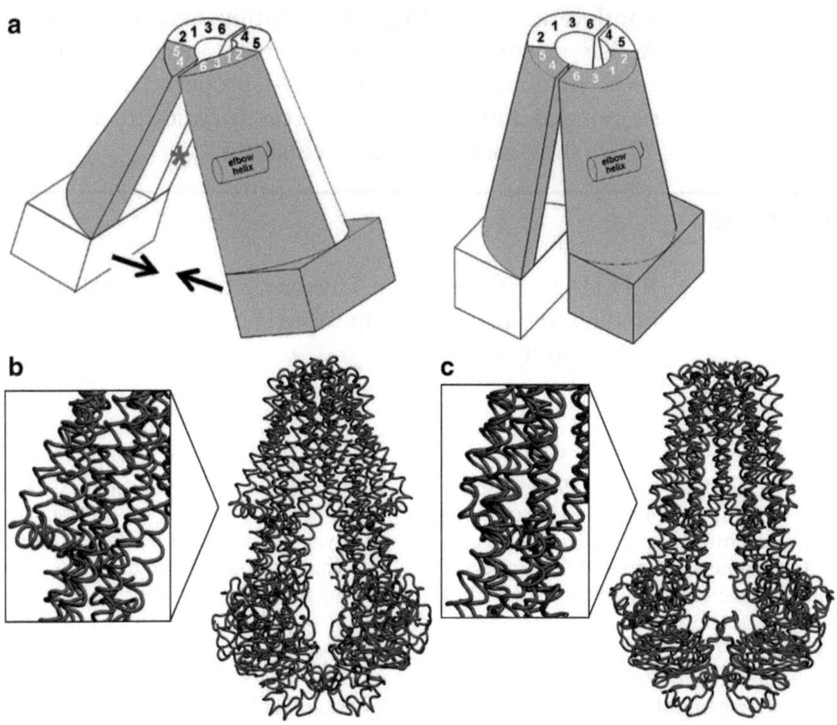

Fig. 8 MsbA conformational transition. (**a**) Cartoon representation of closing domain motion between monomer helices 1–6 [99]. © 2007 National Academy of Sciences. (**b**) Displacement along the open conformer's lowest frequency normal mode reproduces the closed crystal structure with 5.7 Å Cα RMSD. (**c**) Normal mode flexible fitting [100] of the open structure into the closed state using 20 modes successfully captures the helical register (4.0 Å RMSD). Monomers are colored *red* and *blue* for each aligned dimer structure. By comparison, a Gaussian network switching simulation with the model of Hills et al. using the closed contact map results in a 6.4 Å final RMSD after relaxation from the open conformer (Color figure online)

contacts. On top of the generic, nonnative interactions, Gaussian potentials were applied for alpha carbons separated by two or more bonds within a 10.5 Å cutoff:

$$V_G(r_{ij}) = -\varepsilon \exp\left(-(r_{ij} - r_{0ij})^2 / 2w^2\right), \tag{6}$$

where $\varepsilon = 5.2$ kJ/mol, $w = 0.0866$ nm, and each r_{0ij} was binned at 0.54, 0.76, or 0.97 nm. After 4 ns the simulation relaxed to within 6.4 Å Cα RMSD of the closed crystal structure, comparable to normal mode calculations performed on a Cα elastic network (Fig. 8) [100]. The nonbond contacts were parameterized to stabilize native structure while still allowing reasonable (2–4 Å) RMS fluctuations. Application to dramatic functional transitions such as these demonstrates the utility of combining Gō-like terms with a general force field.

In conclusion, substantial variations on the Gō model originally proposed for protein folding are finding increasing application in molecular processes of current interest including pulling, translocation, fly-casting, glycosylation, and crowding [101–106]. As has recently been observed from advances in normal mode conformational analysis [107, 108], proper separation of the intrinsic bonded and nonbonded forces is critical to constructing and interpreting useful CG representations.

Acknowledgments

R.D.H is grateful to the University of New England for startup funding, the Brooks Group for MD parameters, and Roy Johnston for providing coordinates.

References

1. Taketomi H, Ueda Y, Go N (1975) Studies on protein folding, unfolding and fluctuations by computer simulation. 1. Effect of specific amino acid sequence represented by specific inter-unit interactions. Int J Pept Protein Res 7:445–459
2. Bryngelson JD, Onuchic JN, Socci ND et al (1995) Funnels, pathways, and the energy landscape of protein folding: a synthesis. Proteins 21:167–195
3. Dill KA, Chan HS (1997) From Levinthal to pathways to funnels. Nat Struct Biol 4:10–19
4. Noel JK, Onuchic JN (2012) The many faces of structure-based potentials: from protein folding landscapes to structural characterization of complex biomolecules. In: Dokholyan NV (ed) Computational modeling of biological systems. Springer, New York, NY, pp 31–54
5. Hills RD Jr, Brooks CL III (2009) Insights from coarse-grained Go models for protein folding and dynamics. Int J Mol Sci 10:889–905
6. Takada S (2012) Coarse-grained molecular simulations of large biomolecules. Curr Opin Struct Biol 22:130–137
7. Adelman JL, Dale AL, Zwier MC et al (2011) Simulations of the alternating access mechanism of the sodium symporter Mhp1. Biophys J 101:2399–2407
8. Best RB, Chen YG, Hummer G (2005) Slow protein conformational dynamics from multiple experimental structures: the helix/sheet transition of arc repressor. Structure 13:1755–1763
9. Daily MD, Phillips GN Jr, Cui Q (2011) Interconversion of functional motions between mesophilic and thermophilic adenylate kinases. PLoS Comput Biol 7:e1002103
10. Grubisic I, Shokhirev MN, Orzechowski M et al (2010) Biased coarse-grained molecular dynamics simulation approach for flexible fitting of X-ray structure into cryo electron microscopy maps. J Struct Biol 169:95–105
11. Hyeon C, Jennings PA, Adams JA et al (2009) Ligand-induced global transitions in the catalytic domain of protein kinase A. Proc Natl Acad Sci USA 106:3023–3028
12. Lammert H, Schug A, Onuchic JN (2009) Robustness and generalization of structure-based models for protein folding and function. Proteins 77:881–891
13. Okazaki K, Koga N, Takada S et al (2006) Multiple-basin energy landscapes for large-amplitude conformational motions of proteins: structure-based molecular dynamics simulations. Proc Natl Acad Sci USA 103:11844–11849
14. Ratje AH, Loerke J, Mikolajka A et al (2010) Head swivel on the ribosome facilitates translocation by means of intra-subunit tRNA hybrid sites. Nature 468:713–716
15. Wang Y, Tang C, Wang E et al (2012) Exploration of multi-state conformational dynamics and underlying global functional landscape of maltose binding protein. PLoS Comput Biol 8:e1002471
16. Brown S, Fawzi NJ, Head-Gordon T (2003) Coarse-grained sequences for protein folding and design. Proc Natl Acad Sci USA 100:10712–10717
17. Favrin G, Irback A, Wallin S (2002) Folding of a small helical protein using hydrogen bonds and hydrophobicity forces. Proteins 47:99–105
18. Honeycutt JD, Thirumalai D (1990) Metastability of the folded states of globular proteins. Proc Natl Acad Sci USA 87:3526–3529

19. Irback A, Sjunnesson F, Wallin S (2000) Three-helix-bundle protein in a Ramachandran model. Proc Natl Acad Sci USA 97: 13614–13618
20. Miller MA, Wales DJ (1999) Energy landscape of a model protein. J Chem Phys 111:6610–6616
21. Takada S, Luthey-Schulten Z, Wolynes PG (1999) Folding dynamics with nonadditive forces: a simulation study of a designed helical protein and a random heteropolymer. J Chem Phys 110:11616–11629
22. Karanicolas J, Brooks CL III (2002) The origins of asymmetry in the folding transition states of protein L and protein G. Protein Sci 11:2351–2361
23. Yue K, Fiebig KM, Thomas PD et al (1995) A test of lattice protein folding algorithms. Proc Natl Acad Sci USA 92:325–329
24. Oakley MT, Wales DJ, Johnston RL (2011) Energy landscape and global optimization for a frustrated model protein. J Phys Chem B 115:11525–11529
25. Clementi C, Nymeyer H, Onuchic JN (2000) Topological and energetic factors: what determines the structural details of the transition state ensemble and "en-route" intermediates for protein folding? An investigation for small globular proteins. J Mol Biol 298:937–953
26. Noel JK, Whitford PC, Sanbonmatsu KY et al (2010) SMOG@ctbp: simplified deployment of structure-based models in GROMACS. Nucleic Acids Res 38:W657–W661
27. Sulkowska JI, Cieplak M (2008) Selection of optimal variants of Go-like models of proteins through studies of stretching. Biophys J 95:3174–3191
28. Noel JK, Whitford PC, Onuchic JN (2012) The shadow map: a general contact definition for capturing the dynamics of biomolecular folding and function. J Phys Chem B 116:8692–8702
29. Garcia LG, Pereira de Araujo AF (2006) Folding pathway dependence on energetic frustration and interaction heterogeneity for a three-dimensional hydrophobic protein model. Proteins 62:46–63
30. Capraro DT, Gosavi S, Roy M et al (2012) Folding circular permutants of IL-1beta: route selection driven by functional frustration. PLoS One 7:e38512
31. Hills RD Jr, Brooks CL III (2008) Subdomain competition, cooperativity, and topological frustration in the folding of CheY. J Mol Biol 382:485–495
32. Sobolev V, Sorokine A, Prilusky J et al (1999) Automated analysis of interatomic contacts in proteins. Bioinformatics 15:327–332
33. Clementi C (2008) Coarse-grained models of protein folding: toy models or predictive tools? Curr Opin Struct Biol 18:10–15
34. Hills RD Jr, Kathuria SV, Wallace LA et al (2010) Topological frustration in beta alpha-repeat proteins: sequence diversity modulates the conserved folding mechanisms of alpha/beta/alpha sandwich proteins. J Mol Biol 398:332–350
35. Karanicolas J, Brooks CL III (2003) Improved Go-like models demonstrate the robustness of protein folding mechanisms towards non-native interactions. J Mol Biol 334:309–325
36. Periole X, Allen LR, Tamiola K et al (2009) Probing the free energy landscape of the FBP28 WW domain using multiple techniques. J Comput Chem 30:1059–1068
37. Ivankov DN, Garbuzynskiy SO, Alm E et al (2003) Contact order revisited: influence of protein size on the folding rate. Protein Sci 12:2057–2062
38. Chan HS, Zhang Z, Wallin S et al (2011) Cooperativity, local-nonlocal coupling, and nonnative interactions: principles of protein folding from coarse-grained models. Annu Rev Phys Chem 62:301–326
39. Enciso M, Rey A (2011) Improvement of structure-based potentials for protein folding by native and nonnative hydrogen bonds. Biophys J 101:1474–1482
40. Kim J, Keyes T (2008) Influence of Go-like interactions on global shapes of energy landscapes in beta-barrel forming model proteins: inherent structure analysis and statistical temperature molecular dynamics simulation. J Phys Chem B 112:954–966
41. Zarrine-Afsart A, Wallin S, Neculai AM et al (2008) Theoretical and experimental demonstration of the importance of specific nonnative interactions in protein folding. Proc Natl Acad Sci USA 105:9999–10004
42. Hills RD Jr, Brooks CL III (2008) Coevolution of function and the folding landscape: correlation with density of native contacts. Biophys J 95:L57–L59
43. Meireles L, Gur M, Bakan A et al (2011) Pre-existing soft modes of motion uniquely defined by native contact topology facilitate ligand binding to proteins. Protein Sci 20:1645–1658
44. Tama F, Brooks CL III (2006) Symmetry, form, and shape: guiding principles for robustness in macromolecular machines. Annu Rev Biophys Biomol Struct 35:115–133
45. Naganathan AN, Orozco M (2011) The protein folding transition-state ensemble from

a Go-like model. Phys Chem Chem Phys 13:15166–15174
46. Cho SS, Pincus DL, Thirumalai D (2009) Assembly mechanisms of RNA pseudoknots are determined by the stabilities of constituent secondary structures. Proc Natl Acad Sci USA 106:17349–17354
47. Feng J, Walter NG, Brooks CL 3rd (2011) Cooperative and directional folding of the preQ1 riboswitch aptamer domain. J Am Chem Soc 133:4196–4199
48. Sosnick TR, Pan T (2004) Reduced contact order and RNA folding rates. J Mol Biol 342:1359–1365
49. Hills RD Jr, Lu L, Voth GA (2010) Multiscale coarse-graining of the protein energy landscape. PLoS Comput Biol 6:e1000827
50. Kamagata K, Kuwajima K (2006) Surprisingly high correlation between early and late stages in non-two-state protein folding. J Mol Biol 357:1647–1654
51. Naganathan AN, Munoz V (2005) Scaling of folding times with protein size. J Am Chem Soc 127:480–481
52. Zou T, Ozkan SB (2011) Local and non-local native topologies reveal the underlying folding landscape of proteins. Phys Biol 8:066011
53. Sieradzan AK, Scheraga HA, Liwo A (2012) Determination of effective potentials for the stretching of Ca-Ca virtual bonds in polypeptide chains for coarse-grained simulations of proteins from ab initio energy surfaces of N-methylacetamide and N-acetylpyrrolidine. J Chem Theory Comput 8:1334–1343
54. Peter C, Kremer K (2009) Multiscale simulation of soft matter systems: from the atomistic to the coarse-grained level and back. Phys Chem Chem Phys 5:4357–4366
55. Monticelli L, Kandasamy SK, Periole X et al (2008) The MARTINI coarse-grained force field: extension to proteins. J Chem Theory Comput 4:819–834
56. Periole X, Cavalli M, Marrink SJ et al (2009) Combining an elastic network with a coarse-grained molecular force field: structure, dynamics, and intermolecular recognition. J Chem Theory Comput 5:2531–2543
57. Winger M, Trzesniak D, Baron R et al (2009) On using a too large integration time step in molecular dynamics simulations of coarse-grained molecular models. Phys Chem Chem Phys 11:1934–1941
58. Hess B, Kutzner C, van der Spoel D et al (2008) GROMACS 4: algorithms for highly efficient, load-balanced, and scalable molecular simulation. J Chem Theory Comput 4: 435–447
59. Kwiecinska JI, Cieplak M (2005) Chirality and protein folding. J Phys Condens Matter 17:S1565–S1580
60. Skrbic T, Micheletti C, Faccioli P (2012) The role of non-native interactions in the folding of knotted proteins. PLoS Comput Biol 8: e1002504
61. Okazaki K, Sato T, Takano M (2012) Temperature-enhanced association of proteins due to electrostatic interaction: a coarse-grained simulation of actin-myosin binding. J Am Chem Soc 134:8918–8925
62. Bellesia G, Jewett AI, Shea JE (2010) Sequence periodicity and secondary structure propensity in model proteins. Protein Sci 19:141–154
63. Chikenji G, Fujitsuka Y, Takada S (2006) Shaping up the protein folding funnel by local interaction: lesson from a structure prediction study. Proc Natl Acad Sci USA 103:3141–3146
64. Munoz V, Lopez EM, Jager M et al (1994) Kinetic characterization of the chemotactic protein from Escherichia coli, CheY. Kinetic analysis of the inverse hydrophobic effect. Biochemistry 33:5858–5866
65. Kathuria SV, Day IJ, Wallace LA et al (2008) Kinetic traps in the folding of beta alpha-repeat proteins: CheY initially misfolds before accessing the native conformation. J Mol Biol 382:467–484
66. Allen LR, Krivov SV, Paci E (2009) Analysis of the free-energy surface of proteins from reversible folding simulations. PLoS Comput Biol 5:e1000428
67. Mohazab AR, Plotkin SS (2009) Structural alignment using the generalized Euclidean distance between conformations. Int J Quantum Chem 109:3217–3228
68. Accary JB, Teboul V (2012) Time versus temperature rescaling for coarse grain molecular dynamics simulations. J Chem Phys 136: 094502
69. Lewis RJ, Muchova K, Brannigan JA et al (2000) Domain swapping in the sporulation response regulator SpoOA. J Mol Biol 297:757–770
70. Raman S, Lange OF, Rossi P et al (2010) NMR structure determination for larger proteins using backbone-only data. Science 327:1014–1018
71. Barducci A, Bonomi M, Derreumaux P (2011) Assessing the quality of the OPEP coarse-grained force field. J Chem Theory Comput 7:1928–1934
72. Bereau T, Deserno M, Bachmann M (2011) Structural basis of folding cooperativity in

model proteins: insights from a microcanonical perspective. Biophys J 100:2764–2772
73. Davtyan A, Schafer NP, Zheng W et al (2012) AWSEM-MD: protein structure prediction using coarse-grained physical potentials and bioinformatically based local structure biasing. J Phys Chem B 116:8494–8503
74. Enciso M, Rey A (2012) Simple model for the simulation of peptide folding and aggregation with different sequences. J Chem Phys 136:215103
75. Golas E, Maisuradze GG, Senet P et al (2012) Simulation of the opening and closing of Hsp70 chaperones by coarse-grained molecular dynamics. J Chem Theory Comput 8:1750–1764
76. Liwo A, Kazmierkiewicz R, Czaplewski C et al (1998) United-residue force field for off-lattice protein structure simulations. III. Origin of backbone hydrogen bonding cooperativity in united-residue potentials. J Comput Chem 19:259–276
77. Sobolewski E, Oldziej S, Wisniewska M et al (2012) Toward temperature-dependent coarse-grained potentials of side-chain interactions for protein folding simulations. II. Molecular dynamics study of pairs of different types of interactions in water at various temperatures. J Phys Chem B 116:6844–6853
78. de Jong DH, Periole X, Marrink SJ (2012) Dimerization of amino acid side chains: lessons from the comparison of different force fields. J Chem Theory Comput 8:1003–1014
79. Singh G, Tieleman DP (2011) Using the Wimley-White hydrophobicity scale as a direct quantitative test of force fields: the MARTINI coarse-grained model. J Chem Theory Comput 7:2316–2324
80. Hall BA, Chetwynd AP, Sansom MS (2011) Exploring peptide-membrane interactions with coarse-grained MD simulations. Biophys J 100:1940–1948
81. Ward AB, Guvench O, Hills RD Jr (2012) Coarse grain lipid-protein molecular interactions and diffusion with MsbA flippase. Proteins 80:2178–2190
82. Seo M, Rauscher S, Pomes R et al (2012) Improving internal peptide dynamics in the coarse-grained MARTINI model: toward large-scale simulations of amyloid- and elastin-like peptides. J Chem Theory Comput 8:1774–1785
83. Bereau T, Deserno M (2009) Generic coarse-grained model for protein folding and aggregation. J Chem Phys 130:235106
84. Cheung MS, Finke JM, Callahan B et al (2003) Exploring the interplay between topology and secondary structural formation in the protein folding problem. J Phys Chem B 107:11193–11200
85. Miyazawa S, Jernigan RL (1996) Residue-residue potentials with a favorable contact pair term and an unfavorable high packing density term, for simulation and threading. J Mol Biol 256:623–644
86. Kim YC, Hummer G (2008) Coarse-grained models for simulations of multiprotein complexes: application to ubiquitin binding. J Mol Biol 375:1416–1433
87. Givaty O, Levy Y (2009) Protein sliding along DNA: dynamics and structural characterization. J Mol Biol 385:1087–1097
88. Hyeon C, Thirumalai D (2005) Mechanical unfolding of RNA hairpins. Proc Natl Acad Sci USA 102:6789–6794
89. O'Brien EP, Christodoulou J, Vendruscolo M et al (2012) Trigger factor slows co-translational folding through kinetic trapping while sterically protecting the nascent chain from aberrant cytosolic interactions. J Am Chem Soc 134:10920–10932
90. Chu X, Wang Y, Gan L et al (2012) Importance of electrostatic interactions in the association of intrinsically disordered histone chaperone Chz1 and histone H2A.Z-H2B. PLoS Comput Biol 8:e1002608
91. Whitford PC, Noel JK, Gosavi S et al (2009) An all-atom structure-based potential for proteins: bridging minimal models with all-atom empirical forcefields. Proteins 75:430–441
92. Chu JW, Voth GA (2007) Coarse-grained free energy functions for studying protein conformational changes: a double-well network model. Biophys J 93:3860–3871
93. Maragakis P, Karplus M (2005) Large amplitude conformational change in proteins explored with a plastic network model: adenylate kinase. J Mol Biol 352:807–822
94. Noel JK, Schug A, Verma A et al (2012) Mirror images as naturally competing conformations in protein folding. J Phys Chem B 116: 6880–6888
95. Whitford PC, Miyashita O, Levy Y et al (2007) Conformational transitions of adenylate kinase: switching by cracking. J Mol Biol 366:1661–1671
96. Singh JP, Whitford PC, Hayre NR et al (2012) Massive conformation change in the prion protein: using dual-basin structure-based models to find misfolding pathways. Proteins 80:1299–1307
97. Okazaki K, Takada S (2008) Dynamic energy landscape view of coupled binding and protein conformational change: induced-fit

versus population-shift mechanisms. Proc Natl Acad Sci USA 105:11182–11187
98. Koga N, Takada S (2006) Folding-based molecular simulations reveal mechanisms of the rotary motor F-1-ATPase. Proc Natl Acad Sci USA 103:5367–5372
99. Ward A, Reyes CL, Yu J et al (2007) Flexibility in the ABC transporter MsbA: alternating access with a twist. Proc Natl Acad Sci USA 104:19005–19010
100. Tama F, Miyashita O, Brooks CL III (2004) Flexible multi-scale fitting of atomic structures into low-resolution electron density maps with elastic network normal mode analysis. J Mol Biol 337:985–999
101. Bacci M, Chinappi M, Casciola CM et al (2012) Role of denaturation in maltose binding protein translocation dynamics. J Phys Chem B 116:4255–4262
102. Chen J (2012) Towards the physical basis of how intrinsic disorder mediates protein function. Arch Biochem Biophys 524: 123–131
103. Lee W, Zeng X, Rotolo K et al (2012) Mechanical anisotropy of ankyrin repeats. Biophys J 102:1118–1126
104. Shental-Bechor D, Arviv O, Hagai T et al (2010) Folding of conjugated proteins. Annu Rep Comput Chem 6:263–277
105. Wang Q, Cheung MS (2012) A physics-based approach of coarse-graining the cytoplasm of Escherichia coli (CGCYTO). Biophys J 102: 2353–2361
106. Whitford PC, Ahmed A, Yu Y et al (2011) Excited states of ribosome translocation revealed through integrative molecular modeling. Proc Natl Acad Sci USA 108: 18943–18948
107. Bray JK, Weiss DR, Levitt M (2011) Optimized torsion-angle normal modes reproduce conformational changes more accurately than cartesian modes. Biophys J 101:2966–2969
108. Lu M, Ma J (2011) Normal mode analysis with molecular geometry restraints: bridging molecular mechanics and elastic models. Arch Biochem Biophys 508:64–71

Chapter 8

A Tutorial on Building Markov State Models with MSMBuilder and Coarse-Graining Them with BACE

Gregory R. Bowman

Abstract

Markov state models (MSMs) are a powerful means of (1) making sense of molecular simulations, (2) making a quantitative connection between simulation and experiment, and (3) driving efficient simulations. A Markov model can be thought of as a map of the conformational space a molecule explores. Instead of having towns and cities connected with roads labeled with speed limits, a Markov model has conformational states and probabilities of transitioning between pairs of these states. This tutorial describes how to build Markov models and a few of the basic analyses that can be performed with the MSMBuilder software package.

Key words Molecular dynamics, Master equation, Protein dynamics

1 Introduction

Molecular dynamics simulations are a powerful way of understanding molecular systems, particularly those that are difficult to probe experimentally due to factors like conformational heterogeneity. However, fully realizing the potential of such simulations requires methods for (1) analyzing simulation datasets to extract understanding, (2) making quantitative predictions of experiments, and (3) driving efficient simulations.

Markov state models are an attractive option for fulfilling the aforementioned objectives [1, 2]. A Markov model consists of a set of states—each of which contains rapidly mixing conformations—and a transition probability matrix, where the entry in row i and column j denotes the probability of jumping from state i to state j in a short time interval—called the lag time of the model—given that the system is currently in state i. Intuitively, it is useful to think of states as corresponding to local minima in the free energy landscape that ultimately determines a molecular system's structure and dynamics. The transition probabilities can then be thought of as being related to the rate for transitioning across the barrier

Dennis R. Livesay (ed.), *Protein Dynamics: Methods and Protocols*, Methods in Molecular Biology, vol. 1084, DOI 10.1007/978-1-62703-658-0_8, © Springer Science+Business Media New York 2014

separating two states. These models are called Markov state models because it is assumed that the current state a system occupies is sufficient to specify the chance that it will transition to any other state during the next time interval; that is, the choice of the next state is independent of the system's history.

In the physical sciences, Markov models are often referred to as discrete-time master equation models [3]. Indeed these models have a long history and there are many well-developed methods for analyzing them and using them to make predictions. However, their application to the analysis of molecular simulations has been limited by the difficulties inherent in identifying a valid set of states—often referred to as a state decomposition.

The primary challenge to identifying a valid state decomposition is finding a set of kinetically relevant states. Ideally, Markov models would be created using a purely kinetic clustering of a simulation dataset. Unfortunately, one cannot generally analytically derive the transition probability between two arbitrary conformations because the free energy landscapes of most biomolecules are too vast and rugged. Therefore, numerical methods are required for building Markov models.

Recently, great progress has been made in the development of robust methods for constructing Markov models [2, 4–9]. The key insight is that geometric distances can be used as surrogates for kinetic distances on very short scales. For example, two conformations differing from one another by a very small RMSD (say 1 Å) are also very likely to be kinetically close (i.e., they are likely to be separated by only a small free energy barrier, allowing them to interconvert rapidly). Therefore, one can build Markov models by first grouping very similar conformations together based on geometric criteria into what are often call microstates and, second, using kinetic information to then group rapidly mixing microstates into larger aggregates called macrostates.

In the remainder of this chapter, I will describe how to build Markov models from molecular dynamics simulations using the MSMBuilder software (version 2.6.0 [7]). To make this concrete, I will demonstrate how to build a Markov model for the alanine dipeptide (Fig. 1) using the data distributed with MSMBuilder. The major steps are

- Setup the appropriate input files.
- Perform a geometric clustering to build a microstate model appropriate for making a quantitative connection with experiment.
- Use BACE [10] to build a macrostate model (i.e., coarse-graining of the microstate model) appropriate for gaining understanding.

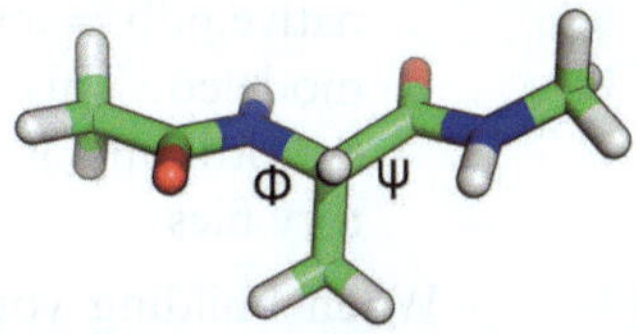

Fig. 1 A representative structure of the alanine dipeptide. This molecule essentially has two degrees of freedom: the Φ and Ψ dihedral angles. Carbons are *green*, hydrogens are *white*, nitrogens are *blue*, and oxygens are *red*

For readers interested in more details regarding the theory behind Markov models and how they can be used to gain understanding, model experiments, and drive efficient simulations, I recommend refs. 1, 2, 11, 12.

2 Methods

1. Download MSMBuilder (currently available at https://simtk.org/home/msmbuilder) and install it by following the instructions provided in the file Tutorial.pdf. The $ sign in the commands provided in this tutorial just denote the beginning of each line in a terminal window.

   ```
   $ tar xzvf msmbuilder2.6.0.tar.bz
   $ cd msmbuilder/
   $ open Tutorial.pdf
   ```

   Installing MSMBuilder will place a number of scripts in your PATH environment variable, allowing you to run them without specifying the entire path. You can learn about the options available with any of these scripts by running with the -h option (e.g., "Cluster.py -h" will print a description of the options available for the clustering script).

2. Move to the MSMBuilder Tutorial directory.

   ```
   $ cd Tutorial/
   ```

   This directory contains all the files required to build a Markov model for the alanine dipeptide (Fig. 1). The key files are:

   - AtomIndices.dat = an array of atom indices (numbered starting at 1) to base the initial geometric clustering of the data on.
   - XTC = a directory of trajectories. Each subdirectory (RUN00, RUN01, ...) contains a series of trajectory files (frame0.xtc, frame1.xtc, ...) in the Gromacs xtc format. Together, the files in one RUN directory form one continuous trajectory (i.e., frame1.xtc is a continuation of frame0.xtc, frame2.xtc is a continuation of frame1.xtc, and so forth).

- native.pdb = a representative structure of the system being modeled. This file is used to determine information like the ordering and names of the atom/residues in the trajectory files.

When building your own Markov models, you will need to create these inputs, though they need not have the same names. You can create an AtomIndices file by hand or use the CreateAtomIndices.py script. Your raw trajectory data should be organized as in the XTC directory. MSMBuilder can also read dcd files organized in the same way (as in the DCD directory provided in the Tutorial directory).

3. Convert the trajectories into MSMBuilder's internal lh5 format (*see* **Note 1** for an explanation of MSMBuilder data formats).

   $ ConvertDataToHDF.py -s native.pdb -i XTC

   You should see output something like

   ...
   21:03:21 - ['XTC/RUN97/frame0.xtc'], length 501, converted to Trajectories/trj97.lh5
   21:03:22 - ['XTC/RUN98/frame0.xtc'], length 501, converted to Trajectories/trj98.lh5
   21:03:22 - ['XTC/RUN99/frame0.xtc'], length 501, converted to Trajectories/trj99.lh5
   21:03:22 - Finished data conversion successfully.
   21:03:22 - Generated: ProjectInfo.yaml, Trajectories/

   The -s option tells MSMBuilder that the simulation trajectories are of the system given in native.pdb. The -I option points the script to a directory of trajectories called XTC. Other options can be seen by running the script with the -h option.

   This command will create a directory called Trajectories containing files called trj0.lh5, trj1.lh5, etc. Each of these files represents one of the input trajectories (e.g., trj0.lh5 is created by concatenating all the trajectory files in XTC/RUN00/and then converting to the MSMBuilder format). This command will also use information from the trajectories and the representative conformation specified (in this case, native.pdb) to create a file called ProjectInfo.yaml that contains basic information about the dataset, like the number and length of trajectories.

4. Build a microstate model by clustering the data with the hybrid k-centers/k-medoids algorithm [7]. *See* **Note 2** for a brief description of this algorithm.

   $ Cluster.py rmsd hybrid -d 0.035
   You should see output something like
   ...

21:06:23 - Sweep 9, swapping medoid 104 (conf 16110) for conf 220...
21:06:23 - Reject. New f = 0.017534, Old f = 0.017479
21:06:23 - Sweep 9, swapping medoid 105 (conf 20355) for conf 47209...
21:06:24 - Reject. New f = 0.017507, Old f = 0.017479
21:06:24 - Sweep 9, swapping medoid 106 (conf 4206) for conf 6730...
21:06:24 - Reject. New f = 0.017473, Old f = 0.017479
21:06:24 - Saving Data/Gens.lh5
21:06:24 - Since stride = 1, Saving Data/Assignments.h5
21:06:24 - Since stride = 1, Saving Data/Assignments.h5.distances

The rmsd option tells MSMBuilder to cluster the data based on the root-mean-square deviation (RMSD) between conformations. You can specify which atoms to base the RMSD calculations on with the -a option but, by default, MSMBuilder will look for a set of indices in a file called AtomIndices.dat. The hybrid option tells MSMBuilder to use the hybrid k-centers/k-medoids algorithm. Finally, the -d option tells MSMBuilder to continue breaking the dataset into progressively smaller states until no state has a radius (i.e., greatest distance between the cluster center and any other data point) greater than 0.035 nm.

Running this command will create a directory called Data containing a number of important outputs. The key files are:

- Assignments.h5 = a matrix where the element in row i and column j denotes the microstate that trajectory i is in at time slice j. Microstates are numbered from 0 to $n - 1$, where n is the total number of microstates. Trajectories are padded with −1's to make them all the same length. These values are ignored in all analyses performed by MSMBuilder.
- Assignments.h5.distances = a matrix specifying the distance from each conformation to the cluster center it is assigned to. This matrix is organized in the same manner as Assignments.h5.
- Gens.lh5 = an MSMBuilder trajectory file containing the structures of the cluster centers.

Note that the number of states you obtain and their definitions may differ slightly from the results presented here due to a stochastic component in the clustering algorithm. *See* **Notes 2–4** for more information on clustering.

5. Test whether or not the model satisfies the Markov assumption and choose a lag time by checking the model's relaxation timescales (also called implied timescales [13]). *See* **Note 5** for a

brief discussion of how these relaxation timescales are calculated and interpreted.

$ CalculateImpliedTimescales.py -l 1,25 -i 1 -o Data/ImpliedTimescales.dat

You should see output something like

...
21:10:48 - Selected component 0 with population 1.000000
21:10:48 - Log-Likelihood of intial guess for reversible transition probability matrix: -192136.084242
21:10:49 - Log-Likelihood after 56 function evaluations; -192134.684426
21:10:49 - Result of last maximization run (run 1): Converged (|f_n-f_(n-1)| ~ = 0)
21:10:49 - Log-Likelihood of final reversible transition probability matrix: -192134.684426
21:10:49 - Likelihood ratio: 4.05445504626
21:10:49 - Saved output to Data/ImpliedTimescales.dat

Together, the -l and -i options tell MSMBuilder to calculate the relaxation timescales of transition matrices estimated with a series of lag times (or observation intervals). The -l option tells the script to calculate relaxation timescales for lag times from 1 to 25 steps and the -i options tells the script to use lag times at 1 step intervals (i.e., relaxation timescales will be calculated for lag times of 1, 2, 3, ... 25 steps). Conformations were stored once a picosecond in the original trajectory data, so each step corresponds to 1 ps. The -o option specifies where to store the results.

The output of this script is a flat text file with two columns. The first is the lag time (in units of steps) and the second is the relaxation timescales.

If the model is Markovian, then these timescales should be invariant with respect to the lag time. Typically, the relaxation times will show a strong dependence on the observation interval at short lag times but will become invariant at longer lag times. For the final model, we will use the lag time where the relaxation timescales first appear to become invariant (this lag time is often called the Markov time). A good Markov time should be significantly shorter than the global relaxation timescale for the system. For example, if you are modeling protein folding for a system that folds on a μs timescale, then your Markov time should be much faster than 1 μs to resolve how the process occurs.

You can plot the results with

$ PlotImpliedTimescales.py -d 1. -i Data/ImpliedTimescales.dat

You should get a plot that looks something like Fig. 2. The three slowest relaxation times have leveled out by 3 ps, so

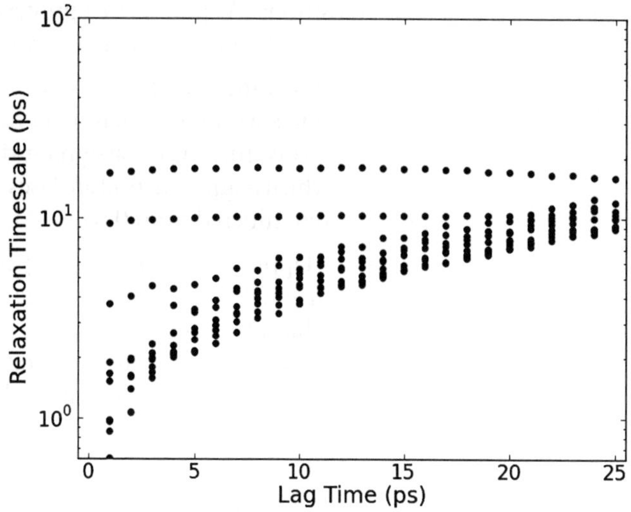

Fig. 2 Relaxation times of the microstate model for alanine dipeptide

we'll choose that as our Markov time. A brief discussion of other approaches to model validation is given in **Note 5**. *See* **Note 6** for a discussion of what steps to take if the relaxation timescales do not level off.

6. Estimate a transition probability matrix at the desired lag time (in this case, 3 ps).

   $ BuildMSM.py -l 3 -o MicroMSM_3ps

   You should see output something like

   ...
   21:13:56 - Log-Likelihood of final reversible transition probability matrix: -187592.661034
   21:13:56 - Likelihood ratio: 1.26278269254
   21:13:56 - Ergodic trimming discarded: 0.000000 % of your data
   21:13:57 - Wrote: MicroMSM_3ps/tProb.mtx
   21:13:57 - Wrote: MicroMSM_3ps/tCounts.mtx
   21:13:57 - Wrote: MicroMSM_3ps/Mapping.dat
   21:13:57 - Wrote: MicroMSM_3ps/Assignments.Fixed.h5
   21:13:57 - Wrote: MicroMSM_3ps/Populations.dat

   The -l option tells MSMBuilder to construct a model with a lag time of three steps (which, in this case, is equivalent to 3 ps). The -o option tells the script to create a directory called MicroMSM_3ps and store the model there.

   This command will count the number of transitions observed between each pair of states and then use a maximum likelihood method [7] to estimate the transition probability matrix that is most likely to have generated the observed transitions. As part of this process, microstates with insufficient data will be discarded. The key output files are:

- Assignments.Fixed.h5 = a new assignments file where microstates with insufficient data have been discarded. In practical terms, this means assigning data in these states to state −1, which is ignored in all analyses. The n remaining states are then renumbered from 0 to $n - 1$.
- Mapping.dat = this file specifies the mapping from the original microstates to the states in the current model. This file is needed at this stage in case some states are discarded (e.g., due to poor statistics). In later steps when we coarse-grain our model, this file will specify which microstates are grouped together. The integer on line i is the state microstate i has been assigned to.
- Populations.dat = the equilibrium probability of each state. The floating point number on line i is the probability the system is in state i.
- tCounts.mtx = a matrix of transition counts where the element in row i and column j denotes the number of transitions from state i to state j. By default, these are not the raw counts, as observed in the original trajectories. Rather, they are the counts after enforcing detailed balance with a maximum likelihood method (*see* **Note 7** for more on the motivation and alternative methods for enforcing detailed balance).
- tProb.mtx = the transition probability matrix estimated from the counts.

This microstate model is useful for modeling experiments because of its high temporal and spatial resolution. For example, you can setup an initial condition (consisting of a vector specifying the initial probability of being in each state) and then model how this ensemble relaxes to equilibrium by repeatedly multiplying by the transition probability matrix. However, microstate models will generally be too complicated (e.g., have too many states) to easily understand. Therefore, it is often useful to create coarse-grained models with fewer states. *See* **Note 8** for further discussion of the motivations for coarse-graining.

7. Use the Bayesian agglomerative clustering engine (BACE) [10] to define coarse-grained models with all possible numbers of macrostates.

```
$ BuildMSM.py -l 1 -o MicroMSM_1ps
$ BACE_Coarse_Graining.py -c MicroMSM_1ps/tCounts.mtx -f
```

You should see output something like

```
...
21:21:01 - Iteration 95, merging 7 states
21:21:01 - Iteration 96, merging 6 states
21:21:01 - Iteration 97, merging 5 states
21:21:01 - Iteration 98, merging 4 states
21:21:01 - Iteration 99, merging 3 states
```

The first command builds a model with a lag time of one step (in this case, 1 ps). This model may not be Markovian, however, it is useful for finding valid state decompositions as it contains the maximum possible data.

The second command uses the BACE algorithm to coarse-grain the model by identifying macrostates (aggregates of microstates) that best preserve the behavior of the original microstate model. BACE works by iteratively merging the two most kinetically similar states (i.e., most rapidly interconverting). The -c option tells MSMBuilder where the model to coarse-grain is (specifically, its transition count matrix). The -f option forces the script to use dense matrices. By default, the script uses whatever format (sparse or dense) the matrix it is coarse-graining was stored in (usually MSMBuilder saves sparse matrices). However, operating on dense matrices is faster and, therefore, is recommended if you have sufficient memory. The model used in this Tutorial has a small enough number of states that using dense matrices is reasonable. However, dense matrices may not be an option for many real-world scenarios where microstate models will have too many states for a dense representation of the transition probability matrix to fit in memory.

BACE stores its output in a directory called Output_BACE (unless you specify an alternative directory name with the -o option). The main outputs are:

- bayesFactors.dat = the Bayes factors during each iteration of BACE. The first column is the number of states (N) at the current stage and the second column is the Bayes factor (which, in this case, can be thought of as the cost of going from $N + 1$ to N states).
- mapN.dat = a mapping from microstates to N macrostates. If, for example, you started with 100 microstates, then you will have mappings called map99.dat to map2.dat.

See **Note 9** for a brief discussion of alternative coarse-graining methods and *see* **Note 10** for alternative implementations of these methods (and other Markov model methods).

8. Choose how many macrostates to use for further analysis by examining the Bayes factors (i.e., costs) as a function of the number of macrostates (Fig. 3). In general, you want to choose the number of macrostates to minimize the amount of

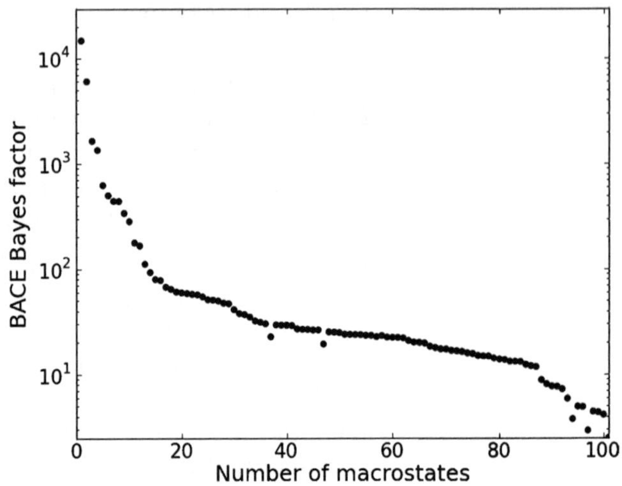

Fig. 3 BACE Bayes factors capture the cost of progressively merging the two most similar states as the algorithm iteratively coarse-grains the initial microstate model

information lost (relative to the original model). Choosing M macrostates is a good idea if there is a large increase in the Bayes factor when going from M macrostates to $M - 1$ macrostates. If you want a coarse-grained model that is still useful for making a quantitative connection with experiment, you would want to identify a jump where the Bayes factors are relatively small (and, usually, the number of macrostates is large). For example, with my output, 88 states would be a reasonable choice in this case because the Bayes factor increases by 0.2–0.5/macrostate as the algorithm coarse grains from 101 to 88 states but then suddenly jumps by ~3 when creating 87 states. Note, however, that you may obtain somewhat different output due to the stochastic nature of the clustering algorithm. To obtain a more qualitative model that is simpler to understand, you want to choose a much smaller number of states near a jump in the Bayes factor. For example, five states would be reasonable in this case because there's a large jump in the Bayes factor when going from 5 to 4 states (*see* Fig. 3).

For the purpose of this tutorial, we'll build five macrostates to get a simple model that one could actually understand (e.g., by looking at representative structures for each state and the transition probabilities between them).

9. Build a complete macrostate model (in this case with five states) using the state definition from BACE. Note that this is not the final macrostate model as we still have to choose an appropriate lag time.

```
$ BuildMSM.py -l 1 -a MicroMSM_1ps/Assignments.Fixed.h5 -m Output_BACE/map5.dat -o BACE_5state_1ps
```

This command is similar to ones we have used already but with two additional options. The -a option points MSMBuilder to an initial set of assignments to states. Previously, we left the default value for this option (Data/Assignments.h5) because we wanted to estimate transition matrices from the original clustering. Now, we want to estimate transition matrices from our microstate model (which may have discarded some poorly sampled states from the initial clustering), so we point the script to the microstate assignments. The -m option points the script to a file defining a set of macrostate definitions (i.e., a mapping from microstates to macrostates). The outputs (as described previously) are written to a directory called BACE_5state_1ps.

10. Test whether the macrostate model satisfies the Markov property and choose a Markov time, again using the relaxation timescales (or implied timescales).

    $ CalculateImpliedTimescales.py -l 1,25 -i 1 -a BACE_5state_1ps/Assignments.Fixed.h5 -o BACE_5state_1ps/ImpliedTimescales.dat -e 4
    $ PlotImpliedTimescales.py -d 1. -i BACE_5state_1ps/ImpliedTimescales.dat

    The main new option is -e, which instructs MSMBuilder to only calculate the first four relaxation times. The relaxation times are a function of the eigenvalues of the transition probability matrix. Since there are only five states, there are only four eigenvalues with dynamic information (the fifth eigenvalue is always 1). Therefore, it is nonsensical to ask MSMBuilder for the default number of eigenvalues (which is 10).

    A good macrostate model should preserve the slowest relaxation timescales from the microstate model it was built from. For example, Fig. 4 shows that this macrostate model preserves the two slowest timescales. As at the microstate level, we must choose an appropriate lag time based on where the relaxation timescales appear to become invariant. As it turns out, three steps is still a reasonable choice in this case. In many real world applications, however, macrostate models will have larger Markov times than microstate models.

11. Build the final macrostate model at the desired lag time (in this case, 3 ps).

    $ BuildMSM.py -l 3 -a MicroMSM_1ps/Assignments.Fixed.h5 -m Output_BACE/map5.dat -o BACE_5state_3ps

12. Get representative PDBs for each state.

    $ SaveStructures.py -a BACE_5state_3ps/Assignments.Fixed.h5 -c 3 -o BACE_5state_3ps/3RandomConfs

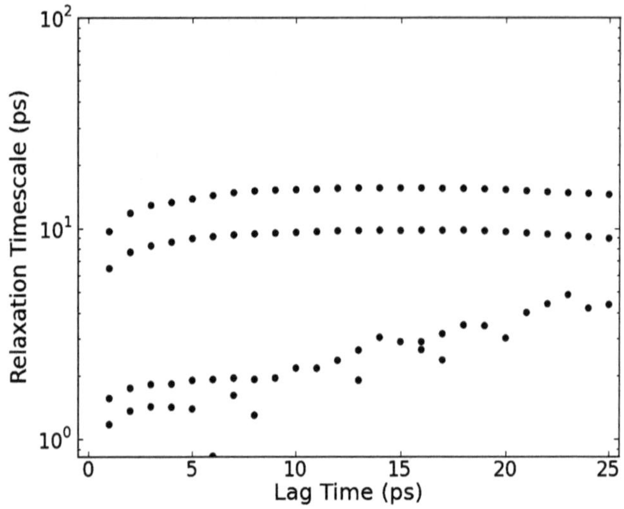

Fig. 4 Relaxation timescales of the macrostate model with six states

By default, this script will store random conformations from every state (specifying a positive integer with the -s option would just select random conformations from that particular state). The -a option points MSMBuilder to the appropriate assignments file, in this case our final macrostate model. The -c option specifies that we want three random conformations from each state and the -o option directs MSMBuilder where to store these conformations. The output files are named StateX-Y.pdb, where *X* is the state number and *Y* is the structure number (in this case 0, 1, or 2 since we asked for 3 random structures).

You can visualize these conformations with your favorite molecular viewing software or use them as a basis for analyzing your model (e.g., calculating properties of each state like the RMSD to some reference structure or the radius of gyration).

13. (Tutorial only) Visualize phase space. Normally, this is not an option because the phase spaces of proteins and other molecules are too high dimensional. However, the alanine dipeptide only has two degrees of freedom—the Φ and Ψ dihedral angles. Therefore, we can visualize our macrostate model with the following command.

$ python PlotDihedrals.py BACE_5state_3ps/Assignments.Fixed.h5

The output should look something like Fig. 5. With this dataset, BACE does not have sufficient statistics to distinguish between the α_L and γ states in a five state model. This is not terribly surprising since these are the highest free energy states and, therefore, are sampled least by the constant temperature

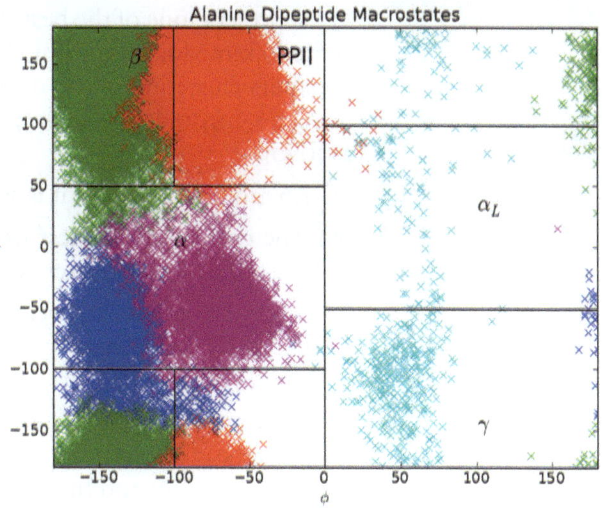

Fig. 5 State decomposition from BACE compared to a reference set of state definitions. The reference state definitions (*black boxes*) are taken from ref. 20 and each "x" represents the projection of one of the conformations sampled by the simulations onto the Φ and Ψ dihedral angles. Each of these points is colored according to the macrostate it is assigned to

simulations included with MSMBuilder. With more sampling, BACE would be able to distinguish these states. With the current sampling, BACE is able to distinguish the two basins in the α state.

3 Notes

1. *MSMBuilder formats*: MSMBuilder stores trajectories and many other large pieces of data in the hdf5 format. The hdf5 format is handy as it can handle a variety of data types and metadata. It also allows for rapid compression/decompression. Trajectory files are stored as .lh5 files (which stands for "lossy hdf5"). Major advantages of this format are (1) use of data compression included with hdf5, (2) additional compression by truncating coordinates, and (3) random access. Random access is useful as many other formats (like Gromacs' xtc format) require code to always read in conformations from the first one (e.g., to get the tenth conformation one first has to read in conformations 1–9). Operations like subsampling (e.g., selecting every tenth conformation) are much faster with random access.
2. *The hybrid clustering method*: The best way to perform an initial clustering of a given data set is still an open question and has been the subject of a number of studies [14, 15].

At present, one of the best available methods is the hybrid k-centers/k-medoids algorithm [7]. This method first performs a k-centers clustering of the data and then refines the clustering using a k-medoids update step. The k-centers clustering works as follows:

(a) An arbitrary data point is selected as the first cluster center.

(b) The data point furthest from any existing cluster center is chosen as a new cluster center.

(c) Every data point that is closer to the new cluster center than the center it is currently assigned to is reassigned to the new cluster center.

(d) Repeat **steps b** and **c** until either (1) the desired number of clusters is created or (2) the maximum distance between any data point and the cluster center it is assigned to (often referred to as the cluster radius) falls below some cutoff distance.

After k-centers, the hybrid method then performs a fixed number of k-medoids updates. The k-medoids update stage works as follows:

(a) For each state, select *N* random conformations.

(b) For each randomly selected conformation in each state, calculate the average distance between that conformation and all the other conformations in that state.

(c) For each state, select the random conformation that is closest to every other conformation in that state (on average) as the new cluster center (i.e., this conformation will replace the previous center, not create a new one).

(d) Assign every data point to the cluster center it is closest to.

(e) Accept the new set of cluster centers if they do not increase the maximum cluster radius from the k-centers stage.

(f) Repeat **steps a** through **e** for some number of iterations.

This hybrid algorithm performs well because the k-centers stage helps to avoid states with large internal free energy barriers and the k-medoids stage helps ensure that the states fall in the densest regions of phase space.

3. *Other clustering options*: You can use the help option to find more information on the other clustering methods available. For example, the command "Cluster.py -h" will provide information on some generic options, including the different distance metrics available. Once you have selected a distance metric (like rmsd), you can obtain information on the clustering algorithms compatible with that distance metric using commands like "Cluster.py rmsd -h". Finally, once you have chosen a clustering algorithm (like the hybrid one), you can

obtain information on the options available with "Cluster.py rmsd hybrid -h".

4. *Subsampling during clustering*: It is often useful to cluster a subsampling of one's data and then to assign the entire dataset to the resulting clustering [5]. Doing so helps to reduce the impact of outliers and reduce the time required to build a model. The degree of subsampling can be set with the Cluster.py script's -S option (e.g., "-S 10" will cluster every tenth conformation). After the clustering step, one can then assign all of the data to the cluster centers identified using Assign.py (e.g., "Assign.py rmsd").

5. *Testing the Markov assumption*: In general, one can test whether a model satisfies the Markov assumption using the Chapman--Kolmogorov equation. The basic idea is that if a model with a lag time of τ is Markovian, then taking two steps with this model should be equivalent to taking one step with a model with a lag time of 2τ. The relaxation times of a model (also called implied timescales) are one way of checking that this property is satisfied [13].

   The relaxation timescales (also called implied timescales) of a model are a function of the eigenvalues of the transition probability matrix. Specifically,

$$\kappa(\tau) = \frac{-\tau}{\ln[\mu(\tau)]}$$

   where κ is a relaxation timescale, τ is the lag time, and $\mu(\tau)$ is an eigenvalue of the transition probability matrix estimated at the given lag time. If the model is Markovian, then

$$\kappa(2\tau) = \frac{-2\tau}{\ln[\mu(2\tau)]} = \frac{-2\tau}{\ln[\mu(\tau)^2]} = \frac{-2\tau}{2\ln[\mu(\tau)]} = \kappa(\tau)$$

   In words, if the model is Markovian, then it should satisfy the Chapman–Kolmogorov equation and the relaxation timescales should be invariant with respect to the lag time.

   Some authors prefer to check the Chapman–Kolmogorov equation on a state-by-state basis [2], arguing that the relaxation timescales can appear reasonable even if a subset of states violate the Markov assumption. However, there are typically too many states to perform such checks for every state so, in practice, one ends up just spot-checking a few states. My preference is to perform a single, more global check using the relaxation timescales. Some states may not be perfectly Markovian, but the invariance of the relaxation timescales strongly suggests the overall behavior of the model is reasonable.

6. *Trouble-shooting relaxation timescales that do not become invariant*: One of the most common problems encountered when

building Markov models is that the relaxation timescales don't appear to become invariant, indicating the state decomposition is inconsistent with Markovian behavior. The most common source of such issues is the presence of large internal free energy barriers. These typically arise from an overly coarse state decomposition that is grouping together kinetically distinct conformations. A good first step is to try generating a new microstate model with smaller states. The other common source of relaxation timescales that do not level off is insufficient statistics. A good rule of thumb is that most every state should have at least ten samples. If your timescales aren't becoming invariant and you can't generate more states without breaking this rule, then you should run additional simulations and try again.

7. *Detailed balance*: At equilibrium, microscopic systems must obey a property called detailed balance (or reversibility). The basic idea is that the number of transitions from state i to state j must equal the number of transitions in the opposite direction. If not, there would be a net flow in one direction and the starting state would lose all of its probability density. The simplest and most robust way to enforce this symmetry is to set $\hat{C}_{ij} = \hat{C}_{ji} = (C_{ij} + C_{ji})/2$, where C_{ij} is the number of transitions observed from state i to state j and $\hat{C}_{ij}$ is the symmetrized counts. This is perfectly valid if one's simulations have reached equilibrium but not otherwise. The maximum likelihood method [7] for symmetrizing the transition count matrix is more appropriate for data that has not reached equilibrium but may fail to converge if the data is too far from equilibrium.
8. *Objectives for coarse-graining*: In general, there are two motivations for coarse-graining a microstate model. First, in the two-stage process of model building presented here, one typically goes out of their way to divide phase space into as many microstates as possible while ensuring each has sufficient statistics. The purpose of doing so is to avoid states with large internal free energy barriers as these will lead to non-Markovian behavior and greatly limit the utility of the model. Therefore, one often would like to build moderately coarse-grained models—typically called mesoscale models—that are still quantitatively predictive but have far fewer states than the original microstate model. However, such models will often still be too complicated to understand. A second common objective is to build macrostate models with very few states. Such models are often only qualitatively correct. However, the small number of states makes it possible to visualize the model and gain an intuition for the system being studied. New hypotheses can be made based on such models and then tested, first with more quantitative models and finally via experiment.

9. *Alternative coarse-graining methods*: Perron Cluster Cluster Analysis (PCCA) [16, 17] and a more robust version called PCCA + [18] are the most common methods for coarse-graining Markov models. Without going into detail, these methods use the eigenspectrum of a models transition probability matrix to find a coarse-graining that best captures the slowest events. These methods work well when there is sufficient data. However, they have a number of known deficiencies, such as poor handling of insufficiently sampled states. A number of new methods have been developed for overcoming these limitations, like the BACE method used here and another method called SHC [19]. PCCA and PCCA + are both available through the PCCA.py script and SHC is coming soon.
10. *Other software packages*: Another software package—called EMMA—is also available for building Markov models [9].

Acknowledgments

I am grateful to the Miller Institute for Basic Research in Science for funding and the rest of the MSMBuilder team for all of their hard work.

References

1. Bowman GR, Huang X, Pande VS (2010) Network models for molecular kinetics and their initial applications to human health. Cell Res 20:622–630
2. Prinz J-H et al (2011) Markov models of molecular kinetics: generation and validation. J Chem Phys 134:174105
3. van Kampen NG (2007) Stochastic processes in physics and chemistry. Elsevier Science & Technology Books, Amsterdam, The Netherlands
4. Bowman GR, Huang X, Pande VS (2009) Using generalized ensemble simulations and Markov state models to identify conformational states. Methods 49:197–201
5. Bowman GR, Beauchamp KA, Boxer G, Pande VS (2009) Progress and challenges in the automated construction of Markov state models for full protein systems. J Chem Phys 131:124101
6. Noé F, Schütte C, Vanden-Eijnden E, Reich L, Weikl TR (2009) Constructing the equilibrium ensemble of folding pathways from short off-equilibrium simulations. Proc Natl Acad Sci U S A 106:19011–19016
7. Beauchamp KA et al (2011) MSMBuilder2: modeling conformational dynamics at the picosecond to millisecond scale. J Chem Theory Comput 7:3412–3419
8. Chodera J, Singhal N, Pande V, Dill K (2007) Automatic discovery of metastable states for the construction of Markov models of macromolecular conformational dynamics. J Chem Phys 126:155101
9. Senne M, Trendelkamp-Schroer B, Mey ASJS, Schütte C, Noé F (2012) EMMA: a software package for Markov model building and analysis. J Chem Theory Comput 8: 2223–2238
10. Bowman GR (2012) Improved coarse-graining of Markov state models via explicit consideration of statistical uncertainty. J Chem Phys 137:134111
11. Pande VS, Beauchamp K, Bowman GR (2010) Everything you wanted to know about Markov State Models but were afraid to ask. Methods 52:99–105
12. Noé F, Fisher S (2008) Transition networks for modeling the kinetics of conformational

change in macromolecules. Curr Opin Struct Biol 18:154–162
13. Swope WC, Pitera JW, Suits F (2004) Describing protein folding kinetics by molecular dynamics simulations. 1. Theory †. J Phys Chem B 108:6571–6581
14. Cossio P, Laio A, Pietrucci F (2011) Which similarity measure is better for analyzing protein structures in a molecular dynamics trajectory? Phys Chem Chem Phys 13:10421–10425
15. Kellogg EH, Lange OF, Baker D (2012) Evaluation and optimization of discrete state models of protein folding. J Phys Chem B 116:11405–11413
16. Noé F, Horenko I, Schütte C, Smith JC (2007) Hierarchical analysis of conformational dynamics in biomolecules: transition networks of metastable states. J Chem Phys 126:155102
17. Deuflhard P, Huisinga W, Fischer A, Schütte C (2000) Identification of almost invariant aggregates in reversible nearly uncoupled Markov chains. Lin Algebra Appl 315:39–59
18. Deuflhard P, Weber M (2005) Robust Perron cluster analysis in conformation dynamics. Lin Algebra Appl 398:161–184
19. Yao Y et al (2009) Topological methods for exploring low-density states in biomolecular folding pathways. J Chem Phys 130:144115
20. Jha AK et al (2005) Helix, sheet, and polyproline II frequencies and strong nearest neighbor effects in a restricted coil library. Biochemistry 44:9691–9702

Chapter 9

Analysis of Protein Conformational Transitions Using Elastic Network Model

Wenjun Zheng and Mustafa Tekpinar

Abstract

In this chapter, we demonstrate the usage of a coarse-grained elastic network model to analyze protein conformational transitions in the NS3 helicase (NS3hel) of Hepatitis C virus (HCV). This analysis allows us to identify and visualize collective domain motions involved in the conformational transitions and predict the order of structural events during the transitions. It is highly efficient and applicable to many multi-domain protein structures which undergo large conformational changes to fulfill their functions. This method is made available through a Web server (http://enm.lobos.nih.gov).

Key words Conformational transition, Coarse-grained model, Elastic network model, Normal mode analysis, Reaction coordinate, Transition pathway, Helicase

1 Introduction

Protein conformational dynamics, which is involved in many protein functions, spans a wide range of temporal scales (from femto-seconds to seconds) and spatial scales (from atomic fluctuations to domain motions). The functions of a large number of protein complexes are thought to invoke conformational transitions between a series of biochemical states, some of which were captured by X-ray crystal structures and electron microscopy. However, it remains very challenging to experimentally probe or computationally model the transient intermediates of these transitions, which determine the kinetic mechanism of these protein complexes. To obtain details of protein dynamics, all-atom molecular dynamics (MD) [1] has been widely employed to simulate protein conformational fluctuations and transitions. Nevertheless, such simulations are limited to nanoseconds to microseconds, which fall short of the typical time scale of protein kinetics (milliseconds to seconds). To overcome the time-scale barrier for MD simulations, a variety of coarse-grained models [2] have been developed. Of particular interest is the elastic network model (ENM)

Dennis R. Livesay (ed.), *Protein Dynamics: Methods and Protocols*, Methods in Molecular Biology, vol. 1084,
DOI 10.1007/978-1-62703-658-0_9,

[3–5], which represents a protein structure as a network of C_α atoms with neighboring ones connected by springs with a uniform force constant [6]. The normal mode analysis (NMA) of ENM often yields a handful of low-frequency modes that capture the large-scale conformational changes observed between two protein crystal structures [7, 8]. ENM has formed the basis of several computational methods for modeling conformational transitions between two given protein conformations [9–13].

In this chapter, we will demonstrate ENM-based NMA and a transition pathway modeling method named interpolated-ENM (iENM) [13]. The iENM constructs a pathway by solving the saddle points of a double-well potential built from two ENM potentials based at the beginning and end conformation of a transition [13]. The predicted pathway allows us to deduce the dynamic order of structural events involving various protein parts [13–17]. We will use NS3hel of HCV as an example [18, 19] (*see* **Note 1**).

2 Methods

2.1 Normal Mode Analysis of Elastic Network Model

Given the C_α atomic coordinates of a protein crystal structure from Protein Data Bank (http://www.rcsb.org), we build an elastic network model by connecting all pairs of C_α atoms that are within a cutoff distance R_c (chosen to be 10 Å by default) by using harmonic spring. The total ENM energy is

$$E_{\text{ENM}} = \frac{1}{2} \sum_{d_{ij}^0 < R_c} C\left(d_{ij} - d_{ij}^0\right)^2, \tag{1}$$

where C is the spring force constant which is set to 1 by default, although it can be determined by fitting the crystallographic B factors if needed [20], d_{ij} is the distance between the C_α atoms i and j, and d_{ij}^0 is the value of d_{ij} as given by the crystal structure.

We expand the ENM energy to second order:

$$E_{\text{ENM}} \approx \frac{1}{2} \delta X^T H \delta X = \frac{1}{2} C \sum_{d_{ij}^0 < R_c} \delta X^T H_{ij} \delta X, \tag{2}$$

where $\delta X = X - X_0$, X is a $3N$-dimensional vector representing the C_α atomic coordinates, X_0 gives the equilibrium C_α coordinates in the crystal structure, $H = C \sum_{d_{ij}^0 < R_c} H_{ij}$ is the $3N \times 3N$ Hessian matrix (second derivatives of E_{ENM}), where $H_{ij} = \frac{1}{2} \nabla^2 \left[\left(d_{ij} - d_{ij}^0\right)^2\right]$.

For the Hessian matrix H, we can perform the normal mode analysis (NMA) to obtain $3N$ normal modes. Each mode m has an eigenvalue λ_m and a $3N$-dimensional eigenvector V_m which satisfy

$HV_m = \lambda_m V_m$. The normal modes can be solved using the dsyevr subroutine of a linear algebra package named LAPACK (http://www.netlib.org/lapack/). The lowest six zero modes, corresponding to three translations and three rotations, are removed from the spectrum (mode numbering starts from #1 for the lowest non-zero mode).

To validate NMA, we need to use two different crystal structures of a protein. We first perform NMA using the first structure (see above). Then, we superimpose the second structure on top of the first structure using the PROFIT program (*see* http://www.bioinf.org.uk/software/profit/). Finally, we compare each mode (mode m) with the observed structural changes between the two superimposed structures (represented by a $3N$-dimensional vector δX_{obs}) by calculating the following overlap:

$$I_m = \frac{|\delta X_{\text{obs}}^T V_m|}{|\delta X_{\text{obs}}| \cdot |V_m|}, \tag{3}$$

where $\delta X_{\text{obs}}^T V_m$ is the dot product between vectors δX_{obs} and V_m, $|\delta X_{\text{obs}}|$ and $|V_m|$ represent their magnitude. I_m ranges from 0 to 1, and higher I_m indicates greater involvement of mode m in the observed structural changes δX_{obs}. In addition, the following cumulative overlap is calculated to assess how well the lowest ten modes describe δX_{obs}:

$$C_{10} = \sqrt{\sum_{1 \le m \le 10} I_m^2}. \tag{4}$$

Because $\sum_{1 \le m \le 3N-6} I_m^2 = 1$, C_{10}^2 gives the percentage of the observed structural changes captured by the lowest ten modes.

For an example of the normal mode analysis, *see* **Note 2**.

2.2 Coarse-Grained Modeling of Protein-DNA System

To study how NS3hel translocates along a single-stranded DNA (ssDNA), we need to refine the modeling of protein–DNA interactions and intra-ssDNA interactions. To this end, we modify the ENM as follows:

1. For residue–residue interactions within NS3hel, we use the C_α-based ENM with $R_c = 10$ Å (*see* Eq. 1).
2. For interactions within ssDNA, we use a modified ENM which represents each nucleic acid by a bead located at the C4′ atom, and adds springs between first, second, and third nearest-neighbor (NN) beads with the same force constant k_{DNA}:

$$E_{\text{DNA}} = \frac{1}{2} \sum_{1 \le |j-i| \le 3} k_{\text{DNA}} (d_{ij} - d_{ij,0})^2, \tag{5}$$

where d_{ij} is the distance between the C4′ atom i and j, $d_{ij,0}$ is the value of d_{ij} as given by an NS3hel-ssDNA structure.

3. To accurately represent protein–DNA interactions, we use a structure-based Leonard-Jones 6–12 potential to allow protein-DNA contacts to form/break readily during a transition:

$$E_{\text{prot-DNA}} = \frac{1}{2} \sum_{\substack{i \in \text{prot} \\ j \in \text{DNA}}} k_{\text{prot-DNA}} \frac{d_{i,\min}^2}{36} \left(1 - \frac{d_{i,\min}^6}{d_{ij}^6}\right)^2, \quad (6)$$

where the summation is over residues which form heavy-atom contacts (within 4 Å) with DNA backbone in a NS3hel-ssDNA structure, and $d_{i,\min}$ is the minimal C_α–C4′ distance for residue i, and $k_{\text{prot-DNA}}$ is the force constant. We only consider contacts between protein and DNA backbone in our modeling because functional and structural data suggested that these contacts are sufficient for ensuring that NS3hel maintains a grip on the ssDNA track and undergoes continuous translocation [21]. For NMA, we replace Eq. 6 with its harmonic counterpart $E'_{\text{prot-DNA}} = \frac{1}{2} \sum_{\substack{i \in \text{prot} \\ j \in \text{DNA}}} k_{\text{prot-DNA}}(d_{ij} - d_{ij,0})^2$.

For NS3hel, we choose $k_{\text{DNA}} = 1$ and $k_{\text{prot-DNA}} = 1.3$ based on the fitting of crystallographic B factors. One should check a range of parameter values to make sure that the modeling results are not sensitive to the particular choice of these parameters.

2.3 Interpolated Elastic Network Model (iENM)

We consider an *arbitrary* double-well potential function $F(E_1, E_2)$ with two minima at the beginning and end conformation of a transition. It satisfies: $F(E_1, E_2) \approx E_1$ if $E_1 \ll E_2$, and $F(E_1, E_2) \approx E_2$ if $E_2 \ll E_1$, where E_1 and E_2 are two single-well potentials. The saddle points (SP) of $F(E_1, E_2)$ are solved as follows

$$0 = \nabla F(E_1, E_2) = \frac{\partial F}{\partial E_1} \nabla E_1 + \frac{\partial F}{\partial E_2} \nabla E_2, \quad (7)$$

which is equivalent to solving the following equation (after setting $\lambda = \frac{\partial F}{\partial E_1} \Big/ \left(\frac{\partial F}{\partial E_1} + \frac{\partial F}{\partial E_2}\right)$)

$$0 = \lambda \nabla E_1 + (1 - \lambda) \nabla E_2, \quad (8)$$

where λ is a parameter of interpolation that varies from 1 to 0 (assuming $\frac{\partial F}{\partial E_1} \geq 0$ and $\frac{\partial F}{\partial E_2} \geq 0$). Therefore, the problem of solving SP for the double-well potential function $F(E_1, E_2)$ is converted to the problem of finding the minima of a linearly interpolated potential function $\lambda E_1 + (1 - \lambda) E_2$. Equation 8 gives a set of minimal-energy crossing points between E_1 and E_2 where $E_1 = E_2$ is at minimum.

Based on the above general formulation, we have proposed an iENM protocol [13] using a double-well potential $F(E_{\text{ENM1}}+$

$E_{col}, E_{ENM2} + E_{col})$, where E_{ENM1} and E_{ENM2} are two ENM potential functions (*see* Eq. 1) based at the beginning and end conformation of a transition, and E_{col} is a steric collision energy defined as follows:

$$E_{col} = \frac{1}{2}\sum_{i=3}^{N}\sum_{j=1}^{i-2} C_{col}\theta(R_{col} - d_{ij})(d_{ij} - R_{col})^2, \quad (9)$$

where $R_{col} = 4$ Å, $C_{col} = 10$, and chemically bonded residue pairs $(j = i \pm 1)$ are excluded. The addition of E_{col} penalizes steric collisions between residues whose C_α atoms are within a distance of R_{col}.

After adding the collision energy, the SPs are solved by setting $\nabla F(E_{ENM1} + E_{col}, E_{ENM2} + E_{col}) = 0$ which is equivalent to solving the following SP equation (the SP is represented by X_{SP}):

$$\lambda\nabla E_{ENM1}(X_{SP}) + (1-\lambda)\nabla E_{ENM2}(X_{SP}) + \nabla E_{col}(X_{SP}) = 0. \quad (10)$$

As λ varies from 1 to 0, X_{SP} traces a pathway that connects the beginning and end conformation of a transition. Because this pathway passes all possible SPs, it gives a *universal* minimal-energy path regardless of the detailed form of $F(E_1, E_2)$. iENM outputs the above pathway as the predicted pathway for the given transition.

We solve Eq. 10 by using the following iterative procedure to find the minima of the linearly interpolated potential function $\lambda E_{ENM1} + (1-\lambda)E_{ENM2} + E_{col}$ with the Newton–Raphson method:

1. Initialization: set $n = 0$, $X_{SP,0} = X_1$, which is the C_α coordinates of the beginning conformation.
2. Given $X_{SP,n}$, calculate $\lambda = \lambda_n = -\frac{[\nabla E_{ENM1} - \nabla E_{ENM2}]\cdot[\nabla E_{ENM2} + \nabla E_{col}]}{|\nabla E_{ENM1} - \nabla E_{ENM2}|^2}$ to minimize $|\lambda\nabla E_{ENM1} + (1-\lambda)\nabla E_{ENM2} + \nabla E_{col}|$.
3. Calculate $R_n = \lambda_n\nabla E_{ENM1} + (1-\lambda_n)\nabla E_{ENM2} + \nabla E_{col}$.
4. If $|R_n| < 0.00001$, go to **step 7**.
5. Displace $X_{SP,n}$ by

$$\delta X_{SP} = -[\lambda_n H_1 + (1-\lambda_n)H_2 + H_{col} + \varepsilon I]^{-1} R_n, \quad (11)$$

where H_1, H_2 and H_{col} are the Hessian matrices calculated for E_{ENM1}, E_{ENM2} and E_{col}, I is identity matrix, ε is a small positive number to render the sum of matrices invertible.

6. Go to **step 3**.
7. Calculate $X_{SP,n+1} = X_{SP,n} + \delta X_{SP}$ and

$$\delta X_{SP} \sim -\delta\lambda[\lambda_n H_1 + (1-\lambda_n)H_2 + H_{col} + \varepsilon I]^{-1}[\nabla E_{ENM1} - \nabla E_{ENM2}], \quad (12)$$

where H_1, H_2 and H_{col} are the Hessian matrices calculated for E_{ENM1}, E_{ENM2} and E_{col}, I is identity matrix, ε is a small positive number to render the sum of matrices invertible, and $\delta\lambda$ is chosen so that the magnitude of δX_{SP} is small (i.e., $|\delta X_{SP}|/\sqrt{N}$ <0.1 Å).

8. Stop if $X_{SP,n+1}$ has reached X_2 which is the C_α coordinates of the end conformation, otherwise set $n \leftarrow n + 1$, then go to **step 2**.

The linear equations in Eqs. 11 and 12 are solved using a highly efficient sparse linear equation solver CHOLMOD (http://www.cise.ufl.edu/research/sparse/cholmod/) [22]. In Eq. 12 we compute an incremental structural displacement δX_{SP} based on the force-induced linear responses—δX_{SP} is calculated as a weighted sum of all normal modes of the Hessian matrix $\lambda_n H_1 + (1 - \lambda_n) H_2 + H_{col}$. Because the weight of each mode is inversely proportional to its eigenvalue, the collective motions described by the lowest modes are favorably sampled along the transition pathway.

For an example of iENM application, *see* **Note 3**.

2.4 Assessment of Motional Order Using Reaction Coordinates

The predicted transition pathway allows us to determine the motional order of different parts/domains of a protein. For this purpose, the following reaction coordinate (RC) is defined for an intermediate conformation of a given part S [13]:

$$RC_S = (\delta X_S \bullet \delta X_{S,obs}) / |\delta X_{S,obs}|^2, \tag{13}$$

where δX_S is the displacement vector of part S from the beginning conformation of a transition to a given intermediate conformation, and $\delta X_{S,obs}$ is the observed displacement of part S from the beginning conformation to the end conformation of a transition. RC_S measures the motional progress of part S in the direction of a transition. $RC_S = 0$ at the beginning of a transition, and $RC_S = 1$ at the end of a transition. For two different parts (named S_1 and S_2) in an intermediate conformation, if $RC_{S_1} > RC_{S_2}$, then S_1's motion precedes S_2's motion.

For an example of RC calculations, *see* **Note 3**.

3 Conclusion

We have demonstrated the use of a coarse-grained ENM for analyzing conformational transitions in protein or protein-DNA complex. The predicted order of structural events has been validated using structural data. This method is highly efficient—it takes only 3 min to run the entire ATP cycle of NS3hel using a dual-core workstation. This method will be useful for future simulations of a variety of molecular motors including many monomeric and ring-shaped helicases. Both ENM-based NMA and iENM are available via a Web server at http://enm.lobos.nih.gov.

4 Notes

1. *Introduction to NS3hel*

   We will illustrate the usage of ENM-based NMA and iENM using NS3hel as an example. To unwind double-stranded DNA/RNA, NS3hel assembles on a 3′ end of ssDNA/RNA tail, and actively translocates along ssDNA/RNA in the 3′–5′ direction [23]. Several crystal structures of NS3hel bound with ssDNA and various ATP analogs have been solved [24–29], which correspond to three biochemical states of its work cycle (apo, ATP, and ADP-Pi, where Pi represents inorganic phosphate). The structure of NS3hel consists of three domains (*see* Fig. 1). The ATP binds at the cleft between domains 1 and 2, while the ssDNA binds in a groove between domains 1, 2 and domain 3 (*see* Fig. 1). From the structural data, an inchworm model has emerged for the translocation of NS3hel along ssDNA (complemented by a ratchet action) [24, 28–31] (*see* Fig. 1): first, ATP binding induces a closure

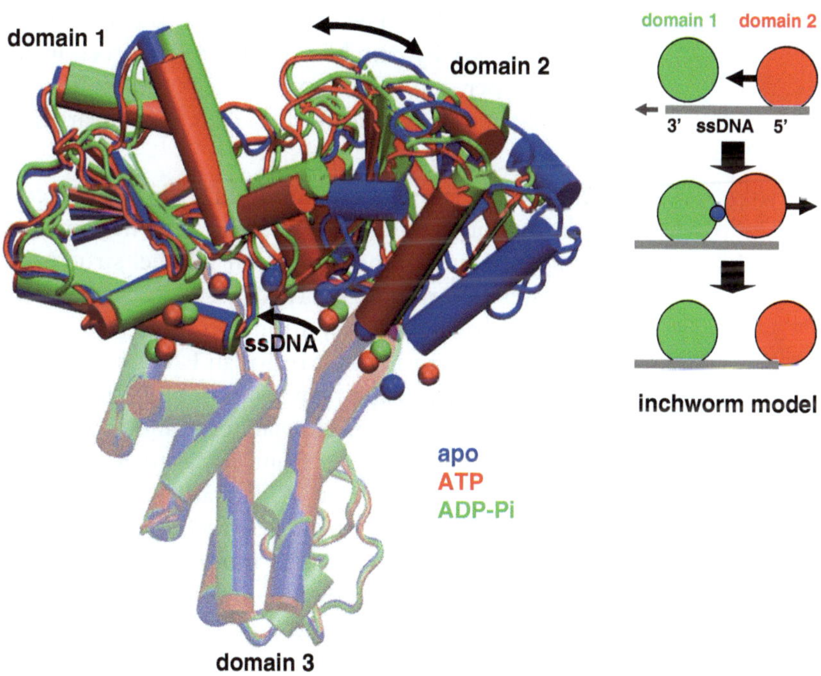

Fig. 1 The structures of NS3hel in three biochemical states (apo, ATP, and ADP-Pi, colored in *blue*, *red*, and *green*, respectively). NS3hel is shown in *cartoon* representation, and ssDNA is shown as a *chain of beads* located at C4′ atoms. The three domains of NS3hel and ssDNA are labeled. The opening/closing motions of domain 2 and the 5′–3′ sliding of ssDNA are marked by *arrows*. The three structures are aligned along domain 3 which is shown as *transparent*. *Inset*: a schematic cartoon illustrates the inchworm model (domain 1, domain 2, ATP, and ssDNA are colored *green*, *red*, *blue*, and *gray*, respectively; the opening/closing motions of domain 2 and the sliding of ssDNA are marked by *arrows*; domain 3 is not shown)

Table 1
Comparison between the lowest ten normal modes and the crystallographically observed conformational changes in HCV NS3hel

PDB_chain id of two NS3hel structures	RMSD (Å)	Mode#	Overlap
3kqk_AD → 3kqu_B	3.26	#3	0.49
		#4	0.57
		#1–10	0.92
3kqu_BM → 3kql_A	1.03	#1	0.61
		#1–10	0.72
3kql_AE → 3kqk_A	3.37	#5	0.45
		#6	0.50
		#1–10	0.82

motion of domain 2 toward domain 1, with domain 1 releasing its grip on ssDNA and sliding along it, while domain 2 maintains its grip on ssDNA; second, following ATP hydrolysis and the release of ADP and Pi, domain 2 opens again as it releases its grip on ssDNA and slides along it, while domain 1 maintains its grip on ssDNA. The net effect is the translocation of NS3hel along ssDNA by one base in the 3′–5′ direction, consuming one ATP per step. The details of the conformational transitions between apo, ATP, and ADP-Pi state remain largely unknown.

2. *Normal mode analysis of ENM*
To validate the use of ENM for NS3hel, we compare the domain motions predicted by NMA of ENM with the observed conformational changes between NS3hel structures in different states (for results, *see* Table 1). Here we will focus on the conformational transition from apo to ATP state.

We have performed NMA for an ENM constructed from an NS3hel-ssDNA structure in apo state (PDB id: 3kqk), and then calculated the overlaps between each mode and the observed conformational changes from 3kqk to an NS3hel-ssDNA-ADP · BeF_3 structure in ATP state [28] (PDB id: 3kqu). Encouragingly, 84 % of the observed conformational changes are captured by the lowest ten modes (with cumulative overlap $C_{10} = 0.92$), among which mode #3 and #4 contribute most (with overlap $I_3 = 0.49$ and $I_4 = 0.57$, respectively). To visualize the domain motions predicted by these two modes, we have deformed the 3kqk structure along the directions given by the eigenvectors of these two modes, and then compared the deformed structures with 3kqk using the VMD program (http://www.ks.uiuc.edu/Research/vmd/).

Mode #3 describes coupled rotations of domains 1 and 2 relative to domain 3, which result in the opening of domains 1–3 interface, the closing of domains 1–2 interface, and the sliding of ssDNA toward its 3′ end (*see* Fig. 2a).

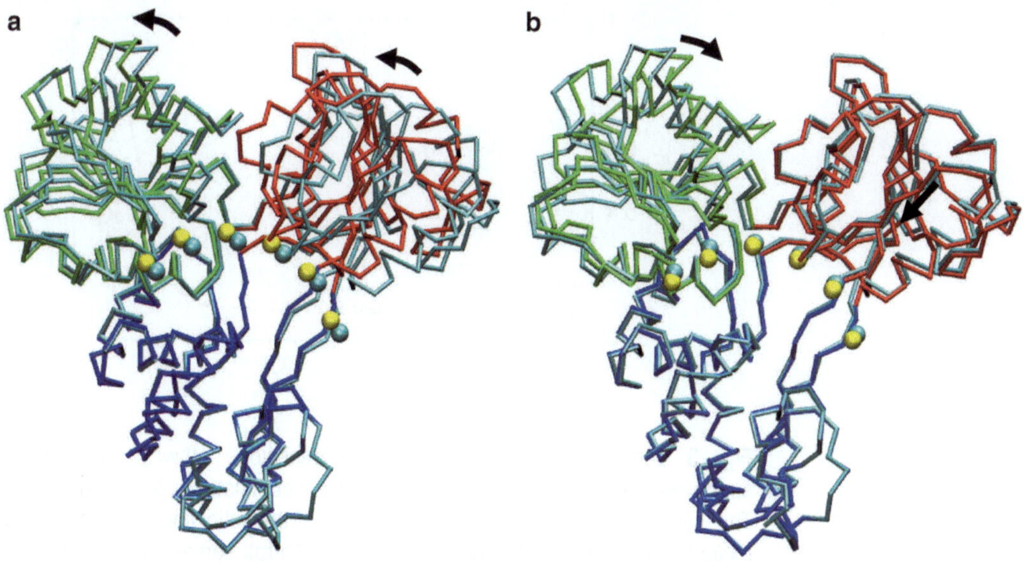

Fig. 2 Conformational changes in NS3hel as predicted by the following ENM-based normal modes: (**a**) mode #3, (**b**) mode #4, which are solved from the apo-state NS3hel-ssDNA structure (PDB id: 3kqk). The original NS3hel-ssDNA structure is colored *cyan*. For the deformed NS3hel-ssDNA structure after the conformational changes, domain 1, 2, and 3 are colored *green*, *red*, and *blue*, respectively, and ssDNA is shown as a *chain of yellow beads* located at C4′ atoms. The two structures of NS3hel are superimposed along domain 3. The domain rotations are shown by *arrows*

Mode #4 describes simultaneous rotations of domains 1 and 2 toward domain 3, which cause the closing of domains 1–2 interface and a shift of ssDNA toward domain 3 (*see* Fig. 2b). Although the lowest ten modes accurately capture the observed conformational changes in NS3hel, they do not correctly predict the translocation of NS3hel along ssDNA which requires domain 1 to slide along ssDNA while domain 2 holds ssDNA. Instead, in both modes #3 and #4, domains 1 and 2 maintain their grip on ssDNA with no sliding between domain 1 and ssDNA (*see* Fig. 2). This is not surprising because domain 1 and ssDNA are linked by elastic springs in ENM which disfavor sliding between them. Therefore, to accurately describe the conformational dynamics underlying NS3hel translocation, one has to modify ENM and account for the anharmonicity of protein–DNA interactions (*see* Section 2.2).

3. *Transition pathway modeling by iENM*
   Next, we use iENM to simulate the conformational transitions in NS3hel between three biochemical states (apo, ATP, and ADP-Pi), which are captured by three crystal structures of NS3hel [28] (PDB ids: 3kqk, 3kqu, 3kql) (*see* Fig. 1). We validate the use of iENM for NS3hel by checking if it correctly predicts the order of inter-domain motions observed among crystal structures of NS3hel in different states. To this end,

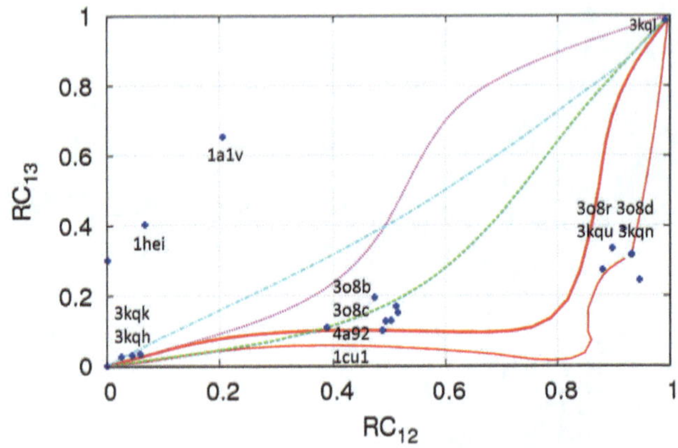

Fig. 3 Comparison between calculated transition pathways and crystal structures of NS3hel using two reaction coordinates (RC_{12} and RC_{13}). The calculated pathways are obtained using the following methods: iENM (*solid lines*), mixed-ENM server (*dashed line*), Morph server (*dot dashed line*), MinActionPath server (*dotted line*). The crystal structures are shown as *points* (PDB&chain ids: 1a1vA, 1cu1A, 1cu1B, 1heiA, 1heiB, 3o8bA, 3o8bB, 3o8rA, 3o8rB, 3o8dA, 3o8dB, 3o8cA, 3o8cB, 3kquA, 3kquB, 3kquC, 3kquD, 3kquE, 3kquF, 3kqlA, 3kqlB, 3kqkA, 3kqkB, 3kqnA, 3kqhA, 3kqhB, 4a92A, 4a92B). The following three pathways are calculated by iENM: 3kqk → 3kql (*thick solid line*), 3kqk → 3kqu (*thin solid line*), 3kqu → 3kql (*thin solid line*)

we have generated pathways for three transitions (apo → ATP, ATP → ADP-Pi, apo → ADP-Pi) using iENM, and then compared the predicted pathways with the crystal structures [24–26, 28, 29] using two reaction coordinates (RC) (*see* Fig. 3): RC_{12} quantifies the progress of motion between domains 1 and 2, and RC_{13} quantifies the progress of motion between domains 1 and 3. Both RCs vary from 0 to 1, where 0 corresponds to the apo state, and 1 corresponds to the ADP-Pi state. The iENM pathway for apo → ADP-Pi transition predicts that the increase of RC_{12} precedes RC_{13} during the transition (*see* Fig. 3), which implies that the domains 1–2 motion precedes the domains 1–3 motion. This order is functionally meaningful, because the domains 1–2 motion occurs upon ATP binding while the domains 1–3 motion occurs during the subsequent transition (ATP hydrolysis) (*see* Fig. 1). This prediction agrees well with the RCs of most crystal structures, which form two intermediate clusters located near the predicted pathway (*see* Fig. 3)—the first cluster includes several apo structures of full-length NS3, and the second cluster includes several crystal structures of NS3hel and full-length NS3 bound with ADP-BeF_3 (corresponding to ATP state).

Two "outlier" structures (*see* Fig. 3) may correspond to off-path intermediates trapped by crystallization conditions. It is remarkable that the iENM pathway for apo → ADP-Pi transition visits the ATP-state structures as intermediates even though these structures are not used in the modeling. For comparison, we have also analyzed the pathways for apo → ADP-Pi transition predicted by alternative methods including Yale Morph server [32], mixed-ENM server [33] and MinActionPath server [12], which do not seem to agree with the crystal structures (*see* Fig. 3).

After validating iENM for exploring conformational transitions in NS3hel, we have used it to simulate the translocation of NS3hel along ssDNA as it undergoes the following three transitions: apo → ATP → ADP-Pi → apo, which comprise the ATP cycle.

For the apo → ATP transition, as predicted by the iENM pathway, domain 2 closes toward domain 1 while it holds the ssDNA (*see* Fig. 4a). Consequently, the ssDNA moves toward its 3′ end by ~4.8 Å as it slides between domains 1 and 3 (*see* Fig. 4a). Therefore, our iENM pathway has reproduced key motions upon ATP binding as postulated by the inchworm model—a closure motion of domain 2 toward domain 1, with domain 1 releasing its grip on ssDNA and sliding along it, while domain 2 maintains its grip on ssDNA (*see* Fig. 1).

For the ATP → ADP-Pi transition, as predicted by the iENM pathway, domains 1 and 2 undergo a coupled rotation, resulting in the movement of ssDNA toward its 5′ end by ~0.8 Å relative to domain 3 (*see* Fig. 4b). Both domains 1 and 2 maintain their grip on ssDNA during this transition, so there is no sliding of ssDNA relative to domains 1 and 2 (*see* Fig. 4b).

For the ADP-Pi → apo transition, as predicted by the iENM pathway, domain 2 opens early to release its grip on ssDNA, which is followed by a small sliding (~1.7 Å) of ssDNA toward its 3′ end and a slight opening of domain 1 (*see* Fig. 4c). Our finding has largely reproduced key motions during ADP-Pi release as postulated by the inchworm model—opening of domain 2 after it releases its grip on ssDNA and slides along it, while domain 1 maintains its grip on ssDNA (*see* Fig. 1).

In sum, our iENM calculations for the above three transitions predict a net translocation of NS3hel along ssDNA in the 3′–5′ direction by ~5.7 Å, which corresponds approximately to 1-base step size.

Acknowledgment

We thank funding support from American Heart Association (grant #0835292N) and National Science Foundation (grant #0952736).

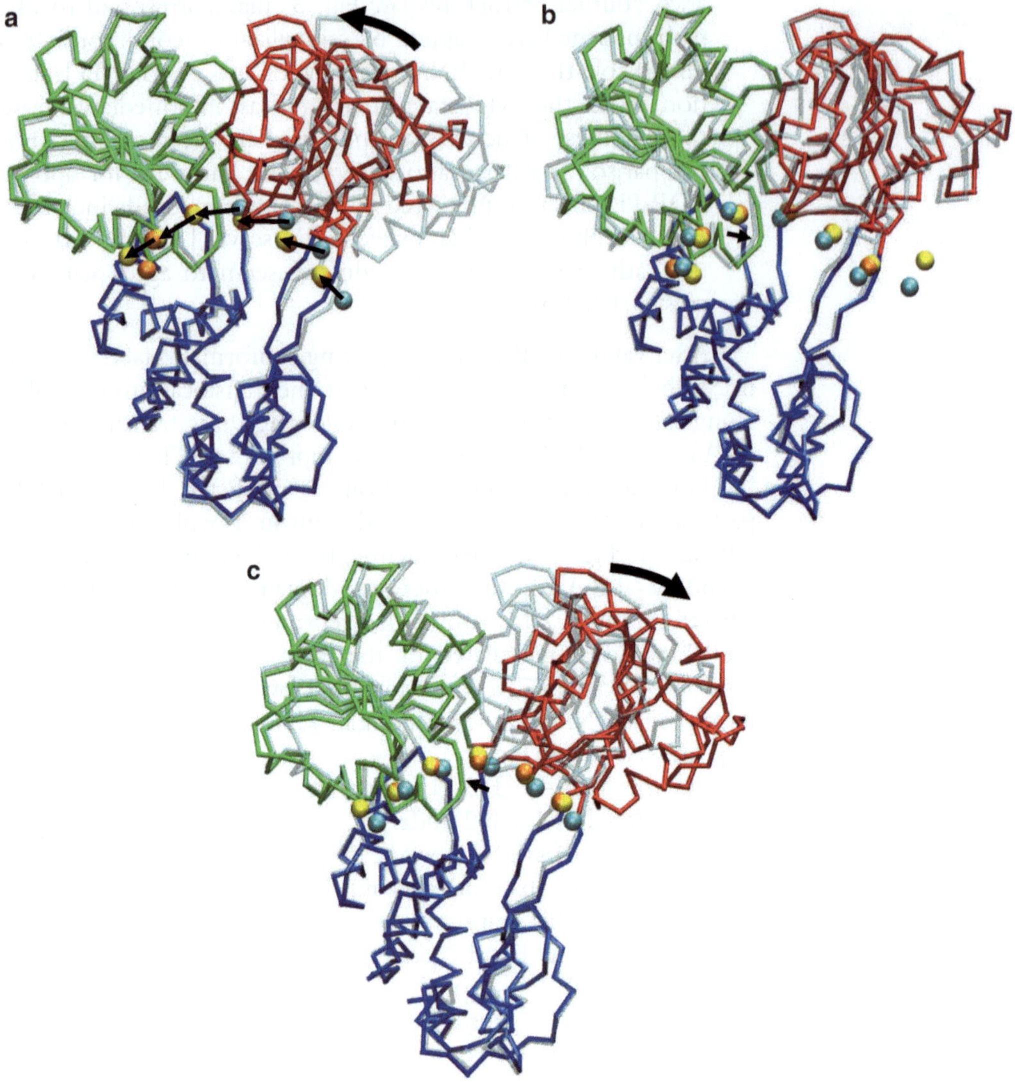

Fig. 4 Snapshots at the beginning and end of iENM pathways for the following transitions in NS3hel: (**a**). apo → ATP transition; (**b**). ATP → ADP-Pi transition; (**c**). ADP-Pi → apo transition. In the end conformation of NS3hel, domain 1, 2, and 3 are colored *green*, *red*, and *blue*, respectively; the beginning conformation of NS3hel is shown as transparent; ssDNA is shown as a chain of beads located at C4′ atoms (for ssDNA, the beginning conformation, end conformation, and target conformation are colored *cyan*, *yellow*, and *orange*, respectively); the movements of domain 2 and ssDNA are marked by *arrows*

References

1. Karplus M, McCammon JA (2002) Molecular dynamics simulations of biomolecules. Nat Struct Biol 9(9):646–652
2. Tozzini V (2005) Coarse-grained models for proteins. Curr Opin Struct Biol 15(2): 144–150
3. Hinsen K (1998) Analysis of domain motions by approximate normal mode calculations. Proteins 33(3):417–429
4. Atilgan AR, Durell SR, Jernigan RL, Demirel MC, Keskin O, Bahar I (2001) Anisotropy of

fluctuation dynamics of proteins with an elastic network model. Biophys J 80(1):505–515
5. Tama F, Sanejouand YH (2001) Conformational change of proteins arising from normal mode calculations. Protein Eng 14(1):1–6
6. Tirion MM (1996) Large amplitude elastic motions in proteins from a single-parameter, atomic analysis. Phys Rev Lett 77(9): 1905–1908
7. Krebs WG, Alexandrov V, Wilson CA, Echols N, Yu HY, Gerstein M (2002) Normal mode analysis of macromolecular motions in a database framework: developing mode concentration as a useful classifying statistic. Proteins 48 (4):682–695
8. Yang L, Song G, Jernigan RL (2007) How well can we understand large-scale protein motions using normal modes of elastic network models? Biophys J 93(3):920–929
9. Miyashita O, Onuchic JN, Wolynes PG (2003) Nonlinear elasticity, proteinquakes, and the energy landscapes of functional transitions in proteins. Proc Natl Acad Sci U S A 100 (22):12570–12575
10. Maragakis P, Karplus M (2005) Large amplitude conformational change in proteins explored with a plastic network model: adenylate kinase. J Mol Biol 352(4):807–822
11. Zheng WJ, Brooks BR, Hummer G (2007) Protein conformational transitions explored by mixed elastic network models. Proteins 69 (1):43–57
12. Franklin J, Koehl P, Doniach S, Delarue M (2007) MinActionPath: maximum likelihood trajectory for large-scale structural transitions in a coarse-grained locally harmonic energy landscape. Nucleic Acids Res 35(Web Server issue):W477–W482
13. Tekpinar M, Zheng W (2010) Predicting order of conformational changes during protein conformational transitions using an interpolated elastic network model. Proteins 78(11): 2469–2481
14. Zheng W (2010) Multiscale modeling of structural dynamics underlying force generation and product release in actomyosin complex. Proteins 78(3):638–660
15. Zheng W (2011) Coarse-grained modeling of conformational transitions underlying the processive stepping of myosin V dimer along filamentous actin. Proteins 79(7):2291–2305
16. Zheng W (2012) Coarse-grained modeling of the structural states and transition underlying the powerstroke of dynein motor domain. J Chem Phys 136(15):155103
17. Zheng W, Auerbach A (2011) Decrypting the sequence of structural events during the gating transition of pentameric ligand-gated ion channels based on an interpolated elastic network model. PLoS Comput Biol 7(1):e1001046
18. Zheng W, Liao JC, Brooks BR, Doniach S (2007) Toward the mechanism of dynamical couplings and translocation in hepatitis C virus NS3 helicase using elastic network model. Proteins 67(4):886–896
19. Zheng W (2010) Computer modeling of helicases using elastic network model. Methods Mol Biol 587:235–243
20. Zheng W (2008) A unification of the elastic network model and the Gaussian network model for optimal description of protein conformational motions and fluctuations. Biophys J 94(10):3853–3857
21. Pyle AM (2008) Translocation and unwinding mechanisms of RNA and DNA helicases. Annu Rev Biophys 37:317–336
22. Chen YQ, Davis TA, Hager WW, Rajamanickam S (2008) Algorithm 887: CHOLMOD, supernodal sparse cholesky factorization and update/downdate. ACM Trans Math Software 35(3):1–14
23. Frick DN (2007) The hepatitis C virus NS3 protein: a model RNA helicase and potential drug target. Curr Issues Mol Biol 9(1):1–20
24. Kim JL, Morgenstern KA, Griffith JP, Dwyer MD, Thomson JA, Murcko MA, Lin C, Caron PR (1998) Hepatitis C virus NS3 RNA helicase domain with a bound oligonucleotide: the crystal structure provides insights into the mode of unwinding. Structure 6(1):89–100
25. Yao N, Reichert P, Taremi SS, Prosise WW, Weber PC (1999) Molecular views of viral polyprotein processing revealed by the crystal structure of the hepatitis C virus bifunctional protease-helicase. Structure 7(11):1353–1363
26. Yao N, Hesson T, Cable M, Hong Z, Kwong AD, Le HV, Weber PC (1997) Structure of the hepatitis C virus RNA helicase domain. Nat Struct Biol 4(6):463–467
27. Cho HS, Ha NC, Kang LW, Chung KM, Back SH, Jang SK, Oh BH (1998) Crystal structure of RNA helicase from genotype 1b hepatitis C virus. A feasible mechanism of unwinding duplex RNA. J Biol Chem 273 (24):15045–15052
28. Gu M, Rice CM (2010) Three conformational snapshots of the hepatitis C virus NS3 helicase reveal a ratchet translocation mechanism. Proc Natl Acad Sci U S A 107(2):521–528
29. Appleby TC, Anderson R, Fedorova O, Pyle AM, Wang R, Liu X, Brendza KM, Somoza JR (2011) Visualizing ATP-dependent RNA translocation by the NS3 helicase from HCV. J Mol Biol 405(5):1139–1153

30. Velankar SS, Soultanas P, Dillingham MS, Subramanya HS, Wigley DB (1999) Crystal structures of complexes of PcrA DNA helicase with a DNA substrate indicate an inchworm mechanism. Cell 97(1):75–84
31. Lee JY, Yang W (2006) UvrD helicase unwinds DNA one base pair at a time by a two-part power stroke. Cell 127(7): 1349–1360
32. Krebs WG, Gerstein M (2000) The morph server: a standardized system for analyzing and visualizing macromolecular motions in a database framework. Nucleic Acids Res 28 (8):1665–1675
33. Zheng W, Brooks BR, Hummer G (2007) Protein conformational transitions explored by mixed elastic network models. Proteins 69 (1):43–57

Chapter 10

Geometric Simulation of Flexible Motion in Proteins

Stephen A. Wells

Abstract

This chapter describes the use of physically simplified analysis and simulation methods—pebble-game rigidity analysis, coarse-grained elastic network modeling, and template-based geometric simulation—to explore flexible motion in protein structures. Substantial amplitudes of flexible motion can be explored rapidly in an all-atom model, retaining realistic covalent bonding, steric exclusion, and a user-defined network of noncovalent polar and hydrophobic interactions, using desktop computing resources. Detailed instructions are given for simulations using FIRST/FRODA software installed on a UNIX/Linux workstation. Other implementations of similar methods exist, particularly NMSim and FRODAN, and are available online. Topics covered include rigidity analysis and constraints, geometric simulation of flexible motion, targeting between known structures, and exploration of motion along normal mode eigenvectors.

Key words Protein flexibility, Geometric simulation, Template, Distance constraint, Hydrogen bond, Hydrophobic tether, Rigidity analysis, Normal mode, FIRST, FRODA

1 Introduction

Geometric simulation is a method for carrying out rapid all-atom simulations of flexible motion in protein structures. It uses a simplified physics which implements only the strongest, most local interactions—covalent bond geometry, steric exclusion of atomic spheres, hydrophobic tethers, and local polar interactions including hydrogen bonds and salt bridges. This allows the simulation to be computationally much cheaper and faster than molecular dynamics (MD) while still being sufficiently detailed to provide informative results. Useful data can typically be generated in a matter of hours on a desktop workstation without the need for high-performance computing facilities or expertise. This means the method can easily be used as an "intuition pump" and hypothesis generator, to visualize the flexibility intrinsic to a protein structure and possibly obtain insight into functional motion and the structure/function relationship. The method is most effective in exploring motion

Dennis R. Livesay (ed.), *Protein Dynamics: Methods and Protocols*, Methods in Molecular Biology, vol. 1084,
DOI 10.1007/978-1-62703-658-0_10,

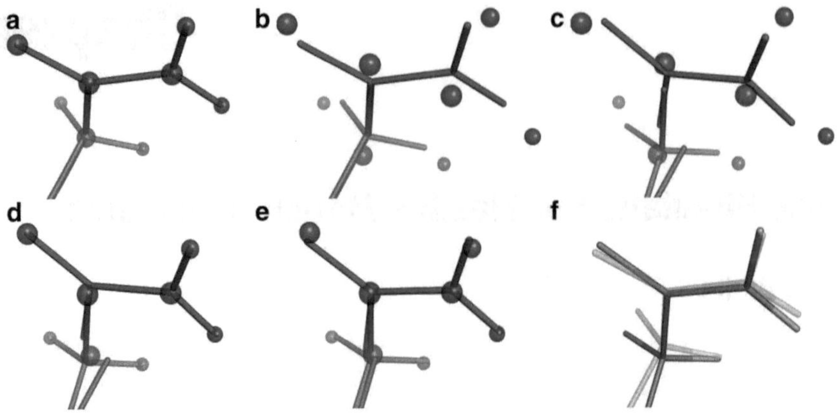

Fig. 1 Template-based geometric simulation. A small portion of a protein structure is shown in *ball-and-stick* form in (panel **a**). The radius of the atomic spheres has been greatly scaled down for clarity. After a perturbation of the atomic positions (panel **b**), geometric templates representing the bonding geometry of rigid clusters are fitted over the atoms (panel **c**). Several rounds of mutual fitting (panels **d** and **e**) reduce the mismatch between atomic positions and template vertices to within an acceptable tolerance, generating a new conformation of the protein structure (panel **f**). Note in particular that rotation of a bond dihedral angle is handled implicitly by the overlap of rigid clusters at a rotatable bond. Adapted by the author from [4]

along a direction of interest, for example by directing motion towards a target structure, or by biasing motion along normal-mode eigenvectors.

The geometric simulation approach described here was originally developed for the study of aluminosilicate zeolite frameworks [1] and has provided informative results on tetrahedral and octahedral frameworks [2, 3]. Subsequently (2005) it was implemented for proteins in a method known as FRODA (Framework Rigidity Optimized Dynamic Algorithm) [4] within the rigidity analysis software FIRST (Floppy Inclusions and Rigid Substructure Topography) [5]. The distinctive feature of the approach is that it avoids the use of a large number of two-body (bond), three-body (bond angle), and four-body (dihedral angle) constraints. Rather, a template represents the covalent geometry of an entire rigid cluster of bonded atoms, with rigid clusters varying in scale from single methyl groups and peptide units up to alpha-helices and entire domains. A harmonic constraint on displacement of an atom from its vertex on the template penalizes all variations from the ideal bonding geometry with respect to all the other atoms in the group. During the course of flexible motion simulation, both atoms and templates are mobile. Figure 1 illustrates the nature of the approach.

The detailed instructions in this chapter explain how to apply FIRST/FRODA to your protein of choice, especially in a biased simulation approach which rapidly explores large-scale motion along directions suggested by elastic network model normal mode analysis. This approach is illustrated schematically in Fig. 2. Several recent studies have shown that such simulations provide

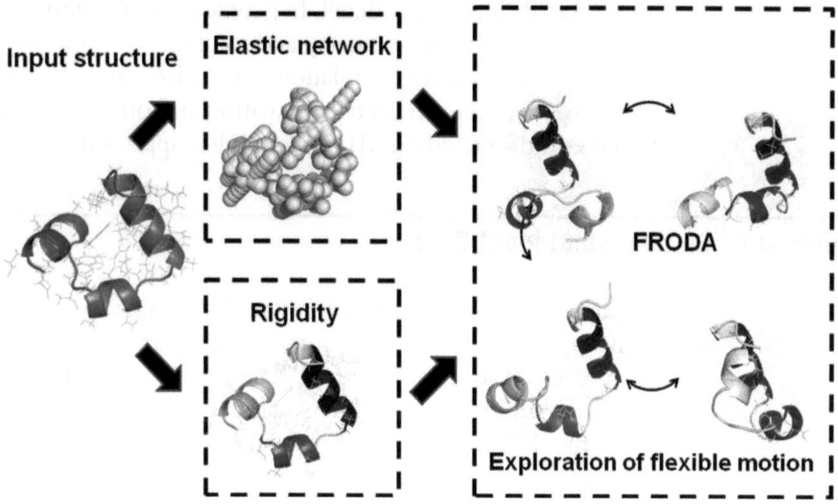

Fig. 2 A combination of analysis and simulation methods leads to rapid simulations of flexible motion. An all-atom input structure (at *left*) undergoes two parallel forms of analysis. A coarse-grained elastic network model generates eigenvectors for flexible motion (*center, above*) while rigidity analysis divides the all-atom structure into rigid clusters of varying size (*center, below*). Geometric simulation by the FRODA algorithm uses information from both analyses to explore large amplitudes of motion. Adapted by the author from [6]

a large amount of biophysical information and insight at minimum computational expense [6–8], and can reconcile apparently disparate crystal-structure and solution-structure data in a consistent picture based on the intrinsic flexibility of a protein [9].

Recently two further pieces of software by other authors have appeared which make use of geometric simulation approaches: FRODAN [10] and NMSim [11]. Broadly speaking, FRODAN is optimized to produce greater amplitudes of motion than FRODA and in particular to explore large-scale structural transitions using targeted simulations. It focuses on small templates rather than larger rigid units, and uses a "momentum" system to explore motion in directions compatible with the constraint network, rather than the normal mode exploration discussed in this chapter. NMSim is more "MD-like" and computationally intensive, being based on repeated cycles of a process including rigidity analysis, normal mode generation, geometric simulation of flexible motion over a small amplitude, and molecular mechanics re-optimisation of the structure. Both of these methods have accessible Web interfaces where structures may be uploaded for simulation: FRODAN is available from the Flexweb Web site at ASU (http://pathways.asu.edu) and NMsim is accessible at http://www.nmsim.de [12]. Users interested in geometric simulation should explore the available methods and use the approach they find most appropriate.

The approach discussed here is conceptually quite distinct from conventional "coarse-graining" approaches to simulation. In coarse-graining, single interaction sites are used to represent clusters of atoms, sacrificing atomic detail. In geometric simulation, by contrast, the all-atom covalent geometry and local steric exclusion are maintained throughout the simulation. This allows seamless transfer of structures between geometric simulation and other all-atom simulation methods, including MD and ab-initio approaches.

2 Prerequisites: Software and Input Data

We assume a Linux/UNIX or MacOS-X command line environment on a desktop workstation. High-performance computing resources are not required. The essential software required is FIRST version 5 or 6, which includes FRODA, and is available from the Flexweb Web site hosted at Arizona State University (flexweb.asu.edu). Academic users can obtain the software free of charge by registering at Flexweb. Invocation of the FIRST executable will be written as "**first**".

The essential input data are an all-atom protein structure in PDB file format. Crystal structures downloaded from the PDB typically require preprocessing (addition of hydrogen atoms; removal of crystal water; selection of alternate sidechain conformations). This will involve a protein visualization and editing tool such as PyMOL, available http://www.pymol.org [13], and a hydrogenation tool such as Reduce, available from http://kinemage.biochem.duke.edu/software/reduce.php [14].

Normal mode eigenvectors for exploration can be generated using any elastic network model implementation. ElNemo software [15] is available online (http://www.igs.cnrs-mrs.fr/elnemo/) and its use is described in Subheading 3.5.

3 Methods

The essential input to FIRST/FRODA is an all-atom (including hydrogens) protein structure in PDB data format (*see* also **Notes 1** and **2**). Subheading 3.1 describes the preprocessing steps typically required in order to set up this input. Subheading 3.2 describes the conduct of rigidity analysis and rigidity dilution in FIRST, and the selection of an appropriate set of constraints for the simulation. Subheading 3.3 describes the invocation of FRODA to simulate flexible motion and discusses the setting of some basic options. Subheading 3.4 describes motion biased towards a known target structure. Subheading 3.5 describes motion biased along a specified vector, and the generation of such vectors using ElNemo. Subheading 3.6 describes a restarting protocol which can be used to continue a previous simulation.

Text to be entered on the command line will be given in bold, thus: "**first foo.pdb -E -2.0 -non -FRODA**". Optional arguments are indicated by square brackets. Scripting commands are given for the popular "bash" shell.

3.1 Preliminary Steps: Structure Preprocessing

A crystal structure downloaded from a source such as the Protein Data Bank typically lacks hydrogen atoms and may contain unwanted entries, such as crystal water oxygens and multiple alternate conformations for side chains. Proper preparation should lead to a structure with hydrogens added, unwanted water molecules and extraneous molecules removed, and a single conformation for each residue (*see* **Note 3**).

Alternate sidechain conformations are labelled by an altLoc field, character 17 in an ATOM record in PDB data format. This field is blank when the conformation is unique and is one of "A," "B," . . . when multiple conformations are given. Structure viewers allow selection on this field; e.g., in PyMOL the command "**remove not (alt ''+A)**" will retain unique and "A" conformations only.

Multiple tools exist to add hydrogen atoms at positions appropriate to the heavy-atom bonding geometry, for example the Reduce tool, part of the comprehensive MolProbity suite. A structure can be uploaded to the MolProbity Web site (http://molprobity.biochem.duke.edu/) for online processing. Reduce can also be downloaded, in which case the command "**reduce foo.pdb > fooH.pdb**" will generate a structure with hydrogens added. In the process of adding hydrogens, certain side groups, such as the carboxamides in asparagine and glutamine, may need to be "flipped" to give a good steric geometry; this flipping is the default behavior of Reduce.

Crystal water oxygens typically have an HOH residue identifier and can be removed from the text file using UNIX tools ("**grep -v HOH foo.pdb**" will give only lines not containing HOH) or deleted while viewing the structure (e.g., in PyMOL, "**remove resn hoh**"). Hydrogenation tools such as Reduce generally do not hydrogenate water oxygens and they should be omitted unless specifically required.

If added hydrogens are initially given an ID number of zero, it is necessary to renumber the atoms sequentially before carrying out any further analysis. PyMOL and other viewers will carry out this renumbering by default simply by loading and saving the structure file.

Further steps assume a prepared structure file, which we will call **myprotein.pdb**.

3.2 Rigidity Analysis, Rigidity Dilution, and Constraint Sets

Rigidity analysis is a necessary step in FIRST before invoking FRODA to simulate flexible motion. This is an opportunity to generate editable lists of covalent and noncovalent constraints found in the input structure. A "rigidity dilution," which examines the rigidity of the structure as hydrogen bonds are progressively eliminated, can provide valuable structural information and assist in the selection of appropriate constraint sets, and in particular of the hydrogen bond energy cutoff value, for simulations using FRODA.

FIRST makes use of four types of constraint when performing rigidity analysis—covalent bonds, hydrogen bonds (including salt bridges), hydrophobic tethers, and stacked ring interactions (rarer in proteins than in RNA/DNA systems)—which are identified based on the geometry of the input structure. For our purposes it is convenient to initially export from FIRST a list of each type of constraint. These lists can then be edited if necessary, and used as input to FIRST for all subsequent steps. To export constraint lists we invoke "**first myprotein.pdb -non -E 0 -covout -hbout -phout -srout**". Taking the options in order: -**non** runs FIRST in noninteractive mode, i.e., without querying the user for text input. -**E 0** ensures that all potential hydrogen bonds with negative bond energy are exported. -**covout** exports covalent bonds to cov.out; -**hbout** exports polar interactions to hbonds.out; -**phout** exports hydrophobic tethers to hphobes.out; and -**srout** exports stacked ring interactions to stacked.out. Editing of the constraint files is discussed in **Notes 4–8**, but is typically not needed for application to ordinary globular proteins. Rename the files for input ("**mv cov.out cov.in; mv hbonds.out hbonds.in; mv hphobes.out hphobes.in; mv stacked.out stacked.in**"). Future invocation of FIRST will use the options "**-covin -hbin -phin -srin**", causing FIRST to read the ***.in** files.

In a "rigidity dilution," bonds are eliminated from the constraint network progressively in order of strength. FIRST produces a graphic illustrating how the rigidity of the protein mainchain changes as constraints are removed. Residues which form part of a rigid cluster in the analysis are shown using a thick colored bar, while flexible parts of the sequence (not part of a multi-residue rigid cluster) are shown as a thin line. Major changes in the plot typically are visible at cutoff values in the approximate range of 0 to −3 kcal/mol [16]. Perform a dilution using the following options: "**first myprotein.pdb -non -E 0 -dil 1 -covin -hbin -phin -srin**". This runs a dilution ("**-dil 1**" option) from a cutoff of 0 kcal/mol downwards. An example of such a graphic, colloquially termed a "stripy plot," is shown in Fig. 3.

This dilution is mainly useful in distinguishing regions of the protein which are rich in constraints, and remain relatively rigid until far down the plot, from regions that are poor in constraints and become flexible early on. The most constrained regions often correlate with the "folding core" of proteins [16]. The dilution provides useful information in choosing hydrogen bond energy cutoff values for flexible motion simulation; in particular, relative motion of two portions of the structure will be impossible if they form part of one rigid cluster! For the purposes of flexible motion

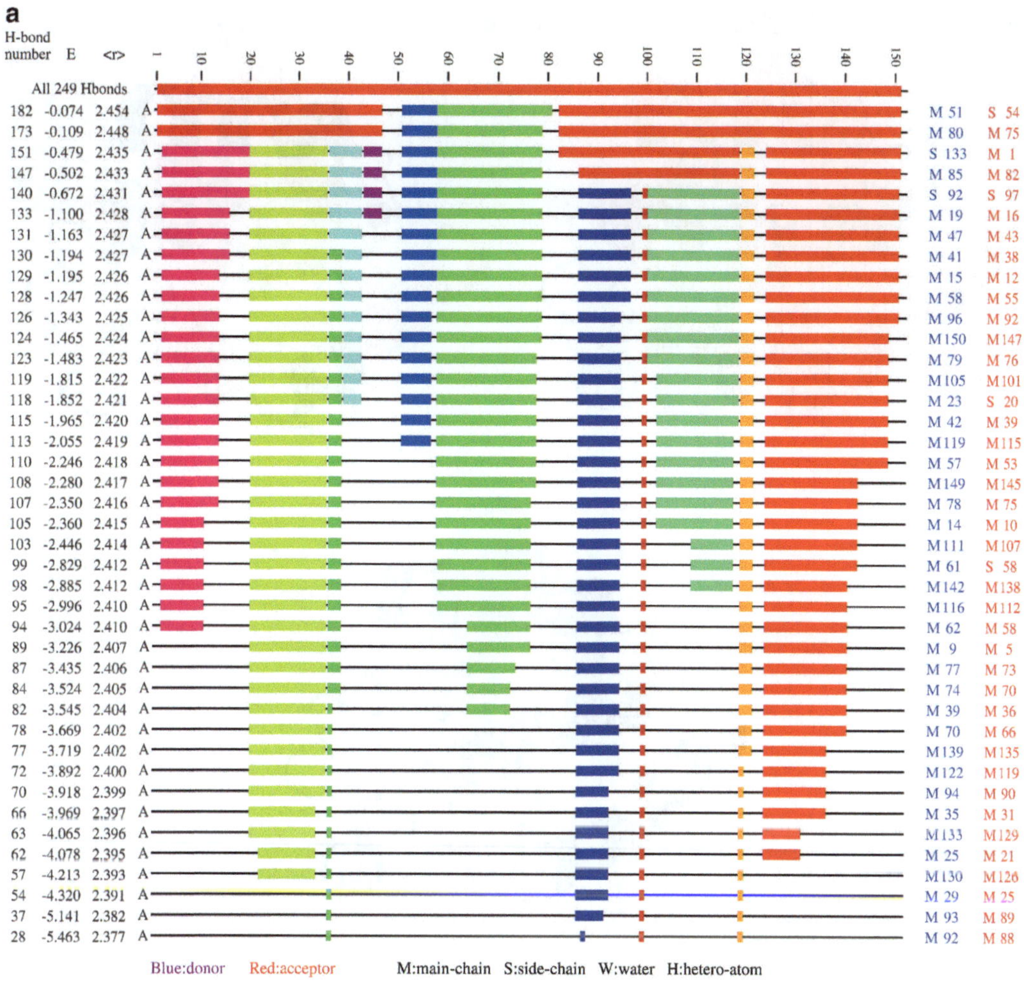

Fig. 3 Rigidity and flexibility in myoglobin: an illustrative analysis of the carbonmonoxymyoglobin structure obtained from PDB entry 1DRW. Rigidity analysis in FIRST using the "pebble game" generates a rigidity dilution or "stripy plot" (panel **a**). Numerical columns at *left* show the number of hydrogen bonds in the constraint network, the energy cutoff value in kcal/mol for the inclusion of bonds as constraints, and the mean coordination of atoms in the protein. Each line in the plot shows a linear representation of the protein backbone, with residues numbers indexed on the scale above. *Thin black lines* indicate flexible regions, *while thick colored lines* indicate rigid clusters. A line is plotted whenever removal of the next weakest hydrogen bond causes a change in the rigid cluster decomposition of the backbone. The side panels show a three-dimensional representation of the protein structure. Flexible portions are shown in *black* with large rigid clusters shown in *color*; the central prosthetic heme group is shown as *spheres*. Hydrophobic tether interactions are shown as *green dashed lines* and hydrogen bonds as *red dashed lines*. At an early stage in the dilution (panel **b**), using a cutoff energy of -0.1 kcal/mol, the bulk of the protein forms a single large rigid cluster, with a few independently rigid helices and flexible loops. Midway through the dilution (panel **c**), using a cutoff energy of -2.0 kcal/mol, flexibility is more widespread, with several helices forming independent rigid clusters. At a late stage (panel **d**), using a cutoff energy of -4.0 kcal/mol, only a few large rigid clusters persist and the structure is largely flexible

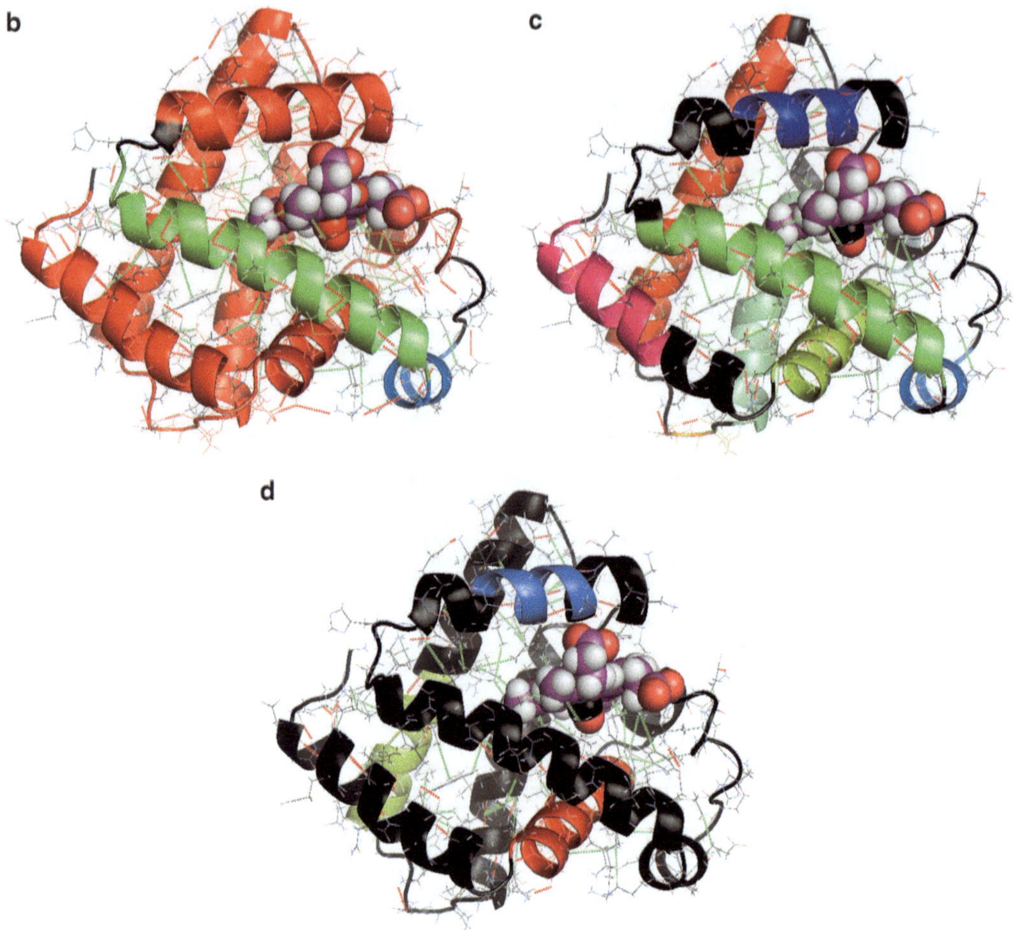

Fig. 3 (continued)

simulation using FIRST/FRODA, experience suggests that *lower (more negative) cutoffs are usually appropriate*, i.e., *on the order of −2/−3 kcal/mol*, even though at these cutoffs the rigidity dilution typically shows the structure as largely or entirely flexible [7].

3.3 Generating Motion in FRODA: Invocation, Basic Settings and Outputs

The use of completely unguided FRODA simulations is deprecated but harmless, and provides an opportunity to discuss a few basic command line options. In a folder containing the input files (**myprotein.pdb, cov.in, hbonds.in, hphobes.in, stacked.in**), run FIRST with the following options:

"first myprotein.pdb -non -E $ECUT -covin -hbin -phin -srin -FRODA -mobRC1 -freq 100 -totconf 1000 -step 0.1 [-body]"

where **$ECUT** is of course replaced by our chosen cutoff in kcal/mol, e.g., −2. -**FRODA** invokes the FRODA module. The default behavior of FRODA is to keep the largest rigid cluster (RC1) immobile; the -**mobRC1** option makes the entire structure mobile and is *strongly recommended* (*see* also **Note 9**). **-freq 100** calls for every 100th conformation generated to be written out as a .pdb data file, while **-totconf 1000** calls for FRODA to stop at the 1000th conformation; both settings can of course be given any desired value. **-step 0.1** causes random perturbations of atomic positions of up to 0.1 Å during conformation generation. **-body**, if given, causes perturbations to be applied on a cluster-by-cluster basis rather than atom-by-atom and favors mobility when large rigid clusters are present.

During a FRODA run, every new conformation generated is reported with a statement of how many fitting cycles FRODA required to satisfy all constraints, and the all-atom RMSD relative to the input conformation:

CONF. 1 FOUND at 5 cycles, ALLATOM 0.0128035;
CONF. 2 FOUND at 2 cycles, ALLATOM 0.0162872;
CONF. 3 FOUND at 2 cycles, ALLATOM 0.0197066;

Note that this is a raw rather than a fitted RMSD. Generation of output structure files is reported:

CONF. 50 FOUND at 2 cycles, ALLATOM 0.176697;
Writing conformation to file 1dwr-h_froda_00000050.pdb

In the event of FRODA being unable to restore all constraints after the random perturbation within a defined maximum number of cycles (100 by default), failure is reported and the conformation is restored to the previous (last known good) state:

FAILED AFTER MAX CYCLES: 101, RECOVERING.

The maximum number of fitting cycles before failure can be controlled by a parameter in FRODA ("**-maxfitc** *N*").

Among the output files generated by FIRST is a script for PyMOL (here **myprotein_RCD.pml**) which will display rigid-cluster information, show hydrogen bonds and hydrophobic tethers, and load the FRODA output files (**myprotein_froda_*.pdb**) into a movie object. The cluster of conformations produced by a random exploration with FRODA gives a visual illustration of the flexibility intrinsic in the structure. However, this random exploration is inefficient in its exploration of conformational space compared to bias-directed or targeted motion.

3.4 Targeting to a Known Structure, Whole or Partial

Given two or more crystal structures of our protein, we can attempt to generate a trajectory of flexible motion from one to another. FRODA can use either complete or partial targets. In a complete target structure there is a one-to-one correspondence of each atom in the input structure (**myprotein.pdb**) to an atom in the target structure (say **mytarget.pdb**). The target structure is specified using a **-target** flag. The user should provide a "directed step size" value using a **-dstep** flag; the perturbation applied to each atom at the start of a FRODA step will be the sum of a random component and a directed component which moves the atom towards its target position. The directed step size should not be large, as a small persistent bias is less likely to lead to jamming or getting stuck in a dead end; values around 0.01 Å are recommended, thus "**-dstep** 0.01". An additional recommended flag is **-propto** which makes the directed step size proportional to the distance to target, so atoms with further to go move faster. This gives a smoother motion. A targeted trajectory halts as specified by the **-totconf** flag, as for random runs. An optional additional halting criterion can be given based on the RMSD between the current conformation and the target structure using a **-dtol** flag (distance tolerance). Small differences of bonding geometry between the input and target structures mean that an RMSD to target less than 0.5 Å is unlikely. A typical all-to-all targeted motion simulation would use a command line like this:

"first myprotein.pdb -non -covin -hbin -phin -srin -E -3.0 -target mytarget.pdb -FRODA -mobRC1 -step 0.1 -dstep 0.01 -propto -freq 100 -totconf 5000 -dtol 0.5"

Conformer generation during targeted motion includes a report of the RMSD to target:

CONF. 3098 FOUND at 5 cycles, ALLATOM 3.84063; TARGET 0.87093;
CONF. 3099 FOUND at 5 cycles, ALLATOM 3.84077; TARGET 0.87082;
CONF. 3100 FOUND at 5 cycles, ALLATOM 3.84115; TARGET 0.870738;

Figure 4 shows a sequence of frames from a simulated transition between apo and carbonmonoxy forms of hemoglobin. An important feature of the geometric simulation, visible in the figure, is that it permits sidechains to rotate naturally during targeted motion, producing physically reasonable covalent geometry at all stages.

A key point in targeted motion is not to include constraints on the input structure which are incompatible with the transition towards the target structure. An initial comparison of the structures, and editing out of unwanted constraints, is highly recommended. This is discussed further in the **Notes 10** and **11**.

A partial target can be given, in which case only those atoms in the input structure which match a target atom are given a directed bias. Atom matching is done based on the PDB ID field, so some care should be taken over preparing the target structure and ensuring that PDB IDs match. Partial targeting must be signalled to FRODA using the **-fancy** flag, thus:

"first myprotein.pdb -non -covin -hbin -phin -srin -E -3.0 -target myPartialTarget.pdb -fancy -FRODA -mobRC1 -step 0.1 -dstep 0.01 -propto -freq 100 -totconf 5000 -dtol 0.5"

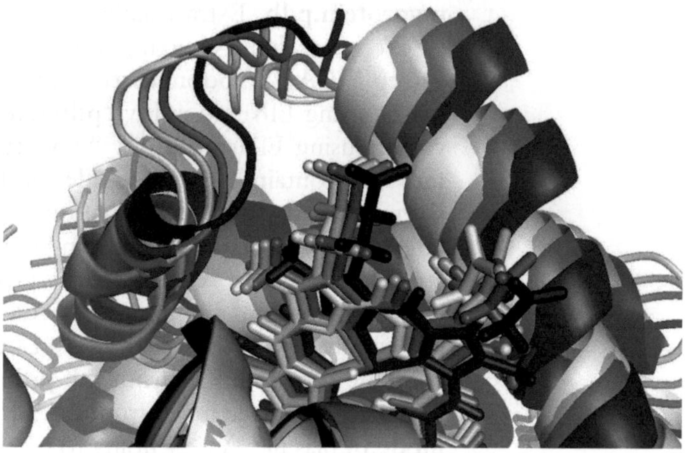

Fig. 4 Stages in a directed motion simulation from apohemoglobin to carbonmonoxyhemoglobin, showing a region around the heme binding site of one chain; the simulation was carried out on the full tetrameric structure. The protein is shown in *cartoon* form with the input, obtained from PDB entry 1A3N. shown in *white* and the target, obtained from PDB entry 1BBB, shown in *black*. Two intermediate frames are shown in *shades* of *gray*. The heme group is shown in *stick* form and the rotation of a side chain is visible; a linear interpolation between start and end points does not capture such local motions

3.5 Exploration of Normal Modes

In this application, the perturbation applied to atoms is composed of two components, one random and one directed. The directed component is specified by a vector provided by the user, describing the bias to be applied to each residue, in a file called "**mode.in**". This file is whitespace-delimited text, containing *one line per residue* of the input structure. Each line consists of an integer, which is the PDB ID of the alpha carbon atom of the residue, followed by three real numbers which are the x, y and z components of a Cartesian vector. This vector specifies the direction of motion for the residue. The bias eigenvector should have unit magnitude, that is, the total sum of the squares of all $3N$ vector components for an N-residue protein should be equal to 1. An example showing the top lines from a **mode.in** file is as follows:

2 0.006566 0.059299 0.025345
11 0.0038335 0.058965 0.020255
30 0.00058834 0.073323 0.026298
41 0.011948 0.073435 0.022475

FRODA itself has no particular requirements on the source or nature of the bias vector, and any form of user-constructed bias may be used. A typical bias vector would be a normal mode eigenvector, or weighted sum of eigenvectors, obtained by coarse-grained elastic network modeling.

Modes can be generated using ElNemo software via the following steps. Create a **Modes**/directory containing the structure file

myprotein.pdb. Extract alpha carbon atoms from protein structure to create an ElNemo input file: "**grep ATOM myprotein.pdb | grep " CA " > pdbmat.structure** ". Generate elastic network matrix using ElNemo utility "**pdbmat**", and then generate eigenvectors using ElNemo utility "**diagstd**". The file **pdbmat.eigenfacs** now contains the eigenmodes of the elastic network. The first six modes are trivial combinations of rigid-body motions with effectively zero frequency and can be neglected. Modes 7 upwards are nontrivial. Each eigenvector is given as one line per residue, with each line containing three Cartesian vector components. Simple text processing tools such as UNIX **cut** and **paste** can be used to combine the atom IDs from the **pdbmat.structure** file with the vectors of a selected mode from the **pdbmat.eigenfacs** file to give a **mode.in** bias file. A C++ utility to extract any desired set of modes is available from the author on request.

The command line options for mode exploration are **-modei**, which triggers reading of mode.in, and a **-dstep** setting giving a bias step size. A key feature of mode exploration is this: *both positive and negative directed step sizes are used.* A positive directed step size will bias motion parallel to the mode eigenvector, while a negative step size will bias motion *antiparallel* to the mode eigenvector.

Mode exploration generally has an initial phase of very rapid conformer generation [6], during which motion in the bias direction is consistent with all structural constraints and FRODA finds new conformers in very few fitting cycles. At larger amplitudes, conflicts will emerge between the bias direction and the constraints (e.g., stretching of tethers, steric clashes) and the rate of generation slows:

CONF. 3151 FOUND at 64 cycles, ALLATOM 12.9298;
CONF. 3152 FOUND at 33 cycles, ALLATOM 12.9316;
CONF. 3153 FOUND at 33 cycles, ALLATOM 12.9334;
CONF. 3154 FOUND at 33 cycles, ALLATOM 12.9353;
CONF. 3155 FOUND at 33 cycles, ALLATOM 12.937;

and may even come to a halt entirely:

FAILED AFTER MAX CYCLES: 100, RECOVERING.
CONF. 3202 FOUND at 101 cycles, ALLATOM 12.9952;
FAILED AFTER MAX CYCLES: 100, RECOVERING.
CONF. 3203 FOUND at 101 cycles, ALLATOM 12.9952;
FAILED AFTER MAX CYCLES: 100, RECOVERING.
CONF. 3204 FOUND at 101 cycles, ALLATOM 12.9952;

The following bash script allows trajectories to be automatically generated over multiple energy cutoff values and multiple bias modes. The execution directory contains the input structure and constraint files, and also contains a **Modes**/directory, which contains a set of mode input files labelled as **modeMM.in** where *MM* is an index—e.g., **mode07.in**, **mode08.in** etc.

```
#!/bin/bash
PROT="myprotein.pdb"
cutlist="1.0 2.0 3.0" # positive values; made negative
in options
TOTCONF=2000
FREQ=200
STEP=0.01
DSTEP=0.01
modelist="07 08 09 10" # etc.
Thisdir=‘pwd‘
echo $Thisdir
mkdir Runs
for CUT in $cutlist
do
  mkdir ./Runs/$CUT
  for MODE in $modelist
  do
    for SIGN in pos neg
    do
      mkdir ./Runs/$CUT/Mode$[17]-${SIGN}
      cd ./Runs/$CUT/Mode$[17]-${SIGN}
      pwd
       cp $Thisdir/$PROT .
       cp $Thisdir/cov.in .
       cp $Thisdir/hbonds.in .
       cp $Thisdir/hphobes.in .
       cp $Thisdir/stacked.in .
       cp $Thisdir/Modes/mode$[17].in ./mode.in
       ##Now we have all the input files
       if [["$SIGN" -- "neg"]]
       then
         COMMAND = " -non $PROT -E -$CUT -FRODA
-mobRC1 -freq $FREQ -totconf $TOTCONF -modei -step
$STEP -dstep -$DSTEP -covin -hbin -phin -srin " # all one
line please! Note -$DSTEP
       else
         COMMAND = " -non $PROT -E -$CUT -FRODA
-mobRC1 -freq $FREQ -totconf $TOTCONF -modei -step
$STEP -dstep $DSTEP -covin -hbin -phin -srin " # all one
line please! Note $DSTEP
       fi
       ##and now we have our command line options
         first $COMMAND ## first is your FIRST
executable
      cd $Thisdir
      pwd
    done # SIGN
  done #MODE
done #CUT
```

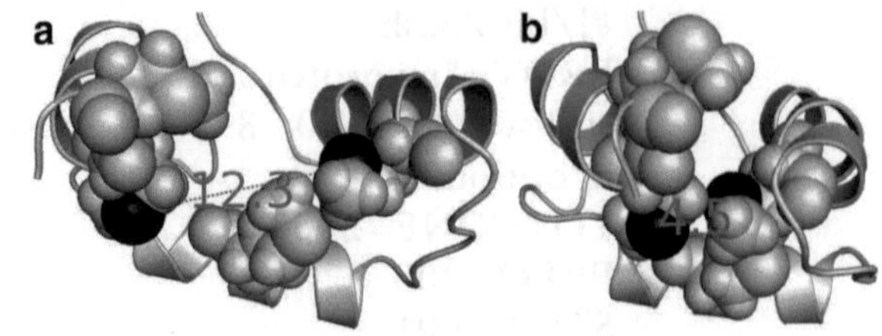

Fig. 5 Exploration of flexible motion captures a significant change in the geometry of a binding site in calmodulin. Panel (**a**) shows part of an input structure based on PDB entry 1CLL. The protein is shown in *cartoon* form. Methionine residues lining a hydrophobic protein-binding site shown are shown as *spheres*. The sulfur atoms of the methionine sidechains are highlighted in *black*. The sulfur–sulfur distance in a specific pair of residues is more than 12 Å. Panel (**b**) shows the same site after simulation of flexible motion along a low-frequency normal mode. The sulfur–sulfur distance is now less than 5 Å, close enough to allow sidechain cross-linking by cisplatin. Adapted by the author from [7]

The logic of this script can be easily adapted to other shells or scripting languages. An initial run using a more limited set of modes and cutoffs may be helpful in finding appropriate values for the **FREQ** and **TOTCONF** values. Figure 5 shows how the flexible motion of a calmodulin structure along a normal mode direction brings key residues close together, permitting cross-linking (ref. 7). (*see* **Note 12**)

3.6 Procedure 3: Restarting Protocol

The user may wish to continue a trajectory from a previously generated endpoint, for example if the number of conformations initially specified with -**totconf** was insufficient. However, simply using a FRODA-generated conformation as input to a new run of FIRST/FRODA is *strongly deprecated*, as it leads to the accumulation of errors. To continue a trajectory we should use the geometry of the initial input, then move the atoms to the restart position and continue generation. This is achieved using the -**restart** option.

For example, an initial command line of "**first myprotein.pdb** ... **[options]** ... -**totconf 2000**" generates a final frame **myprotein_froda_00002000.pdb**. We wish to continue a trajectory from this final frame. Make a directory: "**mkdir Continuation**". Copy all starting files: "**cp myprotein.pdb Continuation/myprotein_more.pdb; cp *.in Continuation/**". Copy the restart position: "**cp myprotein_froda_00002000.pdb Continuation/myrestart.pdb** ". Note that the input and restart positions have been given distinct names for clarity.

In the Continuation directory, run with the desired options, and pass the restart position using the -**restart** flag: "**first myprotein_more.pdb** ... **[options]** ... -**restart myrestart.pdb**".

The conformations produced will be named as **myprotein_more_froda_########.pdb**. Shell scripting can be used to unite the original and continuation frames into a single consistently named trajectory.

4 Notes

1. *On the quality of input structures.* The most important consideration is that the input structure should have good covalent and steric geometry, allowing accurate identification of noncovalent polar and hydrophobic interactions. The best input is a high-quality crystal structure (resolution better than 2 Å). Usable inputs are lower-quality crystal structures (resolution 2–3 Å), NMR consensus structures, and structures from MD or DFT equilibration at low temperatures, or from molecular-mechanics relaxation. Deprecated inputs are poor crystal structures (resolution >3 Å), individual NMR structures from an ensemble, and individual MD frames.
2. *On the quality of output structures.* This is hard to quantify, but roughly analogous to individual members of an NMR structural ensemble. Backbone and sidechain dihedral distributions, and steric clashes, are typically rather worse than in a good X-ray structure, but much better than in an individual frame taken from an MD trajectory. Experience shows that FRODA output structures can be used as inputs to MD or DFT simulations without causing problems, and are good enough for meaningful use in structural biology.
3. *On the importance of preprocessing.* Much time can be saved at the production stage by properly preparing your structure at the very beginning. Nonstandard residues, if present, may have to be hydrogenated by hand. If an explicit buried water is important to your system you should ensure that it is present and properly hydrogenated. Alternate conformations of sidechains are a particular hazard. The presence of multiple sidechain conformations will typically *not* cause fatal errors at the rigidity analysis stage, but results in irresolvable steric clashes in FRODA. Find and remove them early. If you are using bias modes, wait to generate modes until you are certain that your structure is good for production.
4. *On the covalent bond file* **cov.out/cov.in**. Each line of this file contains three integers; the first two are atom IDs from the input structure, and the third is a number of "bars" used in rigidity analysis. This bar number is either 5 or 6. Covalent bonds with a rotatable dihedral angle, such as aliphatic carbon-carbon, have 5 bars; non-rotatable bonds have 6. In proteins the commonest example of the latter are the C—N bonds of the backbone peptide plane. FIRST can recognize the bonding of standard residues, but if the input structure contains (1) nonstandard residues such as drug molecules, and/or (2) bound metal ions, some editing of the covalent bond file may be required. The bonding of the Fe atom of a heme group to coordinating protein nitrogen atoms is a particularly common case.

Covalent bonds can be added in at least three ways. Firstly, CONECT records can be placed in the input PDB data file specifying interatomic bonding. This is generally the best solution for nonstandard residues. Secondly, the cov.in file can be edited in any text editor to add a line "**$ID1 $ID2 5**" to connect atoms ID1 and ID2. Thirdly, FIRST can be run interactively by omitting the -**non** argument ("**first myprotein.pdb -covout**"). In this mode, apparent steric clashes where nonbonded atoms overlap will be queried and the user can choose to declare them as covalent bonds, which will then appear in the **cov.out** file.

5. *On hydrophobic tethers.* These are placed by FIRST between aromatic or aliphatic sidechain carbons at a distance less than 4 Å. -**phout** will save them to **hphobes.out** in the format "**$ID1 $ID2 2**"; that is, hydrophobic tethers are assigned 2 bars in rigidity analysis. The user can eliminate unwanted hydrophobic tethers by deleting lines from **hphobes.in**. In FRODA these atoms pairs will be kept within 4 Å plus a small tolerance of around 0.15 Å. This small tolerance is a key reason why continuation runs (Subheading 3.6) should use the initial input structure and the -**restart** option.
6. *On hydrogen bonds.* These are assigned a (negative) bond energy based on the donor-hydrogen-acceptor geometry by FIRST. Salt bridges (fully polarized noncovalent polar interactions) are rated with a different energy functional reflecting their typical greater bond strength and reduced directionality. In rigidity analysis, hydrogen bonds included in the constraint network are assigned 5 bars. Energies are expressed in kcal/mol, with the strongest interactions having energies in the range −5 to −10 kcal/mol; note that room temperature kT is about 0.6 kcal/mol. The **hbonds.out/hbonds.in** file has the format "**$ID1 $ID2 5 $Energy**". Bonds whose energy lies below the cutoff value are thus included as equivalent to rotatable covalent bonds, while those with energies above the cutoff value are excluded. A single hbonds.in file will therefore do for simulations at many different energy cutoffs.

   A particularly tricky point! The energy functional used in FIRST penalizes too-close approach of hydrogen-bonding atoms. Thus a bond which should naturally relax to a good low-energy bond may initially register with a poor (excessively high, even positive) energy. It will then be excluded by the cutoff energy. The relevant atoms will now register as a severe steric clash which may not be resolvable. This can produce apparently baffling behavior in FRODA, with simulations failing to find any new conformations and jamming persistently from the start. A simple workaround is to find the offending bond in **hbonds.in** and edit its energy to a more sensible, negative value. Molecular-mechanics relaxation of the input structure is a more comprehensive solution.

7. *On energy cutoff values.* Even at the lower end of a dilution plot where the protein registers as largely flexible, the structure in FIRST/FRODA still contains all hydrophobic tethers found in the input structure, and a number of strong hydrogen bonds, especially the backbone-backbone interactions within well-formed alpha helices and beta sheets. These noncovalent constraints will maintain the domain structure of the protein in a FRODA simulation and, together with steric clashes, will limit the amplitude of motion that is possible when exploring a bias direction. It is therefore preferable to use energy cutoff values for FRODA runs that are lower (more negative) than the values at which features appear in the rigidity dilution of the static structure. Thus rigidity analysis papers typically discuss cutoffs in the room temperature range (-0.5 to -1 kcal/mol) whereas mobility is best explored in FRODA at cutoffs of multiple kT (-2 kcal/mol or below).

   A loose justification for this use of lower cutoffs comes from considering the probability of a bond of strength nkT being broken by a thermal fluctuation. This probability will be proportional to $\exp(-n)$. For $n = 1$ this term is around 1/3, whereas once we reach $n = 4$ it has declined to less than 0.01. Thus we do not expect bonds with room-temperature energies to be permanent constraints on motion.

   We should also note that *exact energy cutoff values transfer poorly even between very similar structures* [18] and should not be taken as accurate to better than around a kcal/mol. The analysis of rigidity information should focus more on significant features in the dilution plot, and on the relative flexibility of structural features in the protein [19], than on the rigidity seen at an arbitrarily chosen cutoff value. Since flexible motion simulations exploring normal mode vectors are computationally cheap—typically a few CPU-minutes to explore a single mode and direction—it is best to carry out multiple runs at several different energy cutoffs and compare the results. Likewise, if there are biological reasons to believe certain interactions are particularly significant, a comparative study looking at the difference made by including or excluding those interactions will be informative.

8. *On stacked ring interactions.* These are generally few in number for proteins, being far more significant for nucleic acid structures. Sensible options are either to accept the interactions found by FIRST as part of the constraint network, or to drop stacked ring interactions entirely by creating an empty file called "**stacked.in**" and using the "**-srin**" option in FIRST.

9. *Make all clusters mobile.* The importance of the **-mobRC1** option cannot be overstated. It is particularly easy to forget to include it when concentrating on other options such as **-target** or **-mode**. This will leave your largest rigid cluster frozen in place, usually preventing any useful output. For a particularly amusing experience, try forgetting **-mobRC1** when using the **-restart** option. Catastrophic mismatches result.

10. *Matching constraints for targeted runs.* Clearly a targeted motion will not be possible if the input structure retains non-covalent interactions that must be broken in the target structure. It is therefore advisable to compare the constraint networks of the input and target structures before carrying out simulations. It is easy, and recommended, to match the hydrophobic tether lists from both structures and only to use interactions common to both. The hydrogen bond networks can similarly be compared, though in this case there is the complexity that the bond strengths will not be identical in the input and target structures; the user should retain bonds that are strong in both.

    A useful tactic is to carry out a preliminary targeted run including only covalent interactions—e.g., by the use of blank, dummy hphobes.in and hbonds.in files. This will identify issues that can be resolved before carrying out a production run. For example, some protein side chains include two indistinguishable atoms, such as the oxygen atoms in a carboxylic acid (—COO) group. If the numbering of these atoms does not match between the input and target structures—a fifty-fifty chance—this will cause an unnecessary rotation during targeted motion, with knock-on steric effects on other nearby atoms. This can be seen in a preliminary run where sidechains fail to rotate properly to match the target structure. The problem is fixed by flipping the order of the identical atoms in the target structure.

11. *FRODA or FRODAN for targeted motion?* FRODA's targeted motion capabilities are best suited to cases where the initial and final structures are fairly similar in topology, and the simulation is mostly concerned with maintaining physically reasonable steric and bonding geometry during the transition. In more complex cases, particularly where substantial unfolding and refolding of the protein is involved, or where significant sequence alignment is required, transition generation using FRODAN on the Pathways site at ASU is recommended.

12. *Extracting useful information from simulation output.* A key point to remember is that no simulation of a protein can tell you that "this is what happens." Rather, question which may be answered are "Is this kind of motion plausible for this structure?" and "Do any of these simulated motions offer an explanation for, or understanding of, something the protein does?."

Aim to extract useful parameters from output files, for example, to probe for residues that are initially distant but are brought into proximity by flexible motion, or to identify modes that give large relative motion of domains. Once significant jamming has begun (many new steric contacts; stretching of constraints; many cycles to refit new conformations in FRODA) the simulation has probably moved beyond its domain of validity and caution should be exercised in using and interpreting structures from this regime. For structural biology purposes, the study of conformations generated in the "easy motion" regime is preferable to the use of those obtained in the constraint-conflict regime.

References

1. Wells SA, Dove MT, Tucker MG, Trachenko K (2002) Real-space rigid-unit-mode analysis of dynamic disorder in quartz, cristobalite and amorphous silica. J Phys Condens Matter 14 (18):4645–4657
2. Sartbaeva A, Wells SA, Treacy MMJ, Thorpe MF (2006) The flexibility window in zeolites. Nat Mater 5(12):962–965
3. Wells SA, Sartbaeva A (2012) Template-based geometric simulation of flexible frameworks. Materials 5:415–431 (Special issue "Computer modelling of microporous materials")
4. Wells S, Menor S, Hespenheide B, Thorpe MF (2005) Constrained geometric simulation of diffusive motion in proteins. Phys Biol 2(4): S127–S136. doi:10.1088/1478-3975/2/4/S07
5. Jacobs DJ, Rader AJ, Kuhn LA, Thorpe MF (2001) Protein flexibility predictions using graph theory. Proteins 44(2):150–165
6. Jimenez-Roldan JE, Freedman RB, Romer RA, Wells SA (2012) Rapid simulation of protein motion: merging flexibility, rigidity and normal mode analyses. Phys Biol 9(1):016008. doi:10.1088/1478-3975/9/1/016008
7. Li H, Wells SA, Jimenez-Roldan JE, Romer RA, Zhao Y, Sadler PJ, O'Connor PB (2012) Protein flexibility is key to cisplatin crosslinking in calmodulin. Protein Sci 21(9):1269–1279. doi:10.1002/pro.2111
8. Burkoff NS, Varnai C, Wells SA, Wild DL (2012) Exploring the energy landscapes of protein folding simulations with Bayesian computation. Biophys J 102(3):446a
9. Amin NT, Wallis AK, Wells SA, Rowe ML, Williamson RA, Howard MJ, Freedman RB (2012) High-resolution NMR studies of structure and dynamics of human ERp27 indicate extensive inter-domain flexibility. Biochem J 450:321. doi:10.1042/BJ20121635
10. Farrell DW, Speranskiy K, Thorpe MF (2010) Generating stereochemically acceptable protein pathways. Proteins 78(14):2908–2921
11. Ahmed A, Rippmann F, Barnickel G, Gohlke H (2011) A normal mode-based geometric simulation approach for exploring biologically relevant conformational transitions in proteins. J Chem Inf Model 51(7):1604–1622. doi:10.1021/ci100461k
12. Kruger DM, Ahmed A, Gohlke H (2012) NMSim web server: integrated approach for normal mode-based geometric simulations of biologically relevant conformational transitions in proteins. Nucleic Acids Res 40(Web Server issue):W310–W316. doi:10.1093/nar/gks478
13. Schrodinger, LLC (2010) The PyMOL molecular graphics system, Version 1.3r1
14. Word JM, Lovell SC, Richardson JS, Richardson DC (1999) Asparagine and glutamine: using hydrogen atom contacts in the choice of side-chain amide orientation. J Mol Biol 285(4):1735–1747. doi:10.1006/jmbi.1998.2401
15. Suhre K, Sanejouand YH (2004) ElNemo: a normal mode web server for protein movement analysis and the generation of templates for molecular replacement. Nucleic Acids Res 32 (Web Server issue):W610–W614. doi:10.1093/nar/gkh368
16. Rader AJ, Hespenheide BM, Kuhn LA, Thorpe MF (2002) Protein unfolding: rigidity lost. Proc Natl Acad Sci U S A 99(6):3540–3545. doi:10.1073/pnas.062492699
17. Cozzini P, Kellogg GE, Spyrakis F, Abraham DJ, Costantino G, Emerson A, Fanelli F,

Gohlke H, Kuhn LA, Morris GM, Orozco M, Pertinhez TA, Rizzi M, Sotriffer CA (2008) Target flexibility: an emerging consideration in drug discovery and design. J Med Chem 51 (20):6237–6255

18. Wells SA, Jimenez-Roldan JE, Roemer RA (2009) Comparative analysis of rigidity across protein families. Phys Biol 6(4):046005. doi:10.1088/1478-3975/6/4/046005

19. Heal JW, Jimenez-Roldan JE, Wells SA, Freedman RB, Romer RA (2012) Inhibition of HIV-1 protease: the rigidity perspective. Bioinformatics 28(3):350–357. doi:10.1093/bioinformatics/btr683

Chapter 11

Principal Component Analysis: A Method for Determining the Essential Dynamics of Proteins

Charles C. David and Donald J. Jacobs

Abstract

It has become commonplace to employ principal component analysis to reveal the most important motions in proteins. This method is more commonly known by its acronym, PCA. While most popular molecular dynamics packages inevitably provide PCA tools to analyze protein trajectories, researchers often make inferences of their results without having insight into how to make interpretations, and they are often unaware of limitations and generalizations of such analysis. Here we review best practices for applying standard PCA, describe useful variants, discuss why one may wish to make comparison studies, and describe a set of metrics that make comparisons possible. In practice, one will be forced to make inferences about the essential dynamics of a protein without having the desired amount of samples. Therefore, considerable time is spent on describing how to judge the significance of results, highlighting pitfalls. The topic of PCA is reviewed from the perspective of many practical considerations, and useful recipes are provided.

Key words Protein dynamics, Principal component analysis, PCA, Subspace analysis, Kernel PCA, Independent component analysis, Conformational sampling, Conformational ensemble, Molecular dynamics simulation, Geometric simulation, Essential dynamics, Collective motions, Elastic network

1 Introduction

Protein dynamics is manifested as a change in molecular structure, or conformation as a function of time. To describe accessible motions over a broad range of time scales and spatial scales, protein conformations are best represented by a vector space that spans a large number of dimensions equal to the number of degrees of freedom (DOF) selected to characterize the motions. Many molecular simulation techniques are available to generate trajectories to sample the accessible conformational ensemble characterized by those DOF. The interpretation of a trajectory can lead to better understanding of how proteins perform biological functions. To this end, the process of extracting information from sampled conformations over a trajectory, and checking whether the sampling is a robust representation of an ensemble of conformations accessible

Dennis R. Livesay (ed.), *Protein Dynamics: Methods and Protocols*, Methods in Molecular Biology, vol. 1084,
DOI 10.1007/978-1-62703-658-0_11, © Springer Science+Business Media New York 2014

to the protein, are tasks well suited for statistical analysis. In particular, Principal Component Analysis (PCA) is a multivariate statistical technique (*see* **Note 1**) applied to systematically reduce the number of dimensions needed to describe protein dynamics through a decomposition process that filters observed motions from the largest to smallest spatial scales [1–5]. PCA is a linear transform that extracts the most important elements in the data using a covariance matrix or a correlation matrix (normalized PCA) constructed from atomic coordinates that describe the accessible DOF of the protein, such as the Cartesian coordinates that define atomic displacements in each conformation comprising a trajectory [6]. When all of the atomic displacements have similar standard deviations, a covariance matrix is typically used; otherwise it is prudent to employ the correlation matrix, which normalizes the variables to prevent rare but large atomic displacements from skewing the results. In constructing the covariance matrix or correlation matrix (henceforth C-matrix will be generically used for either matrix type), it is often assumed that the amount of sampling is sufficient, but this always requires many more observations than the number of DOF (variables) used in the matrix. An eigenvalue decomposition (EVD) of the C-matrix leads to a complete set of orthogonal collective modes (eigenvectors), each with a corresponding eigenvalue (variance) that characterizes a portion of the motion, where larger eigenvalues describe motions on larger spatial scales (*see* **Note 2**). When the original (centered) data is projected onto an eigenvector, the result is called a principal component (PC).

While PCA can be performed on any high dimensional dataset, for the analysis of a protein trajectory, a C-matrix associated with a selected set of atomic positions must be constructed. Often, a coarse grained description of the protein motion is made at the residue level by using the alpha carbon atom as a representative point for the position of a residue. In this case, the C-matrix will be a $3m \times 3m$ real, symmetric matrix, where m is the number of residues. Performing an EVD results in $3m$ eigenvectors (modes) and $3m - 6$ non-zero corresponding eigenvalues, provided that at least $3m$ observations are used. When the eigenvalues are plotted against mode index that are presorted from highest to lowest variance, a "scree plot" typically appears as a function of mode index. When such a scree plot forms, a large portion of the protein motions can be captured with a remarkably small number of modes that define a low dimensional subspace. The top set of modes typically has a higher degree of collectivity [7], meaning the PCA modes have many appreciable components distributed quite uniformly. Conversely, a low degree of collectivity indicates there are a small number of appreciable components, although they are not necessarily tied to a localized region of space. When analyzing proteins, 20 modes are usually more than enough (even for large proteins) to define an "essential space" that captures the motions governing biological function, thus achieving a tremendous reduction of dimension.

The process of applying PCA to a protein trajectory is called Essential Dynamics (ED) since the "essential" motions are extracted from the set of sampled conformations [8–10]. Of course, a linear combination of the $3m$ orthogonal PCA modes can be used to describe exact protein motions (at the selected coarse grained level). In practice, the presence of large-scale motions makes it difficult or impossible to resolve small-scale motions because the former has much greater relative amplitude in atomic displacements. Indeed, it is for this reason that the large-scale motions are often the most biologically relevant. Therefore, only a small number of PCA modes having the greatest variances are used to characterize large-scale protein motions. When small-scale motions are of interest, the method of PCA can still be used successfully by applying it to sub-regions of a protein as a way to increase the resolution for describing the dynamics within those sub-regions.

An alternative method to quantify large-scale motions of proteins is to use a Normal Mode Analysis (NMA) [11, 12] derived from an Elastic Network Model (ENM) [13–15]. In the ENM, one typically considers nearby alpha carbon atoms to interact harmonically, where the connectivity is determined from a single structure to extract an elastic network. Typically, the large-scale motions quantified by a small set of lowest frequency modes of vibration are in good agreement with the same corresponding number of PCA modes when direct comparisons of subspaces are made [16–18]. One advantage of performing PCA to obtain the ED of a protein is that information from any selected set of atoms can be used to obtain the PCA modes associated with that subspace. While it is true that ED is often applied to the analysis of alpha carbons, this is not required. The spatial resolution of PCA analysis can be coarser than the resolution of the structures that comprise the trajectory, which, for example, may come from an all-atom based simulation. Another advantage of ED is that statistics from many trajectories may be pooled allowing a great deal of flexibility in the way data from different simulations can be combined. The overall large-scale motions and any number of selected small-scale motions can be determined in a post-simulation phase of research as the nature of the protein motions is being interrogated.

Perhaps the most important difference between NMA and PCA is in the assumption of harmonicity. The premise of NMA requires the molecular motion is confined near the local minimum in the free energy landscape where residues in close proximity (i.e., atomic packing) respond as harmonic pairwise interactions (i.e., springs). Since proteins display a significant amount of anharmonicity in their behavior [19, 20], this assumption is not always suitable [21–23]. PCA makes no assumption of harmonicity, and thus is not limited to harmonic motions. Indeed, because PCA is independent of the model invoked during the simulation to generate the trajectory, the resulting conformational changes that can be explored can deviate far from the harmonic assumption. On the other hand, the limitations of PCA stem from using a linear transform that is based on second moments (covariance), and the fact that subsequent factorization yields eigenvectors that are orthogonal. While a linear transform of the data is always possible, if the variables are not intrinsically linearly related, any nonlinear relationships present will not be properly described. Nonetheless, in practice, standard PCA is similar to the standard ENM approach. In other words, relying on covariance implies higher-order correlated motions related to higher moments are missed.

Nonlinear generalizations of PCA are available such as kernel PCA [24] that can be applied directly, or employed after the most relevant subspace is identified first using standard PCA. A disadvantage of kernel PCA is that the choice of kernel is not obvious because it is problem dependent, although we show below that some common choices work well for protein trajectories. Also problematic is that the reconstruction of data is difficult to interpret because the mapping involves feature space, which is distinctly different than conformational space that has a geometric interpretation despite being of high dimensionality. The reason for employing kernel PCA is to differentiate conformations within an ensemble beyond that possible using standard PCA, which may give insight into structural mechanisms governing protein function. Our work suggests that the simplest PCA, which follows from the C-matrix, offers a validated method to describe the dominate correlations present in atomic motions found in proteins, and it provides an effective dimension reduction scheme that can be used for subsequent analysis to capture nonlinear (or higher order correlations) effects when they are of interest. Nevertheless, in practice it is always important to ensure and test the robustness of the PCA modes.

Keep in mind that individual PCA mode directions are subject to errors related to finite sampling of conformations to construct the empirical C-matrix. The empirical C-matrix should be a good estimate for the actual population C-matrix (infinite samples). In practice, PCA can be strongly influenced by the presence of outliers in a dataset. The main concern is that the outliers may skew the first few mode directions. While there are robust algorithms that are useful in stabilizing PCA in the presence of outliers [25–32], it is often effective to remove identifiable outliers or simply consider a sufficiently long trajectory for which the results are significant. Generating a large number of conformational samples and removal of outliers before the C-matrix is calculated mitigates concerns about robustness of the results. Moreover, this type of intrinsic error does not pose much of a problem as long as biologically relevant motions are described using a superposition of a small set of dominant modes (instead of focusing on one mode). As the mode number increases the core part of this subspace becomes stable against sampling noise. However, only the top several modes tend to be useful.

The choice of which modes to include is often made by examining the scree plot for a visible "kink" (the Cattell criterion) [33, 34], such that all modes up to the kink are important (*see* **Note 3**). Although a kink does not have to exist, it typically does in the study of protein dynamics. In fact, a kink will generally appear for any high dimensional dataset. Hence the name scree (geological debris at the bottom of a cliff) plot has been tied to PCA. Other criteria are commonly used for the choice of essential modes. For example, the top set of modes associated with greatest variances when added should reach some fraction (say 80 %) of the total variance possible given by the trace of the C-matrix. The problem with this method is that some a priori set fraction is arbitrary, and for fractions greater than 50 % one tends to end up with many more modes than are truly relevant to the problem. The scree plot provides an objective criterion. Figure 1a shows the scree plots for PCA of two protein simulations and a random process created from independent and identically distributed variables. Notice there is a rapid decrease in the eigenvalues for the proteins that is not present in the random process.

When PCA is applied to Cartesian coordinates that describe the positions of atoms, an alignment step is necessary prior to the process of constructing the C-matrix because the intent is to capture the internal motions of a protein. The structural alignment step requires the center of masses to coincide as well as a global rotation to optimally align the structures. The authors implemented a quaternion rotation method to obtain optimal alignment defined by the minimum least-squares error for the displacements between corresponding atoms [35]. PCA is not limited to the analysis of a Cartesian coordinate-based C-matrix. Any set of dynamic variables that describe the protein motion can be used. For example, one may choose to use internal dihedral-angle coordinates such as the (Φ, Ψ) angles or interatomic distances, which eliminates the need to optimally align conformations. However, in the former case, it has been realized there is an intrinsic nonlinear effect that is not well described using standard PCA, suggesting kernel PCA should be employed or an alternative internal coordinate system that is naturally linear should be chosen. In the latter case, internal atomic distances offer the possibility of an all-to-all distance C-matrix for the alpha carbons, which has a row dimension equal to the number of structures in the trajectory and a column dimension equal to $m(m-1)/2$, where m is the number of residues considered. A distance based C-matrix can be created, which is

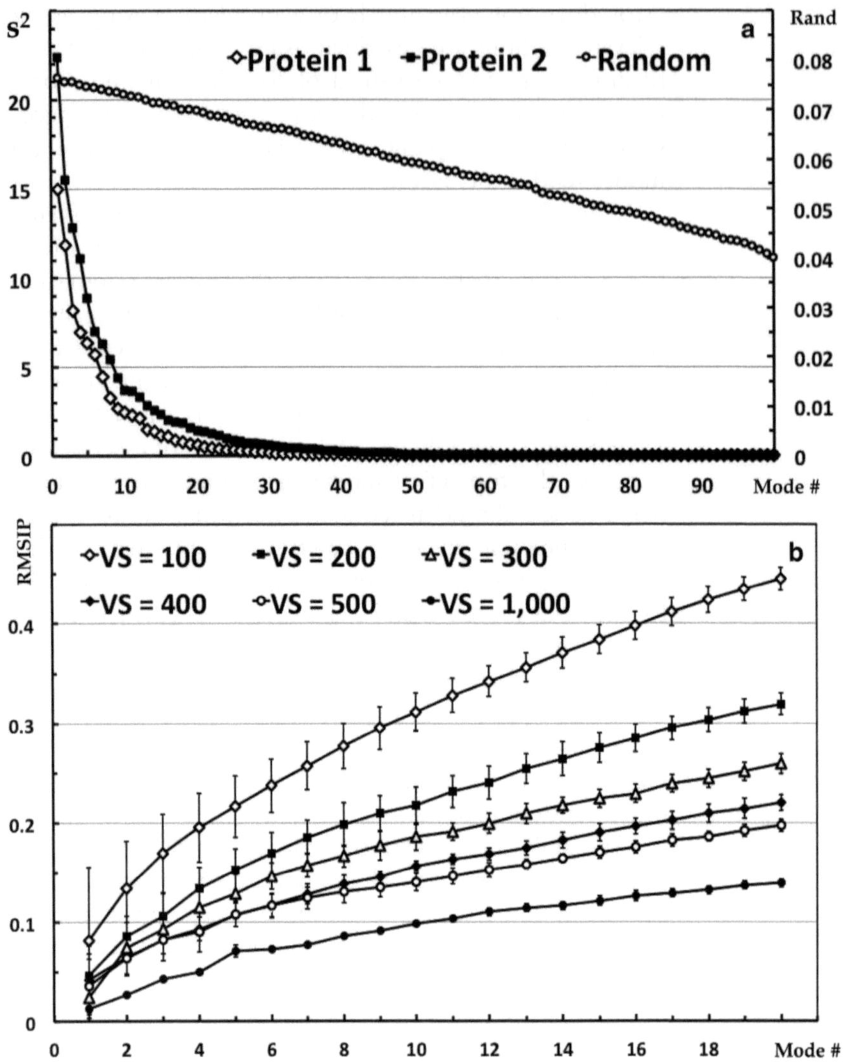

Fig. 1 (**a**) Eigenvalue Scree plot for first 100 modes of two example protein simulations (primary *y*-axis) and a random process (secondary *y*-axis), each having 225 dimensions. The units are angstrom squared (positional variance). (**b**) Average RMSIP scores for a random process in different vector space dimensions as a function of subspace dimension. Error bars show plus and minus one standard deviation

a square matrix with dimension $m(m-1)/2$, and therefore requires much more sampling. In this case, the PCA modes reveal the coordinated changes in distances between all residue pairs. Despite the advantage of working directly with internal coordinates, performing all-to-all distance PCA quickly becomes computationally prohibitive due to the need to diagonalize very large non-sparse matrices. More importantly, the interpretation of the eigenvectors becomes difficult when the number of residues is greater than ten. Nevertheless, this approach has proven useful when studying a small subset of atoms where the interpretation is clear [36, 37].

The task of applying PCA to a conformational ensemble (CE) requires that a CE be generated. There are multiple ways to create a CE including molecular dynamics (MD) and geometrical simulations such as FIRST/FRODA [38–40]. A CE may be generated by experimental methods such as using protein structures from X-ray crystallography or nuclear magnetic resonance (NMR) techniques. For certain applications it is prudent to combine multiple CEs together that define a single dataset. One reason for combining different CEs is to boost statistics, where each CE has the same characteristics. This is convenient, as the simplest way to apply parallel computing occurs when multiple simulations are run simultaneously and independently. However, the CEs that are combined could represent different conditions, such as different temperatures in MD simulation, fixing a different set of distance constraints in geometric simulation or contrasting mutant structures. Clustering different CEs in the subspace defined by the most relevant PCA modes provides insight into the effect of varying conditions. In some cases, a protein may undergo large-scale (anharmonic) conformational changes that bridge two distinct basins of low free energy. The combined CEs will allow these basins to be clearly identified, as well as the paths connecting them. Similarly, different CEs that represent a set of mutant structures, or apo and holo forms of a protein, possibly with different ligands bound, allow one to differentiate the conformations easily by clustering in a small dimensional subspace.

The most appealing and intuitive way to investigate the nature of protein motions is to project the displacement vectors (DV) defined in the original high dimensional space that characterize different conformations onto a pair of PCA modes. It is even possible to project onto higher dimensions as one visualizes multiple PCA modes simultaneously using specialized software such as R or XL-STAT™, which is a plug-in for Microsoft Excel developed by Addinsoft™. Such plots are indispensible for assessing how well certain parts of the subspace are sampled, especially in comparative studies where differentiation in dynamics can have functional consequences. The results of such an analysis show how each state occupies a region of the conformational space defined by the first two PCA modes.

Given that the ED of a protein is characterized using a small vector space defined by PCA modes that reflect different CEs and a combined CE, it becomes necessary to benchmark how similar these subspaces are to one another. When subspaces are sufficiently similar, this implies that the different ensembles capture the same type of protein dynamics. Conversely, when subspaces are dissimilar, different types of motions are being captured, which may have biological consequences tied to the different conditions analyzed. As such, it is necessary to define a measure to quantify the overlap of vector subspaces, as a natural generalization to the concept of a projection (dot product) of one vector onto another. That said, note that a set of n PCA modes forms an orthogonal n dimensional *subspace* (SS) within the full *vector space* (VS) defined by the size of the C-matrix (*see* **Note 4**). Common metrics that quantify SS similarity include cumulative overlap (CO), root mean square inner product (RMSIP), and principal angles (PA) [12, 41–45]. The CO metric quantifies how well one SS is able to capture the PCA modes of the other SS. The RMSIP metric is a single number that quantifies the SS similarity in terms of multiple inner products between the two. The PA method provides a quantification of the optimal alignment between the two SS that is based on the singular value decomposition (SVD) of a matrix of overlaps (inner products) between the two SS. The result is a sorted (monotonically increasing) set of n angles, where n is the dimension of the compared subspaces, that quantify how well the two SS can be aligned.

A final concern with assessing the PCA output is the significance of the results. While PCA is robust when there is sufficient sampling, the questions that remain are: What constitutes sufficient sampling and how trustworthy are the modes? Since PCA relies on the factorization of the C-matrix, the condition number of the C-matrix indicates the numerical accuracy that can be expected within the solution of the associated set of equations. For a given process, more sampling reduces the condition number. Therefore, if the condition number for a C-matrix is high, this could be in indication there is not enough statistics. If possible, the number of independent samples should be at least ten times the number of variables. Two direct measures for sampling significance are known as the Kaiser-Meyer-Olkin (KMO) score given as:

$$\mathrm{KMO} = \left(\sum_j \sum_{k \neq j} r_{jk}^2\right) \Big/ \left(\sum_j \sum_{k \neq j} r_{jk}^2 + \sum_j \sum_{k \neq j} p_{jk}^2\right) \tag{1}$$

and the associated measure of sampling adequacy (MSA) given as:

$$\mathrm{MSA}_j = \left(\sum_{k \neq j} r_{jk}^2\right) \Big/ \left(\sum_{k \neq j} r_{jk}^2 + \sum_{k \neq j} p_{jk}^2\right) \tag{2}$$

where r is the standard correlation coefficient and p is the standard partial correlation coefficient [46]. These statistics can take values between 0 and 1. If all the partial correlations are zero, then the MSA score is 1. The KMO score indicates the amount of partial correlations between the sampled variables and provides an indicator for when applying PCA is appropriate. The MSA provides a metric for each variable. KMO and MSA should ideally be greater than ½. It is worth noting that the MSA scores for each variable are related in a nontrivial way to the protein environment. Specifically, there tends to be a moderate negative correlation between the MSA scores and the residue RMSD.

When comparing essential subspaces, keep in mind that all of the subspace metrics described above depend on both the dimension of the SS and the dimension of the full VS as shown in Fig. 1b.

One way to assess PCA modes is to compare them to the modes of a random process to obtain a baseline for determining the significance of the subspace comparisons as the dimensions for the SS and full VS change. With these baselines, a *Z*-score can be calculated to assess the statistical significance of the scores, for example when using RMSIP:

$$Z = \frac{\mathrm{RMSIP_{obs}} - \mathrm{RMSIP_{rand}}}{\mathrm{stdev(RMSIP_{rand})}} \tag{3}$$

However, the essential SS of a random process has very different characteristics than the essential SS constructed from a protein trajectory as Fig. 1 clearly shows. Randomly shuffling the indices for the components of modes produces a new set of modes that have essentially the same character as the modes determined by PCA on a purely random process. Consequently, any two same-sized proteins share much more in common than would be expected by a random process, making large *Z*-scores not very useful in practice. This is due to the fact that compared to a completely random process all proteins share much more common dynamics because they share common structural features such as a covalent backbone even if their fold topology is very different. What this means in practice is that any of the metrics described above for any two proteins will show much more overlap compared to a random process. In fact, using two different trajectories under the same conditions, we found that the scores for overlap between two identical proteins can be *lower* than the overlap between two *different* proteins when the number of residues is small (<100). This result escalates when using a coarse-grained approach that prunes many discriminating features (to reduce DOF). To obtain a more stringent criterion for *Z*-score determination, the data presented strongly suggests that a comparison to other proteins, possessing the same number of DOF, that define a decoy set should be used to define the random baseline in (1), rather than a generalized random process. However, to the best of our knowledge, baselines from decoys have not been done.

Figure 2 shows the risks of comparisons made for small proteins using a coarse-grained model. For this analysis, four proteins having distinctly different folds were simulated under the same conditions using geometrical simulation and then subjected to PCA as a combined set, where only the first 75 residues were included in the covariance matrix starting from the N-terminus and always remaining within the N-terminal domain. Figure 2b shows the *Z*-scores for the comparisons in Fig. 2a. Here it is critical to note the similarity between the random process and the decoy

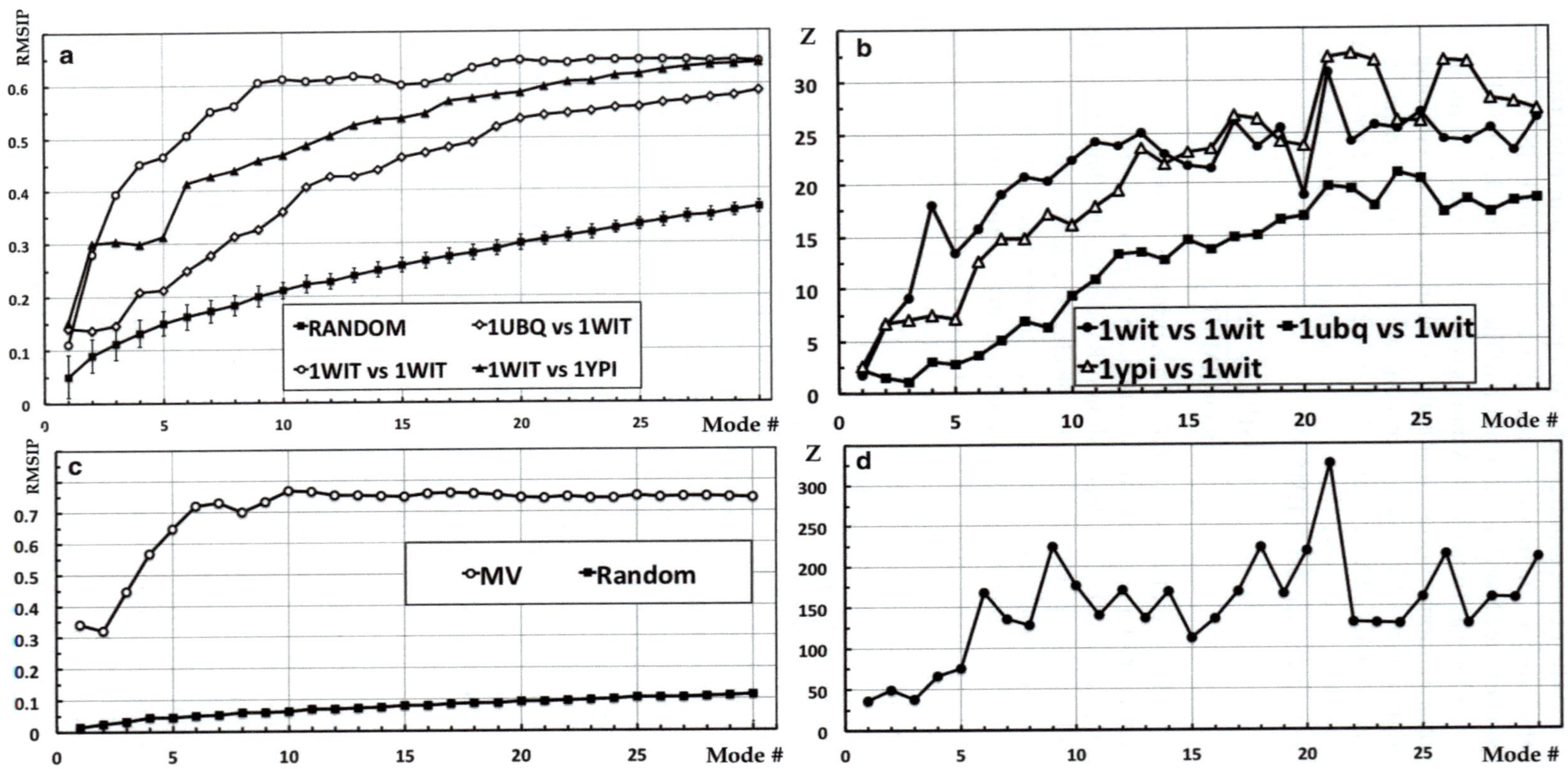

Fig. 2 (**a**) RMSIP scores for inter-comparisons between three proteins each having 75 residues and a random process with 225 DOF. Only the true self-comparison yields a curve that saturates rapidly within a small essential space defined by the first nine modes. The decoy plots have much more in common with the protein dynamics of interest compared to the random process up to the first 30 modes. (**b**) The *Z*-scores for the RMSIP scores shown in panel **a**. (**c**) Comparison of two myosin V (795 residues) CEs run under different simulation conditions and a random process with 2,385 DOF. Again, note the rapid saturation of the RMSIP scores in an essential subspace defined by the first ten modes. (**d**) The *Z*-scores for the RMSIP scores in Panel **c**

comparisons. When 1WIT is compared to itself (using different simulation conditions), RMSIP saturation suggests that the proper essential subspace dimension is nine modes. However, the random process and the decoy comparisons do not reach a saturation point within the first 30 modes. When working with larger proteins, such comparisons are much safer, as shown in Fig. 2c, d with myosin V (MV). The moral here is that extra care must be taken to claim significance of PCA results on small proteins when coarse-graining is used.

Another way to assess how stable the PCA results are can be made by looking for cosine content within the top few PCs. It has been noted that when MD trajectories insufficiently sample conformational space the top few PCs resemble cosine functions with periods equal to half the mode number, which is what occurs for a random diffusion process [47]. The resemblance is determined by finding the correlation between the set of T values of the ith PC and $\cos(2\pi t/bT)$ where $0 < t < T,\ b = i/2$. We note that CEs derived from geometrical simulation do not produce PCs that resemble cosines due to the restriction of conformational space imposed by locking in the distance constraints at the beginning of the simulation. However, when it occurs in MD simulations, this indicates sampling is limited.

Lastly, we note that the contributions of variables to a PC can be assessed to determine if any variables are strongly influencing a particular PC. Additionally, when interpreting the component loadings (eigenvector components multiplied by the square root of the associated eigenvalue), the squared cosine between the variables and a PC can be used to determine if real correlations exist, or if there is only an apparent relation due to the projection onto a low dimensional subspace. To infer a correlation, there should be a clustering on a two-dimensional loading plot, and the squared cosine should be greater than one half. As in all statistical interpretations, the best practice is to examine multiple sources of information. For the case of a single CE analysis, these sources include the KMO scores and the MSA for each variable and/or the condition number of the C-matrix, the scree plot, the collectivity of the PCs, the correlations between the variables and the PCs, RMSD mode plots, two-dimensional scatter plots of observations projected on the PCs, the cosine content of the top few modes, and the squared-cosines for variables. When analyzing multiple CEs, additional sources of information include the PA spectra, the RMSIP scores, and CO scores. Comparisons can be made between each individual CE and a reference CE, constructed by combining all of the CEs together, as well as with an appropriate random process. Each CE may also be directly compared to each other.

2 Methods

2.1 Preliminaries

A dynamic trajectory provides snapshots depicting the protein in multiple configurations called frames (*see* **Note 5**). Denote the trajectory as the set $A_{\mathrm{Raw}} = \{X(t)\}$ where t is a discrete variable referring to a particular frame. The vector X may be composed of a subset of atoms within the protein. Here, we consider the set of alpha carbons. If the protein consists of m residues, then X will be a column vector of dimension $3m$. If A contains n observations, then set A is a matrix of dimension $3m \times n$, since each alpha carbon has (x, y, z) coordinates in Cartesian space. To study internal motions of the protein, it is essential to set the center of mass of each frame at the origin, and to rotate each frame to its optimally aligned orientation relative to a selected reference structure, defined by X_{ref}, which also has its center of mass at the origin. Since this translation and rotation process changes the coordinates of each frame, the transformed A matrix is denoted as A_{Aligned}, which is denoted as $A_{\mathrm{Aligned}} = \{X_{\mathrm{Aligned}}(t)\}$ (*see* **Note 6**). The choice of the reference structure X_{ref} is not critical, and although it is common to use the initial input structure to the simulation, any frame is as good as any other. It is also possible to use an average structure, but the method of averaging needs to be done with care in an iterative fashion [48]. The data in matrix A_{Aligned} is then mean centered (here, this means row centering) and we denote this as A'. The covariance matrix Q associated with the $3m$ variables is then defined as $Q = A'A'^{\mathrm{T}}$, which is always real and symmetric, and has dimension $3m \times 3m$. If $n \geq 3m$, then the EVD of Q will result in $3m - 6$ non-zero eigenvalues, where the six zero eigenvalues correspond to the modes of the trivial degrees of freedom, 3 for translation and 3 for rotation. The same is true for the correlation matrix R, which only differs from Q in that the values of the variances are divided by the associated standard deviations, yielding the value of 1 for each diagonal element. It is prudent to use the correlation matrix when the standard deviations of the variables are strongly skewed. Conversely, correlation based PCA employs normalized variables and this standardization tends to inflate the contribution of variables whose variance is small, and reduce the influence of variables whose dimensions are large. It is therefore not a priori possible to know which approach will provide more insight for any given problem. Both methods should probably be explored (*see* **Note 7**).

In order to construct a C-matrix based on the internal coordinates defined by interatomic distances, it is first necessary to construct the all-to-all distance matrix D for the residues of interest. This is a matrix of dimension $m(m-1)/2 \times n$, where m is the number of residues considered and n is the number of observations (each residue is represented solely by its alpha carbon). Here, the data must be centered so that deviations in all lengths will average out to zero. Therefore, D' is constructed in which each row is centered. The covariance matrix associated with the $m(m-1)/2$ variables is then defined as $Q_D = D'D'^T$. The correlation matrix R_D is Q_D normalized by the variable standard deviations. This type of PCA, called dPCA is interpretable if one restricts the size of the set of atoms to small numbers. For example, if three residues (alpha carbons) are used, then three modes will result from the EVD of Q_D or R_D and the interpretation of the eigenvectors, which are composed of three components reveals correlations (if any) in the fluctuations in the lengths between the three sets of pairs (*see* **Note 8**). This can be useful to interpret fluorescence resonance energy transfer (FRET) experiments [36, 37].

When choosing to work in the sample space, either due to a small number of samples or to implement a non-linear method, one must construct the kernel matrix (K), which is a $n \times n$ square symmetric matrix, where n is the number of observations. Each element of K is formed by computing $K(i, j)$, where i and j represent two observations from the centered data set, using the definition for the specific kernel function of interest, k. Essentially, the kernel function maps N dimensional vectors in $\Re^N$ from the sample space to a new high dimensional (possibly infinite) vector space referred to as feature space. Working in the high dimensional feature space can often detect features that are not apparent in sample space. The "curse of dimensionality" is avoided by constructing the feature space from a collection of inner-products so that the actual mapping function is never calculated. Calculating inner products over the sampled data is not by itself an intensive operation. This method of avoiding the difficulties normally associated with high-dimensional spaces is known as the "kernel trick". It is worth noting that using this approach, only a subset of feature space is being explored, which is limited by the range of the data of the original sample space.

The kernels that can be employed must yield positive-definite symmetric square matrices [24]. When the kernel is defined simply as the inner product of the input data (linear kernel), then the results of the analysis are identical to the standard PCA. Specifically, one will recover the same set of non-zero eigenvalues as that from the covariance matrix based PCA. In this sense, kernel PCA (kPCA) subsumes standard PCA. Additional features may be detected by using other types of non-linear kernels, such as a Gaussian kernel, a Neural Net kernel (i.e., a tanh function), a kernel that maps the data to a set of degree n polynomials (either homogeneous or inhomogeneous), or a mutual information kernel. There are no rigorous guidelines for which kernel to apply to the data of interest and thus the method of kPCA requires intimate knowledge of one's data (or based on trial an error) as well as how a particular kernel might or might not affect the resolution of multiple states. Furthermore, most kernel functions have adjustable parameters that need to be set to obtain the best resolving power within feature space. Unfortunately, there is no a priori formula for parameter optimization because this process is highly dependent on the data used. Lastly, unlike standard PCA where the PCs are generated by taking the dot product of the DVs and the appropriate eigenvector, the process for kPCA is more involved. First, the eigenvectors must be normalized in the sample space to reflect the fact that their magnitude in the feature space is unity, and then the PCs (for the training set) are calculated by determining the sum of the inner products of the normalized eigenvectors with the kernel columns. Having used both standard PCA and kPCA, we note that when the parameters are suitably tuned, the ability of kPCA to discriminate multiple states from a trajectory is impressive.

If kPCA is to be used, we note that an ideal approach for computationally intensive kernels is to first use PCA to reduce the dimension of the data and then apply the kernel methods to the top set of PCs. In this approach, we have found that as few as five PCs may be used as input to kPCA with no substantial loss in numerical accuracy. This filtering process greatly reduces the computational intensiveness of the kPCA (*see* **Note 9**), although it does not reduce the size of the kernel matrix. Many more properties of kPCA can be found in [24].

For completeness, we briefly consider the method of Independent Component Analysis (ICA) [49]. ICA is a method for performing blind source separation, as when one wishes to decompose a mixed signal into two signals or a signal plus noise. The underpinning mathematics of the method is to detect non-Gaussian processes by looking at higher order correlations than second degree. To achieve this, ICA is typically implemented using either kurtosis or an information theoretic quantity like mutual information (FastICA) as a contrast function [50]. To apply ICA, one must first center the data and then whiten it. Whitening is the process of transforming an observed data vector *linearly* so that one obtains a new vector, which is *white*, i.e., its components are uncorrelated and their variances equal unity. In other words, the covariance matrix of a whitened data vector equals the identity matrix. One method for whitening data involves an EVD of the covariance matrix and is given by $\widetilde{x} = \mathrm{E}\mathrm{D}^{-1/2}E^{\mathrm{T}}x$ where x is the centered data, E is the matrix of eigenvectors from the EVD of the covariance matrix, with E^{T} its transpose, and D is the diagonal matrix of eigenvalues from the EVD of the covariance matrix. Once the data has been centered and whitened, the ICA algorithm essentially computes the optimal rotation of the data using higher order statistics (e.g., fourth moments), thereby determining the independent components (ICs). We note that the algorithm can be computationally expensive for high dimensional data when a large number of ICs are to be extracted.

In order to make ICA amenable to large, high-dimensional datasets like protein CEs, PCA is first applied to perform a dimensionality reduction and whitening preprocessing step. Similar results to ICA may be obtained from kPCA by choosing to work with a kernel that maps the data to inner products of degree two polynomials. Such kernels have the property of detecting fourth moments, i.e., kurtosis. Alternatively, we note that one may perform *post hoc* analyses of the PCs derived from either standard PCA or kPCA to determine which ones have the highest amount of kurtosis. Choosing to examine such PCs will allow the investigator to see if non-Gaussianity, as measured by kurtosis, leads to the detection of a biological signal. The real criterion for assessing the usefulness of ICA is determining if the assumptions of the model are met. We find that for investigating native state dynamics, where proteins are described by a large set of DOF and are not undergoing large conformational shifts, ICA does not provide greater insight than what PCA (or kPCA) provides because most of the variables in the CEs are Gaussian.

PCA is a multivariate statistical approach, and there is almost no limit to the variants available to an investigator. For example, one may perform sparse PCA (SPCA) in which one attempts to form linear combinations that are *sparse*, meaning that they are combinations of less than all the variables. This is done in an attempt to make the interpretation of the PCA more manageable as is the case of standard PCA the linear combinations include all the variables and in high dimensional data, rendering an interpretation as non-trivial at best. Typically this is done by using a thresh-holding method such as any component less than *c* is mapped to zero, where c is an *ad hoc* chosen number between 0 and 1 or by solving an optimization criterion as in the case of SPCA [51]. The effect of such a sparsification is the reduction of complexity in interpretation of correlated motions and often better cluster separation. The problem with the approach is that there is no guarantee that the sparse variables are the important ones. Another approach combines PCA and ICA methodologies in a process called Independent Principal Component Analysis (IPCA) [52], based on the assumption that biologically meaningful components can be obtained if most noise has been removed from the associated loading vectors. In IPCA, PCA is used as a preprocessing step to reduce the dimension of the data and to generate the loading vectors. The FastICA algorithm is then applied to the previously obtained PCA loading vectors to generate the Independent Principal Components (IPCs). In this method, the kurtosis measure of the loading vectors is used to order the IPCs. There is also a sparse variant with a built-in variable selection procedure implemented by applying soft-thresholding on the independent loading vectors (sIPCA). Because of the breath of the topic and the system dependent details that depend on the data itself, it is beyond the scope of this article to provide recipes for ICA, SPCA, IPCA, or sIPCA. The interested reader should refer to the references given for more details on the theory and application of those approaches. One distinct advantage of standard PCA is that recipes can be provided to define protocols and best practices that are largely independent of the specific nature of the data.

Before proceeding to describe the recipes for PCA and kPCA, we note that there are numerical considerations that must be addressed to suit the investigation at hand. Full eigenvector decompositions of large non-sparse matrices scale as (O^3) and are thus memory intensive. When the DOF in the covariance matrix are less than 10,000 it is reasonable to perform a full decomposition on a standard computer, however, for larger matrices, one may need to consider numerical approaches such as factoring the C-matrix or kernel matrix or computing only a small number of greatest eigenvalues and corresponding eigenvectors. Additional concerns include the condition number of the C-matrix as this is strongly influenced by the number of observations and is related to the KMO statistic. Typically, the condition number improves as the number of samples increases. If a C-matrix is constructed from a set of observations that is smaller than the number of DOF represented by the matrix, it will almost always be ill-conditioned. Furthermore, in this case, the C-matrix in not invertible and contains many zero eigenvalues. In general, it is good practice to have at least ten times more samples than variables to ensure a reasonable KMO score and that most of the variables will have a MSA score of 0.5 or greater. Another option is to switch the analysis from sample space to feature space by implementing kPCA with an appropriate kernel function.

2.2 Recipe I: Essential Dynamics Using Cartesian Coordinate Based PCA

1. Obtain trajectories (one or more) from dynamic simulation. For illustrative examples, one MD and three geometrical simulation (FRODA) trajectories for myoglobin (PDB ID 1a6n) are considered to explain aspects of PCA. For this purpose, details about the setup of the various simulations are ignored, except when it pertains to methodology. Additional details can be found in [16]. The MD trajectory consists of 2,000 frames after equilibration. One FRODA trajectory has 2,000 frames (100,000 explored conformations), and the other two FRODA trajectories have 10,000 frames. The sampling rate of FRODA is normally set at 1 out of 50 conformations generated. Here, one long trajectory is obtained from sampling every conformation (10,000 explored conformations), meaning it is 10 % as long as the 2,000 frame FRODA trajectory in terms of MC-steps, while the other is obtained from sampling every tenth conformation (100,000 explored conformations), is of equal length.

Table 1
Descriptive statistics for three variables in the MD simulation data

Variable	Minimum	Maximum	Mean	Standard deviation
Var 1	3.456	11.489	7.085	1.610
Var 10	9.568	12.980	11.530	0.707
Var 20	8.390	10.467	9.423	0.301

2. Remove overall translations and rotations by aligning each frame to a reference structure.
   - We use the starting (crystal) structure as our reference, and our quaternion alignment program to optimally align each structure to the reference structure. Only the alpha carbon atoms were included in the alignment process.
3. Choose the set of atoms for the analysis: This forms the data matrix A_{Aligned}.
   - Protein conformations (observations or frames) define columns, and rows describe the (x, y, z) coordinates of the alpha carbon atoms. In this example, all 151 of the alpha carbons are used, giving 453 total DOF (variables).
4. Examine the descriptive statistics for the variables.
   - Table 1 shows some statistics for three selected coordinates (variables) to highlight the nonuniformity of the standard deviations.
5. Examine the KMO for each CE and MSA scores for each coordinate. The MD and FRODA trajectories each with 2,000 samples are compared in Fig. 3. Most coordinates from (MD, FRODA) simulation (do not, do) meet the recommended KMO cutoff criterion of 0.50. We assess how the KMO statistic changes when the number of FRODA samples is increased from 2,000 to 10,000, and investigate how the sampling frequency affects the sampling adequacy in Fig. 3b. The overall KMO statistic remains about the same, and the individual coordinates that had a low KMO statistic did not improve by increasing the number of samples. Even more surprising, the sample rate of 1 leads to a slight improvement of the KMO. Thus, there exists a trade-off between the amount of conformational space that a simulation explores and the statistical sampling adequacy of those states (*see* **Note 10**).
6. Center the variables of A_{Aligned} (row centering): This forms the centered data matrix A'.
7. Construct the covariance matrix of the $\{x, y, z\}$ positions for the atoms using A': $Q = A'A'^{\mathrm{T}}$
   - For comparisons, construct the correlation matrix R.

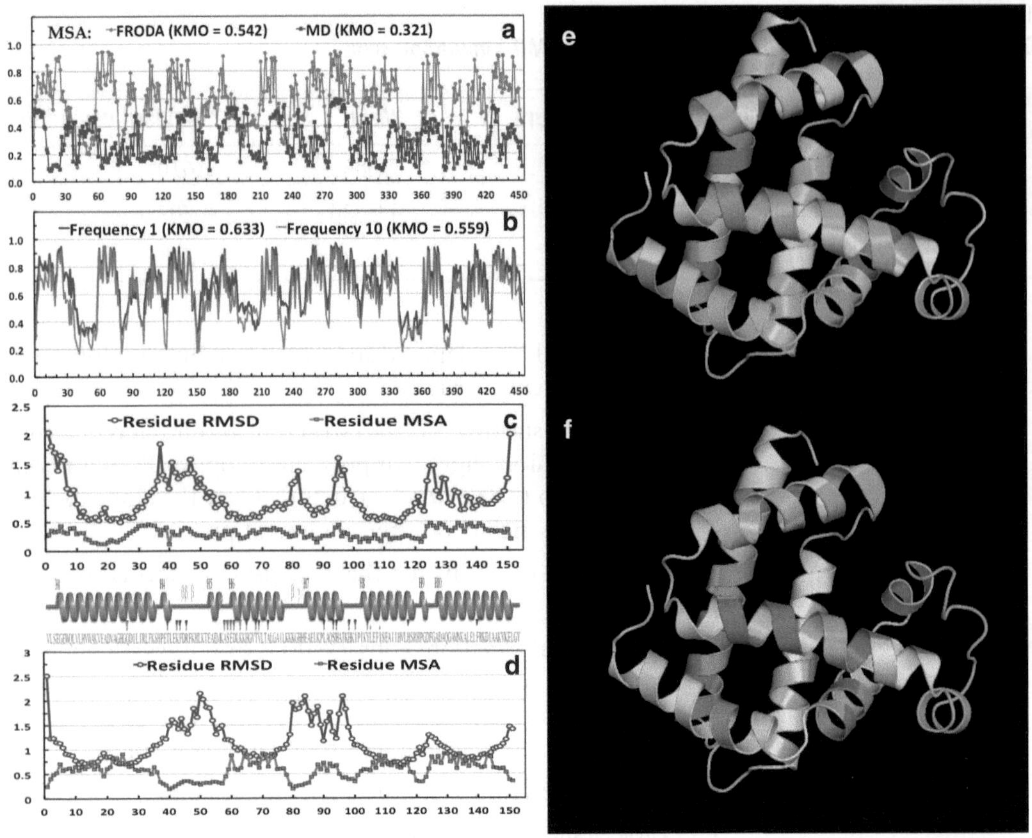

Fig. 3 The Kaiser-Meyer-Olkin MSA for (**a**) the FRODA and MD CEs each with 2,000 frames, and (**b**) for the FRODA CEs each with 10,000 frames. The overall KMO score is shown *parenthetically* in the legend. (**c**) Relationship between residue RMSD and MSA for MD. (**d**) Relationship between residue RMSD and MSA for FRODA. (**e**) Ribbon diagram colored by the MSA scores for MD. (**f**) Ribbon diagram colored by the MSA scores for FRODA

8. Diagonalize Q or R using an EVD.
9. Examine the eigenvalue scree plot to determine the number of eigenvectors to include in the reduced vector space that describes the most relevant features. Figure 4 shows these plots in Panel a along with the conformational and residue RMSDs in Panels b and c.
   - It is not advisable to include all modes up to a preset percent of variance cutoff (*see* **Note 3**). Note that the characteristics of the scree plot depend heavily on whether one is analyzing fluctuations within a single native basin or is analyzing combined trajectories of multiple states. For a single native basin of random motions, many modes will be required to achieve 50 % of the variance. For multiple states/configurations, the first two modes may subsume more than 50 % of the variance. Our example MD plot

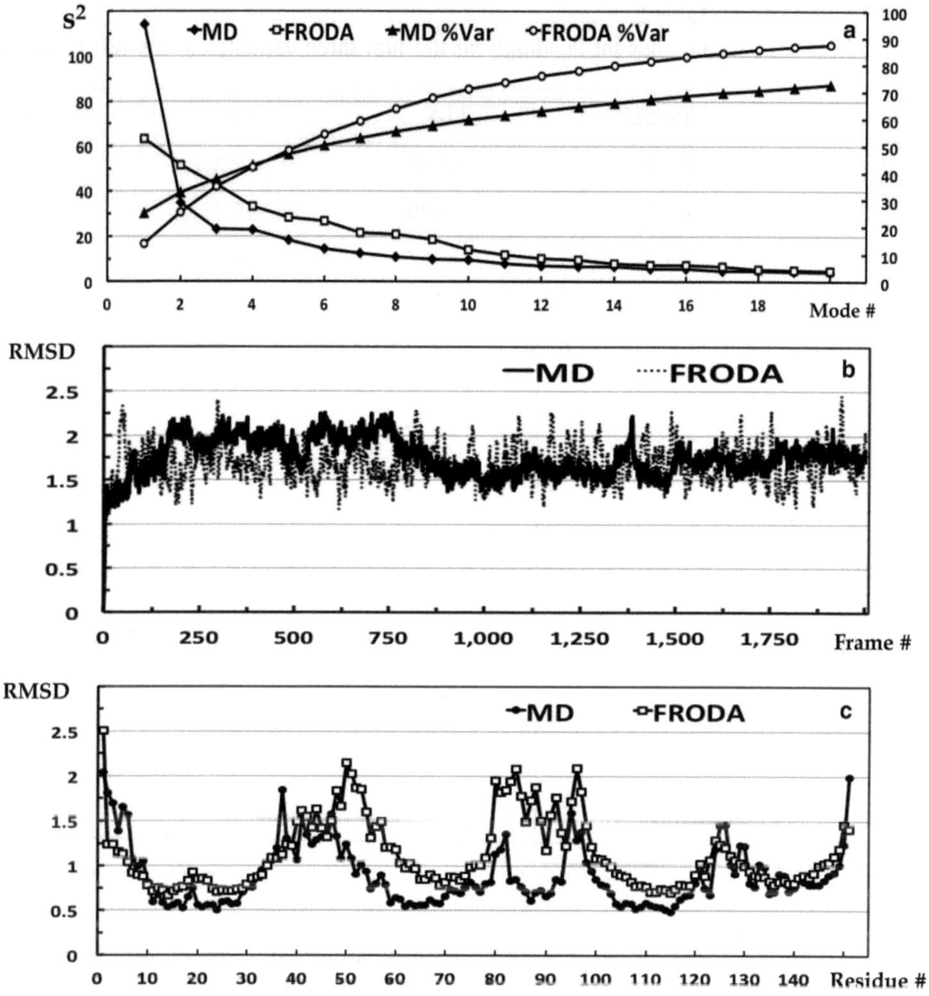

Fig. 4 (**a**) Eigenvalue scree plots for the FRODA and MD CEs showing both the correlation explained in each mode and the cumulative correlations (Since the PCA was based on the correlation matrix). (**b**) The conformation RMSD of the MD and FRODA trajectories. Each value is with respect to the starting structure (crystal structure). (**c**) The residue RMSD for the MD and FRODA trajectories

shows that most of the variance is captured by one mode, because its CE clusters into two conformational states. In contrast, the FRODA plot does not have a dominant mode, but rather shows a monotonically decreasing trend indicative of random fluctuations about the native state of the protein (the input structure).

10. Select the top set of eigenvectors for forming the PCs (Usually 2–20). In our MD example, the top two modes reveal how two distinct states of the protein were sampled. However, at least ten modes are required to define the essential subspaces for a comparison between MD and FRODA CEs (See the RMSIP plots below).

Table 2
Component Loadings for the first three variables in the MD trajectory

Variable	PC1	PC2	PC3
Var 1	0.807	−0.218	−0.056
Var 2	0.890	−0.223	−0.095
Var 3	0.867	−0.254	−0.111

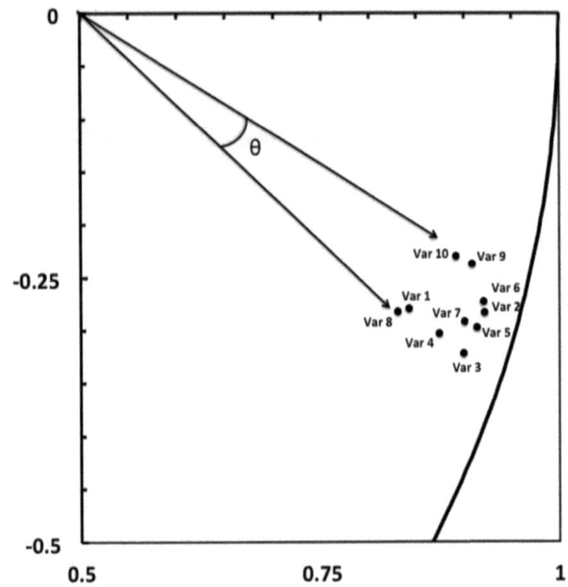

Fig. 5 The correlations between the first ten variables and the top two PCs. Notice how these variables form a tight cluster with small angles between each, indicating that they are correlated on these PCs. The boundary line on *right* is an arc of the unit circle to indicate how close the values are to 1

11. Examine the component loadings, which are the product of the square root of the eigenvalue with the eigenvector. When the correlation matrix is used, they are also the correlation coefficients (cosines) between the variables and factors (PCs). Analogous to Pearson's *r*, the squared component loading (squared cosine) is the percent of variance in that variable explained by the PC. In Table 2, PC1 is clearly capturing the behavior of the first three variables.
    - Scatterplots of the component loadings for the top two factors should be examined. In Fig. 5 the first ten variables (Var 1 to Var 10) are seen to cluster. The angle between the variables on this scatterplot indicates the level of correlation, with (0, 90, 180) degrees indicating a correlation of (1, 0, −1).

Table 3
Squared cosines of the variables

Variable	PC1	PC2	PC3
Var 1	**0.651**	0.048	0.003
Var 2	**0.791**	0.050	0.009
Var 3	**0.752**	0.065	0.012

Table 4
Contribution of the variables to the PCs as percent

Variable	PC1	PC2	PC3
Var 1	0.570	0.137	0.014
Var 2	0.693	0.143	0.039
Var 3	0.659	0.186	0.053

12. Examine the squared cosines of the variables. These values indicate whether a correlation is worthy of interpretation or likely an artifact of projection into a low dimensional subspace. Only the first three are shown in Table 3, and they strongly support the correlations shown in Fig. 5.
13. Examine the contribution of the variables. Here we show only the first three in Table 4, but even from this truncated list, it is clear that the N-terminus residues have a large contribution to the first mode.
14. Examine the eigenvector collectivity (*see* Fig. 6). The top modes tend to be more collective than lower modes indicating that many residues are participating in collective motions. For our example, the FRODA eigenvector collectivity drops off rather steeply suggesting that the top 40 or so modes capture most of the collective motions occurring in the native state. This trend of having a set of highly collective modes highlights the fact that real protein motions tend to be captured by a superposition of PC modes, not a single mode. In contrast, the MD collectivity does not drop off rapidly indicating many more modes are required to capture the dynamics that the MD simulation produced. These results also clearly demonstrate that while PCA modes in totality always form a complete basis set, they are derived from statistics, and will be dependent on the sampling. The top PCA modes reflect biasing in the sampling, which may not necessarily be of biological

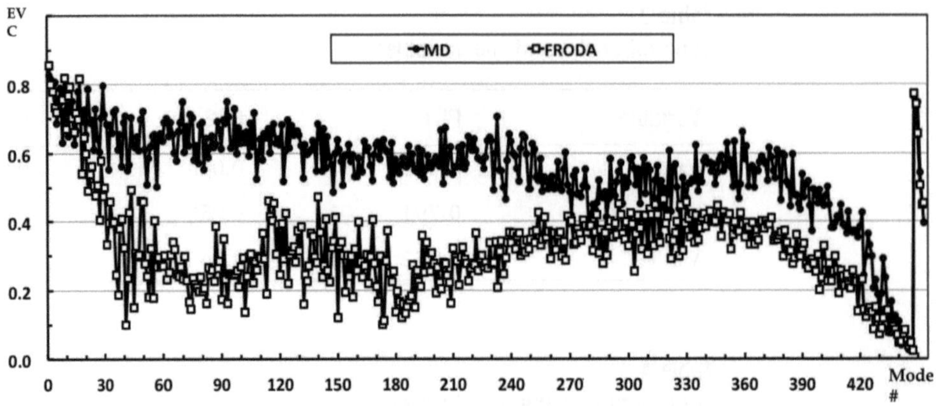

Fig. 6 The eigenvector collectivity (EVC) for the entire set of eigenvectors from both the MD and FRODA PCA. Note that the mode index is plotted with decreasing size of the eigenvalue, so mode index 1 is the top mode. This plot indicates that the collectivity measure should not be of primary concern

importance. It is therefore important to carefully choose what and how to sample so that biological interpretations can be made.

15. Construct the weighted RMSD modes: Here we map the $3m$ components of the eigenvectors to m new variables that capture the squared displacements of each residue to visualize which residues contribute most to the fluctuations of each PCA mode. For each eigenvector i, the new mode N_i has m components, with each component defined by the square root of the sum of the squares of the three variables that contribute to the associated residue, scaled by the square root of the corresponding eigenvalue (*see* **Note 11**). These results are shown in Fig. 7. The mapping equation is given by:

$$N_i = \sqrt{\lambda_i \begin{pmatrix} x_1^2 + y_1^2 + z_1^2 \\ \vdots \\ x_m^2 + y_m^2 + z_m^2 \end{pmatrix}} \tag{4}$$

- Weighting is done by multiplying by the square root of the eigenvalue for the mode, λ_i. This gives units of angstroms.
- It is often useful to compare the RMSD modes to the overall residue RMSD plot from the entire trajectory. Also, one may use the un-weighted RMSD modes to see relative displacements that are hard to see in the weighted plots due to the typical rapid decrease in the eigenvalues with mode index.

16. Construct the DVs for the trajectory, given by $DV_i = X_i - X_{\text{ref}}$ and construct the PCs.

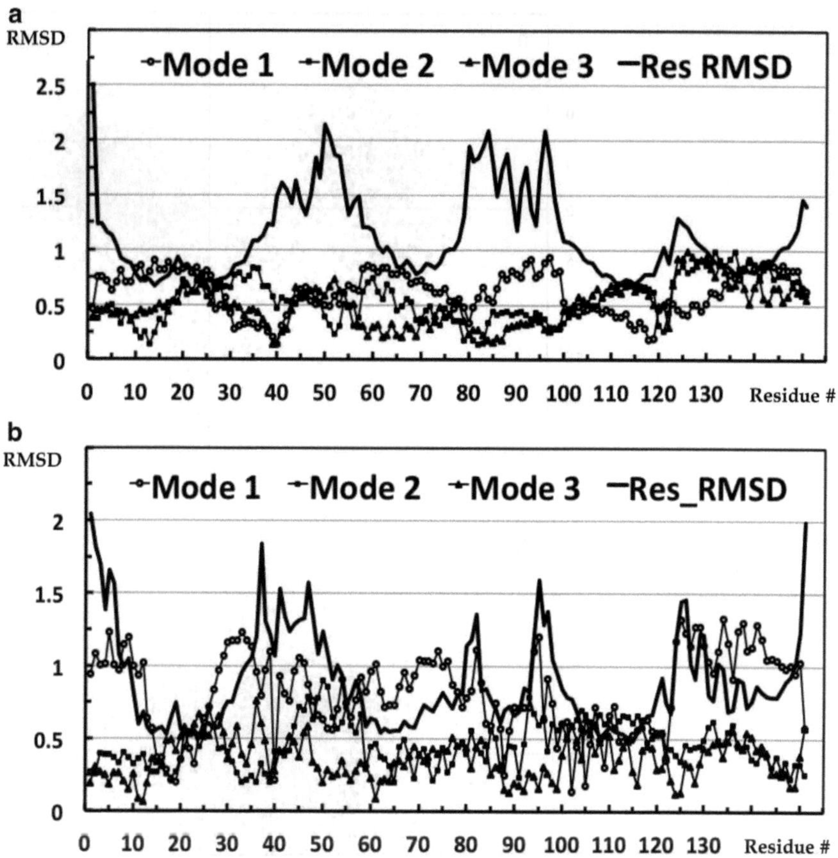

Fig. 7 The RMSD and the top three RMSD modes are compared from (**a**) MD and (**b**) FRODA PCA

- PC_i is formed by taking the inner product between eigenvector i and each DV (Observation) (*see* **Note 12**). Projections can be made on singe modes to view as line graphs. Projections on sets of two PCA modes create scatter plots that show how the simulation explored the configuration space defined by the selected set of modes. In Fig. 8, it is evident that the MD trajectory sampled two states of the protein as seen by the two clusters in the scatterplot of PC1 versus PC2. In contrast, the projection of the FRODA trajectory onto the top two modes shows a uniform distribution.

17. Check the contribution of the observations to the PCs to see if there are particular ones that unduly influence the analysis. Here we show only the first three observations in Table 5 and the values are percentages.
18. We also examine the squared cosines of the observations when determining if an observation belongs to a particular cluster or not. In Table 6, we show values for the first three observations. Values in bold are significant at the 0.01 level.

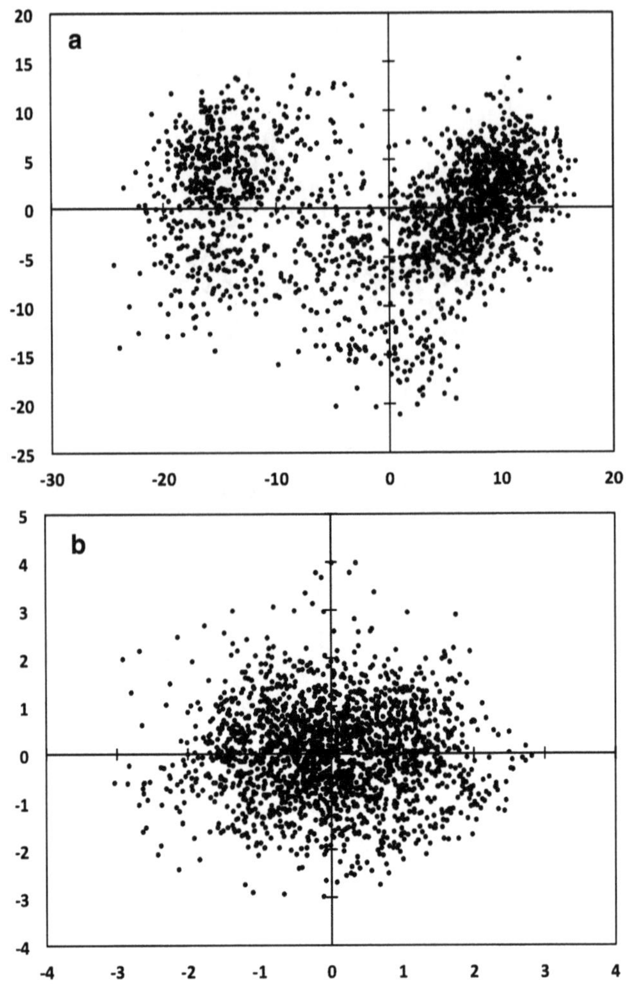

Fig. 8 (**a**) MD and (**b**) FRODA displacement vectors are projected onto their respective top two PCs as a scatter plot

Table 5
Contribution of the observations to the PCs as percent

Observation	PC1	PC2	PC2
Obs 1	0.015	0.529	0.147
Obs 2	0.002	0.329	0.121
Obs 3	0.003	0.485	0.033

19. Since the sampling in the MD simulation was poor for many variables, we check the cosine content of the top two PCs. Comparing PC1 to a half-period cosine, we find a 0.63 correlation and in comparing PC2 to a full period cosine, we find a

Table 6
Squared cosines of the observations

Observation	PC1	PC2	PC3
Obs 1	0.026	**0.285**	0.052
Obs 2	0.005	**0.222**	0.054
Obs 3	0.007	**0.351**	0.016

0.16 correlation. The high cosine content in mode one suggests that the MD simulation should be run longer.

20. When examining two or more sets of PCA modes, determination of how similar the trajectories are to each other may be assessed using the CO, RMSIP or PA metrics.
    - In Fig. 9, we compare the vector space of the top modes from the MD trajectory to that of the FRODA trajectory, each with 2,000 frames. Note that the various metrics for SS comparisons depend on the size of the VS and SS (*see* **Note 13**). As the SS DIM increases, the ability of that SS to capture a given eigenvector increases. Because all the metrics have dependencies on dimensionality, it is best to have a baseline score for random comparisons as a function of the dim(VS) and dim(SS).

2.3 Recipe II: Essential Dynamics Using Internal Distance Coordinate Based PCA

1. Obtain trajectories (one or more) from dynamic simulation.
2. No need to remove overall translations and rotations as internal coordinates are being used.
3. Choose the set of atoms.
    - For a set of N atoms, there will be $N(N-1)/2$ modes. It is recommended that less than ten atoms be selected, because otherwise the interpretation of the resulting modes becomes increasingly difficult.
4. Construct an all-to-all distance matrix D for the residue set chosen for each trajectory.
5. Construct the centered data matrix D' by centering the variables (row center).
6. Construct the covariance (or correlation) matrix, Q_D (or R_D), from D'.
7. Diagonalize Q_D (or R_D) using an EVD.
    - It is best to implement both methods.
8. Examine the eigenvalue scree plot.
9. Select the top set of modes, typically, this is one or two.

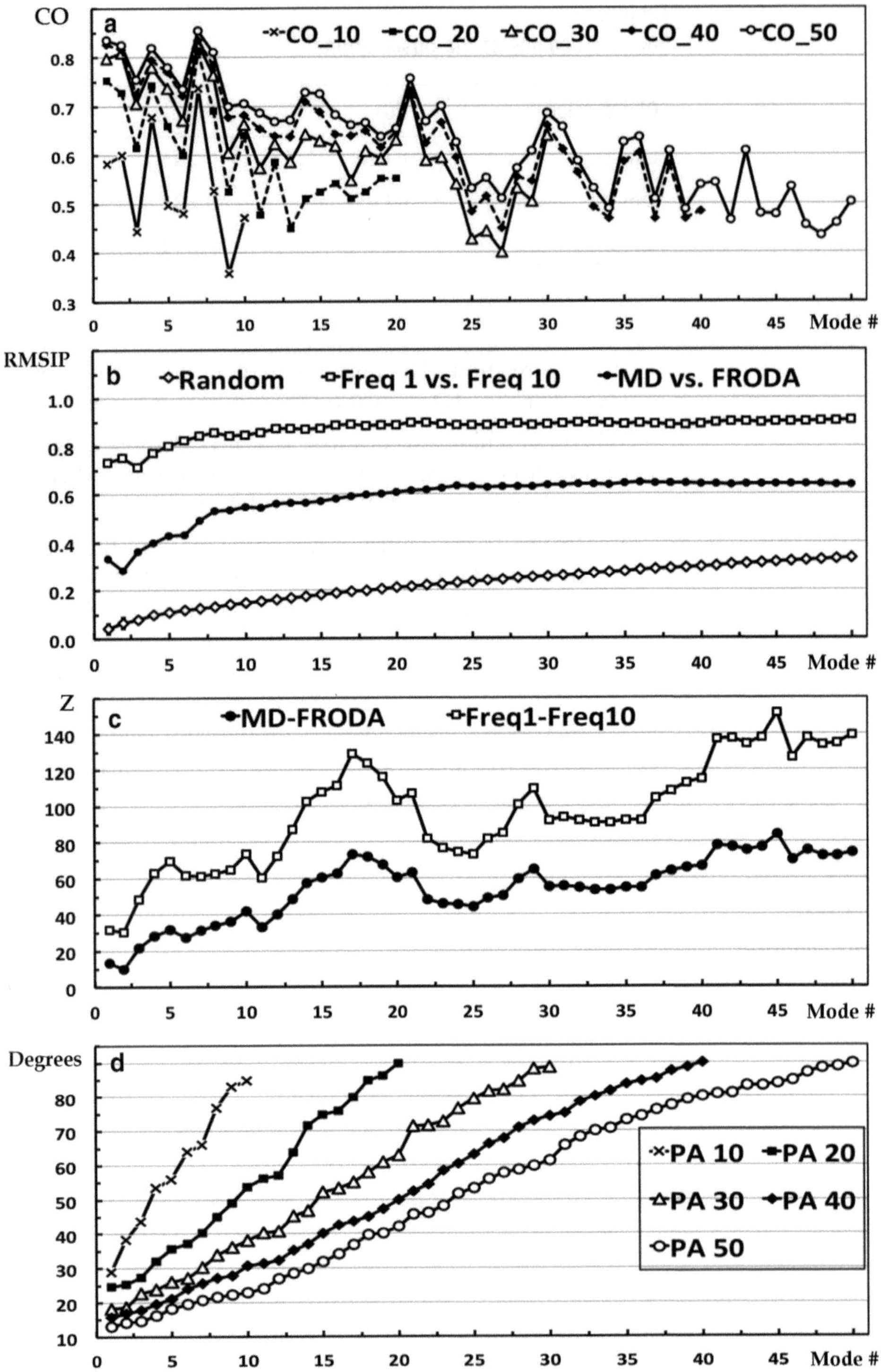

Fig. 9 (**a**) The cumulative overlap (CO) of each MD eigenvector with the entire set of FRODA eigenvectors defining the subspace of indicated size. We do not show the reverse metric, which is not symmetric, but yields similar values. (**b**) The RMSIP scores for the comparisons of random processes with 453 DOF, two FRODA

- Each component of the distance PCA modes indicates how the relative distance between a pair of atoms change. There is no way to map the mode components to individual residues.

10. Construct the weighted distance modes (*see* **Note 14**).
    - Weighting is done by multiplying by the square root of the eigenvalue for the mode, λ_i.
11. Construct the DVs for the trajectory, given by $DV_i = X_i - X_{\text{ref}}$, and construct the PCs.
    - Although there is a physical difference between using internal and Cartesian coordinates, mathematically the same procedures described above in terms of taking inner products and forming projections are identical.
12. When examining two or more sets of PCA modes, determination of how similar the trajectories are to each other may be assessed using the CO, RMSIP or PA metrics.

2.4 Recipe III: Essential Dynamics Using Cartesian Coordinate Based Kernel PCA

1. Obtain trajectories (one or more) from dynamic simulation.
2. Remove overall translations and rotations by aligning each frame to a reference structure.
3. Select the set of atoms for the analysis to define the data matrix, A.
4. Center the variables of A (row center) to define the data matrix A'.
5. Construct the kernel matrix, K, of $\{x, y, z\}$ positions for the atoms using A'.
    - The matrix K has dim $(n \times n)$ where n is the number of observations.
    - Each element (i, j) in the kernel is determined using a chosen kernel function, which has the general form as $K_{i,j} = K(k(x_i, x_j))_{i,j}$. A linear kernel is given as $K(x, y) = (x \cdot y)$, and a homogeneous polynomial is given by $K(x, y) = (x \cdot y)^d = (C_d(x), C_d(y))$ where C_D maps x to the vector $C_D(x)$ with entries that are all possible nth degree ordered products of the entries of x. Another kernal type uses a Gaussian weighting function given by

Fig. 9 (continued) simulations using the same conditions, and the MD and FRODA simulations. *Error bars* on the random process scores indicate plus and minus one standard deviation for 50 iterations. (**c**) The *Z*-scores for the RMSIP scores. (**d**) The PA spectra for the comparisons of the MD and FRODA simulations using the indicated SS DIM

$K(x, y) = \exp\left(-\frac{\|x - y\|^2}{2\sigma^2}\right)$ where the standard deviation, σ, is an adjustable parameter. A neural net kernel is given as $K(x, y) = \tanh(m(x \cdot y) + b)$, and a mutual information kernel is given as $K(x, y) = \mathrm{MI}(x, y)$ where $\mathrm{MI}(X, \Upsilon) = \sum_{y \in \Upsilon} \sum_{x \in X} p(x, y) \log\left(\frac{p(x, y)}{p(x)p(y)}\right)$. These are commonly employed kernels in many fields, and are not necessarily particularly useful for protein dynamics. Nevertheless, because higher order correlations in large datasets can be filtered with these kernels, and as such, we have explored all of them.

6. Diagonalize K using an EVD, and ignore the zero eigenvalues.
7. Examine the scree plot, and from where the kink is, select the top modes.
   - The characteristics of this plot depend heavily on whether one is analyzing fluctuations within a single native basin or is analyzing combined trajectories of multiple states. In kPCA, typically that first few eigenvalues are much larger than the remainder.
8. Determine the eigenvector collectivity. When using kPCA with properly tuned parameters, the top eigenvector often has a collectivity of 0.5 or higher.
9. Select the top set of eigenvectors for forming the kernel principal components (kPCs) (Usually 2–5).
10. Scale the top eigenvectors using the condition $1 = \lambda_n(\alpha_n \cdot \alpha_n)$ where α_n is the nth eigenvector (a column vector) of K and λ_n is the corresponding nth eigenvalue of K.
    - The eigenvectors are derived from the feature space and usually do not have a meaningful interpretation in the sample space.
11. Construct the DVs for the trajectory given by $DV_i = X_i - X_{\mathrm{ref}}$, and then construct the kPCs.
    - Calculate kPC_n using $(\mathrm{kPC})_n(x) = \sum_{i=1}^{M} \alpha_i^n k(x_i, x)$. Note that x is a test vector, and not a training vector (a vector are used to create the kernel). If only the original centered data is to be used, i.e., the data used to construct K, then all the elements of K are already determined. Projections can be made on singe modes to view as line graphs or on two PCA modes create scatter plots that show how the simulation explored the configuration space defined by the selected set of modes.

- We applied PCA and kPCA to the set of four 75 residue proteins to assess the ability of the methods to achieve cluster separation. The results are shown in Fig. 10.

12. When examining two or more sets of kPCA modes, determination of how similar the trajectories are to each other may be assessed using the following metrics. We note that the essential subspaces in kPCA are quite small, comprised of usually five or so modes. This is especially true when standard PCA was used as a preprocessing dimensional reduction step. Additionally, subspace comparisons require that the parent vector spaces have the same dimensionality. Therefore, it is possible to compare the essential subspaces derived from different kernels only when the same number of samples is used.
    - In Fig. 10f, we show that the subspaces for the top modes generated from the different kPCA approaches are quite similar using the RMSIP scores and the first PA. The most dissimilar was the SS derived from the MI kernel.

3 Notes

1. Many statistical packages support PCA and factor analysis (FA). While both methods use EVD, what is being factored is not the same. In PCA there is no underlying model for interpreting the "factors", and second, PCA does not account for error in the measurements, and thus if using the correlation matrix, it places all ones on the diagonal unlike FA, which places the communalities on the diagonal.
2. Here we refer to the spectral decomposition of a matrix as an eigenvalue decomposition (EVD). With square symmetric matrices there is no need to use a singular value decomposition (SVD) since the right and left vectors from the SVD are identical and the singular values are equal to the square root of the eigenvalues from the EVD.
3. There are multiple criteria for choosing modes (eigenvectors) in PCA (or FA). Since no underlying model is being used, the "interpretability" criterion does not apply. Also, the "Eigenvalue Larger than 1" only applies when using the correlation matrix. In protein dynamics, we find that trying to capture a specific amount of variance, say 50 %, does not work well and often over-estimates the essential subspace. The Cattell criterion for mode selection tends to work best and is applied by constructing the eigenvalue scree plot and identifying the "kink". Unlike with FA, there is no harm in doing this subjectively. We suggest that this approach be combined with subspace analysis to identify the saturation point for the RMSIP

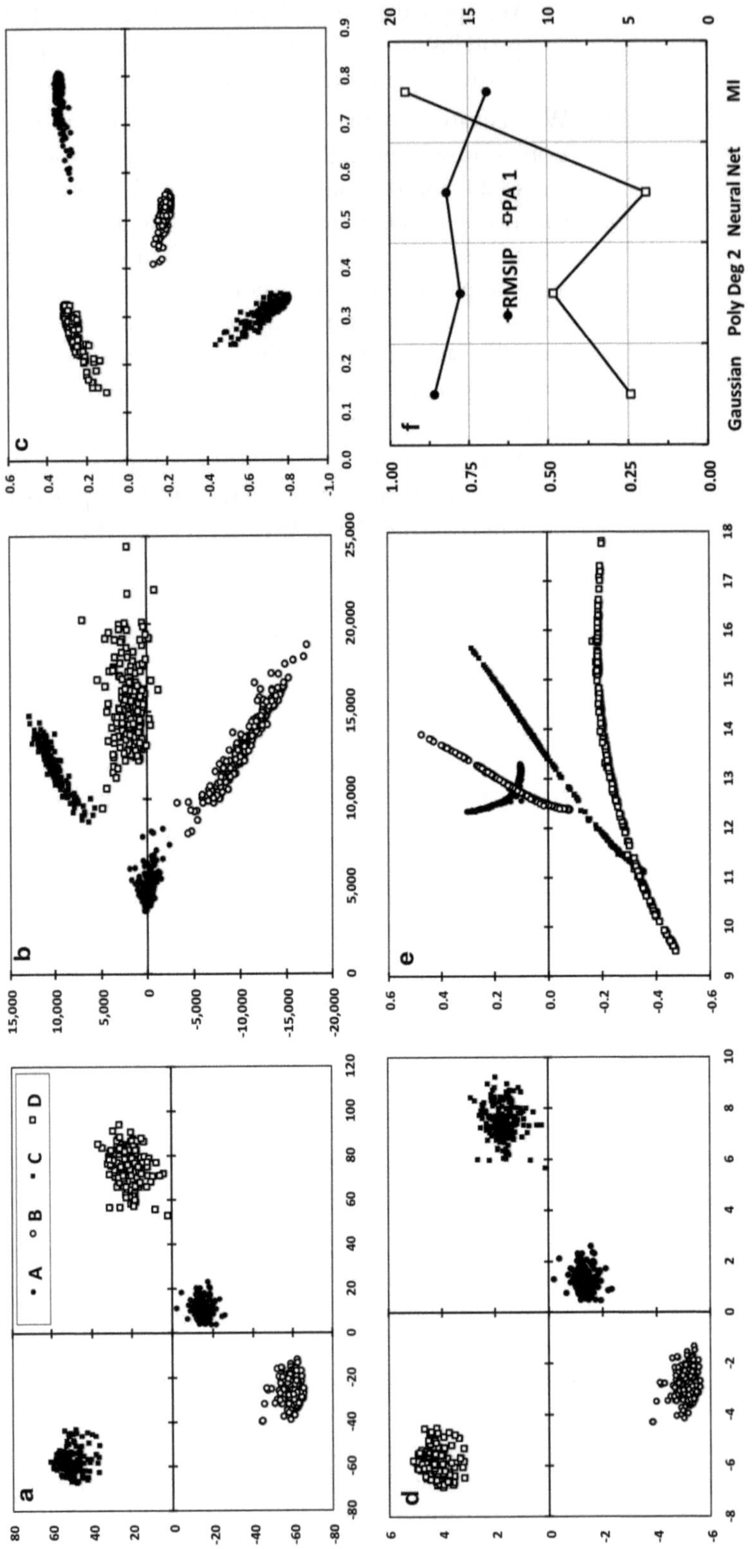

Fig. 10 Cluster separation for the dynamics of four different proteins using different kernels, but all using the same CE containing trajectories involving 2,000 FRODA frames for each of the four proteins. (**a**) Linear kernel

plots, as this is a good indicator of the essential subspace that is invariant to the "noise" in the data.

4. Given a C-matrix that is well conditioned, most common algorithms that perform EVDs (LINPACK, JAMA, etc.) will generate a set of eigenvalues in increasing order and a matching set of eigenvectors. The eigenvectors are orthogonal and normalized to have a magnitude of 1. Thus, any set of N eigenvectors constitutes an N dimensional orthonormal subspace of the parent vector space, defined by the full rank of the C-matrix.
5. There are numerous dynamic simulation packages available to generate the CEs, and many different formats for saving the coordinates from such a simulation. The method of "packing" the coordinates is not critical, but consistency is of utmost importance. This is especially true when a subset of atoms is selected from the main trajectory data. Extreme care should be taken to ensure that the data that is analyzed is in the expected format of the PCA package employed.
6. It is the nature of dynamic simulations to "shake up" the protein. Thus, to analyze the real internal fluctuations in the protein, all the trivial translations and rotations must be eliminated. The way this is normally done is to select a reference structure X_{ref} and a correspondence set (CS), e.g., the set of all alpha carbons. Then every structure from the trajectory is optimally aligned to X_{ref} using the chosen CS. In some cases, where there are very flexible tails, etc., it may be beneficial to exclude these from the CS so as to achieve better overall alignments. It is important to realize that if a subset of atoms is used as a reference, the choice of atoms to use in the CS is nontrivial and does affect the outcome of the PCA.
7. The results from Q and R based PCA are usually quite similar. We have found that if the movement of a small set of mobile atoms defines two or three clusters in the top two PC scatterplots, Q analysis will tend to enhance the separation, while R will tend to lessen it, resulting in subtle differences.
8. PCA based on internal distance coordinates (dPCA) can be very informative when combined with experimental data. In the case where three residues (alpha carbons) are analyzed (e.g., 25, 50,

Fig. 10 (continued) equivalent to standard PCA. (**b**) Homogeneous polynomial kernel of degree two, which is sensitive to fourth order statistics. (**c**) Gaussian kernel with standard deviation set to 50. (**d**) Neural net kernel with no offset and a slope parameter set to 10^{-4}. (**e**) Mutual Information kernel. (**f**) Subspace comparisons of the four kernels in **b–d** using the linear kernel essential space as the reference. The SS DIM in all cases was five. The primary *y*-axis shows RMSIP scores while the secondary *y*-axis shows the principal angle value in degrees

100), the eigenvector components convey how the distance between each alpha carbon pair is correlated (25–50, 35–100, 50–100). Since the information provided is all-to-all pair correlations, it is challenging to interpret the results of dPCA on even ten residues, which yields $10 \times 9/2 = 45$ pairs.

9. We find that performing standard PCA on our datasets and then extracting the top five modes works as an excellent data compressor/filter. These top five PCs are then analyzed with kPCA and additional features can be extracted. In our testing, we did not find a significant difference between using all the raw data or just the top five PCs: The kernels performed about the same in both cases, but in the latter, the computations were completed must faster.
10. How to improve sampling adequacy in locations with low MSA scores is nontrivial, since more sampling in the same way has diminishing returns. We found that the highest MSA scores were obtained when the sampling frequency (in FRODA) was set to one. The key to good MSA scores involves picking structures that are close together so as to enhance correlation in the variables. Sampling with larger time intervals for MD or lower frequencies with FRODA means that there are smaller correlations between the variables and larger partial correlations between sets of variables under the influence of the other variables. On the other hand, a CE consisting of uncorrelated samples is required to ensure statistical significance on representing the real dynamics of the system. Thus a best practices approach would be to sample over different time scales within a combined CE, in order to obtain sets of samples that are very close in conformational space with sets of samples that are more spread out in the conformational space.
11. One may also decide to not take the square root and work in units of variance.
12. PCs can be scaled by multiplying each PC by its corresponding eigenvalue, called a PC score. This has the effect of showing the differences in variance in the modes.
13. We have found that using a 20 dimensional subspace is a good compromise between reducing dimension and capturing the essential subspace. Often the RMSIP plots can be used to determine a saturation point that indicates the size of the essential space. One should always compare to a random process for aid in interpretation.
14. These plots are the most informative results from the dPCA on a single trajectory. Furthermore, when there are few components, the interpretation is straightforward.

References

1. Pearson K (1901) On lines and planes of closest fit to systems of points in space. The London, Edinburgh and Dublin Philosophical Magazine and Journal of Science 2:572
2. Hotelling H (1933) Analysis of a complex of statistical variables into principal components. J Educ Psychol 24:441
3. Manly B (1986) Multivariate statistics—a primer. Chapman & Hall/CRC, Boca Raton, FL
4. Abdi H, Williams LJ (2010) Principal component analysis. Wiley Interdisciplinary Reviews: Computational Statistics 2:433–459
5. Jolliffe IT (2002) Principal component analysis, vol XXIX, 2nd edn, Springer series in statistics. Springer, New York, p 487, p. 28 illus. ISBN 978-0-387-95442-4
6. Balsera MA, Wriggers W, Oono Y, Schulten K (1996) Principal component analysis and long time protein dynamics. J Phys Chem 100:2567–2572
7. Brüschweiler R (1995) Collective protein dynamics and nuclear spin relaxation. J Chem Phys 102(8):3396–3403
8. Berendsen HJ, Hayward S (2000) Collective protein dynamics in relation to function. Curr Opin Struct Biol 10:165–169
9. Amadei A, Linssen AB, de Groot BL, van Aalten DM, Berendsen HJ (1996) An efficient method for sampling the essential subspace of proteins. J Biomol Struct Dyn 13:615–625
10. Amadei A, Linssen AB, Berendsen HJ (1993) Essential dynamics of proteins. Proteins 17:412–425
11. Krebs WG, Alexandrov V, Wilson CA, Echols N, Yu H, Gerstein M (2002) Normal mode analysis of macromolecular motions in a database framework: developing mode concentration as a useful classifying statistic. Proteins 48:682–695
12. Sanejouand TF (2001) Conformational change of proteins arising from normal mode calculations. Protein Eng 14:1–6
13. Atilgan AR, Durell SR, Jernigan RL, Demirel MC, Keskin O, Bahar I (2001) Anisotropy of fluctuation dynamics of proteins with an elastic network model. Biophys J 80:505–515
14. Tirion MM (1996) Large amplitude elastic motions in proteins from a single-parameter, atomic analysis. Phys Rev Lett 77:1905–1908
15. Yang L, Song G, Carriquiry A, Jernigan RL (2008) Close Correspondence between the motions from principal component analysis of multiple HIV-1 protease structures and elastic network modes. Structure 16:321–330
16. David CC, Jacobs DJ (2011) Characterizing protein motions from structure. J Mol Graph Model 31:41–56
17. Van Aalten DMF, De Groot BL, Findlay JBC, Berendsen HJC, Amadei A (1997) A comparison of techniques for calculating protein essential dynamics. J Comput Chem 18 (2):169–181
18. Rueda M, Chacó P, Orozco M (2007) Thorough validation of protein normal mode analysis: a comparative study with essential dynamics. Structure 15:565–575
19. Cui Q, Bahar I (eds) (2005) Normal mode analysis: theory and applications to biological and chemical systems. Chapman and Hall/CRC, Boca Raton, FL, 432 pages
20. Kitao A, Go N (1999) Investigating protein dynamics in collective coordinate space. Curr Opin Struct Biol 9:164–169
21. Ma J (2005) Usefulness and limitations of normal mode analysis in modeling dynamics of biomolecular complexes. Structure 13:373–380
22. Hayward S, Kitao A, Go N (1995) Harmonicity and anharmonicity in protein dynamics: a normal mode analysis and principal component analysis. Proteins 23(2):177–186
23. Hayward S, Kitao A, Go N (1994) Harmonic and anharmonic aspects in the dynamics of BPTI: a normal mode analysis and principal component analysis. Protein Sci 3 (6):936–943
24. Scholkopf B, Smola A, Muller K-R (1999) Kernel principal component analysis. In: Scholkopf B, Burges CJC, Smola AJ (eds) Advances in kernel methods—support vector learning. MIT Press, Cambridge, MA, pp 327–352
25. Sapra S (2010) Robust vs. classical principal component analysis in the presence of outliers. Appl Econ Lett 17:519–523
26. Storer M, Peter M, Roth PM, Urschler M, Bischof H. Fast-robust PCA (2009). Institute for Computer Graphics and Vision Graz University of Technology Inffeldgasse 16/II, 8010 Graz, Austria
27. Gnanadesikan R, Kettenring J (1972) Robust estimates, residuals, and outlier detection with multiresponse data. Biometrics 28:81–124
28. Huber P (1981) Robust statistics. Wiley, New York
29. De La Torre F, Black M (2003) A framework for robust subspace learning. Int J Comput Vis 54:117–142
30. Handling of data containing outliers. Wolfram Stacklies and Henning Redestig CAS-MPG

Partner Institute for Computational Biology (PICB) Shanghai, P.R. China and Max Planck Institute for Molecular Plant Physiology Potsdam, Germany

31. Joint Outliers and Principal Component Analysis. Georgy Gimel'farb, Alexander Shorin, and Patrice Delmas. Dept. of Computer Science, University of Auckland, P.B. 92019, Auckland, New Zealand
32. Kriegel HP, Kröger P, Schubert E, Zimek A (2008) a general framework for increasing the robustness of PCA-based correlation clustering algorithms. Scientific and Statistical Database Management. Lecture Notes in Computer Science, vol 5069. p 418
33. Cattell RB (1966) The scree test for the number of factors. Multivariate Behav Res 1 (2):245–276
34. Cattell RB, Vogelmann S (1977) A comprehensive trial of the scree and KG criteria for determining the number of factors. Multivariate Behav Res 12:289–325
35. Charles David (2012) Essential dynamics of proteins using geometrical simulations and subspace analysis. Ph.D. Dissertation, UNC Charlotte, Department of Bioinformatics and Genomics
36. Jacobs DJ, Trivedi D, David CC, Yengo CM (2011) Kinetics and thermodynamics of the rate limiting conformational change in the myosin V mechanochemical cycle. J Mol Biol 407(5):716–730
37. Trivedi D, David CC, Jacobs DJ, Yengo CM (2012) Switch II mutants reveal coupling between the nucleotide- and actin-binding regions in myosin V. Biophys J 102 (11):2545–2555. doi:10.1016/j.bpj.2012.04.025
38. Wells SA, Menor S, Hespenheide BM, Thorpe MF (2005) Constrained geometric simulation of diffusive motion in proteins. Phys Biol 2: S127–S136
39. Farrell DW, Kirill S, Thorpe MF (2010) Generating stereochemically acceptable protein pathways. Proteins 78:2908–2921
40. Jacobs DJ, Rader AJ, Kuhn LA, Thorpe MF (2001) Protein flexibility predictions using graph theory. Proteins 44:150–165
41. Amadei A, Ceruso MA, Di Nola A (1999) On the convergence of the conformational coordinates basis set obtained by the essential dynamics analysis of proteins' molecular dynamics simulations. Proteins 36:419–424
42. Leo-Macias A, Lopez-Romero P, Lupyan D, Zerbino D, Ortiz AR (2005) An analysis of core deformations in protein superfamilies. Biophys J 88:1291–1299
43. Miao J, Ben-Israel A (1992) On principal angles between subspaces. Linear Algebra Appl 171:81–98
44. Gunawan H, Neswan O, Setya-Budhi W (2005) A formula for angles between subspaces of inner product spaces. Contribut Algebra Geom 46(2):311–320
45. Absil PA, Edelman A, Koev P (2006) On the largest principal angle between random subspaces. Linear Algebra Appl 414(1):288–294
46. Cerny CA, Kaiser HF (1977) A study of a measure of sampling adequacy for factor-analytic correlation matrices. Multivariate Behav Res 12(1):43–47
47. Hess B (2002) Convergence of sampling in protein simulations. Phys Rev E 65:031910
48. Kabsch W (1978) A discussion of the solution for the best rotation to relate two sets of vectors. Acta Crystallogr A 34:827–828
49. Hyvärinen A, Oja E (2000) Independent component analysis: algorithms and applications. Neural Netw 13(4–5):411–430
50. Hyvärinen A (1999) Fast and robust fixed-point algorithms for independent component analysis. IEEE Trans Neural Netw 10(3):626–634
51. Zou H, Hastie T, Tibshirani R (2006) Sparse principal component analysis. J Comput Graph Stat 15(2):265–286
52. Yao F, Coquery J, Lê Cao K (2012) Independent principal component analysis for biologically meaningful dimension reduction of large biological data sets. BMC Bioinformatics 13:24

Chapter 12

A Case Study Comparing Quantitative Stability–Flexibility Relationships Across Five Metallo-β-Lactamases Highlighting Differences Within NDM-1

Matthew C. Brown, Deeptak Verma, Christian Russell, Donald J. Jacobs, and Dennis R. Livesay

Abstract

The Distance Constraint Model (DCM) is an ensemble-based biophysical model that integrates thermodynamic and mechanical viewpoints of protein structure. The DCM outputs a large number of structural characterizations that collectively allow for Quantified Stability–Flexibility Relationships (QSFR) to be identified and compared across protein families. Using five metallo-β-lactamases (MBLs) as a representative set, we demonstrate how QSFR properties are both conserved and varied across protein families. Similar to our characterizations on other protein families, the backbone flexibility of the five MBLs are overall visually conserved, yet there are interesting specific quantitative differences. For example, the plasmid-encoded NDM-1 enzyme, which leads to a fast spreading drug-resistant version of *Klebsiella pneumoniae*, has several regions of significantly increased rigidity relative to the other four. In addition, the set of intramolecular couplings within NDM-1 are also atypical. While long-range couplings frequently vary significantly across protein families, NDM-1 is distinct because it has limited correlated flexibility, which is isolated within the active site S3/S4 and S11/H6 loops. These loops are flexibly correlated in the other members, suggesting it is important to function, but the others also have significant amounts of correlated flexibility throughout the rest of their structures.

Key words β-Lactamase, Distance constraint model, Protein flexibility, Allostery, Quantitative stability–flexibility relationships

1 Introduction

Feynman famously stated nearly 50 years ago that "Everything that living things do can be understood in terms of the jigglings and wigglings of atoms" [1]. Since then, our appreciation of how these dynamical properties play out in proteins over a wide spectrum of timescales has been greatly expanded [2]. However, the experimental and computational methods to interrogate and compare these jigglings and wigglings across many different proteins remain elusive [3]. To that end, we have developed a powerful and

Dennis R. Livesay (ed.), *Protein Dynamics: Methods and Protocols*, Methods in Molecular Biology, vol. 1084,
DOI 10.1007/978-1-62703-658-0_12,

computationally efficient Distance Constraint Model (DCM) [4–7] that allows us to efficiently compare protein stability and flexibility properties across many different proteins, which we refer to as Quantitative Stability–Flexibility Relationships (QSFR) [8, 9].

Starting with our initial QSFR comparisons in 2006 of a mesophilic and thermophilic RNase H pair [10], to our most recent comparisons of larger protein datasets [11–13], we have observed fairly consistent trends. That is, backbone flexibility is largely conserved across protein families, whereas intramolecular couplings (i.e., allostery) can be quite variable. Put otherwise, backbone flexibility is, in large part, engrailed by structure, whereas allostery is highly sensitive to small perturbations. While these qualitative observations are consistent and robust, many quantitative variations therein do occur. For example, single point mutations in lyosozyme have a frequent, large, and long-range effect on stability, backbone flexibility and intramolecular couplings [14, 15]. In contrast, while they are indeed sensitive, allosteric couplings do quantitatively reflect similarities within protein structures. For example, we demonstrated that allosteric response in *Escherichia coli* and *Salmonella typhimurium* CheY orthologs was more similar than either was to the response in *Thermotoga maritima*, which is the naïve expectation based on evolutionary relationships [12].

Our most recent QSFR comparisons have focused on a dozen class-A β-lactamase (BL) structures, which is the enzyme that is primarily responsible for conferring resistance to penicillin and related antibiotics. There are four different BL classes, and the class-A enzymes are the most clinically relevant. Interestingly, the TEM-1 class-A enzyme has been shown to have an extremely rigid backbone [16], which is consistent with our results [17]. Moreover, our backbone flexibility predictions are well conserved across the family, whereas allosteric couplings are overall quite variable. Similar to our earlier work on CheY, we further demonstrate that the systematic variations within the flexibility properties parallel the evolutionary history of the family. Large and systematic differences in both quantities occur between evolutionary out-groups, whereas properties are more conserved between close homologs.

While the class-A family is currently the most clinically relevant group of BL enzymes, the metallo-β-lactamases (MBLs) pose a looming pandemic threat. The MBLs (also referred to as class-B) are mechanistically distinct from the other BLs, and frequently have activity against carbapenems, which represent our last lines of antibiotic defense [18]. For example, the plasmid-encoded New Delhi metallo-β-lactamase (NDM-1) has led to a number of patient deaths due to its incredible broad-spectrum activity [19]. In this report, we describe how QSFR properties vary across five MBL enzymes, which is discussed as a representative of all of our comparative studies, while also revealing several specific and interesting details about NDM-1 that could lead to future antimicrobial strategies.

2 A Brief Summary of the Distance Constraint Model

From conception, the DCM has been designed to optimally balance computational efficiency with biophysical accuracy. To keep the method fast, it is fundamentally based on a free energy decomposition (FED) scheme where component enthalpies and entropies can be placed into look-up tables and accounted for quickly when present. However, FED descriptions of large, highly cross-linked macromolecules like proteins typically fail because of entropy nonadditivity [20, 21]. As such, the critical component of the DCM is robustly accounting for nonadditivity within entropic components [22]. The DCM models structure as a graph, where each chemical interaction is described by an edge. When present, an edge lowers the enthalpy, and also lowers the entropy when independent. When a constraint is placed into a flexible region, it is said to be *independent* because it removes a degree of freedom (DOF). Conversely, when a constraint is placed into a region that is already rigid, it is said to be *redundant* and does not further reduce the entropy because it is placed into a region that is already rigid, meaning all DOF have already been removed. For large atomic systems, a constraint is determined to be or not to be independent by a fast graph rigidity algorithm, called the Pebble Game [23, 24], which provides a complete and rigorous mechanical description of the molecular network [25].

To account for thermal fluctuations, the DCM generates an ensemble of rigidity graphs where weak chemical interactions are allowed to fluctuate on and off. A Gibbs ensemble of rigidity graphs is modeled, each weighted by its free energy using the FED scheme described above. In the standard way, appropriate derivatives of the partition function provide a complete thermodynamic description of the protein. Subsequently, the partition function is used to weight the rigidity/flexibility descriptions of the protein, thus providing a feedback cycle that integrates mechanical and thermodynamic viewpoints. Put otherwise, thermodynamic characterizations are improved by distinguishing between independent and redundant constraints, whereas the calculated Boltzmann weights are used to appropriately average the mechanical properties. An important consequence of this approach is that the DCM appropriately models cooperativity because network rigidity is used as an underlying interaction that accounts for enthalpy–entropy compensations. That is, competition emerges between an enthalpically stabilized rigid structure with many redundant constraints and a flexible entropically stabilized unfolded state [26, 27].

In typical usage, the model is parameterized by reproducing experimental heat capacity curves [6, 7]. Our current minimal DCM (mDCM) has three parameters (u_{sol}, v_{nat}, δ_{nat}) that, respectively, describe the energy of forming a hydrogen bond to solvent,

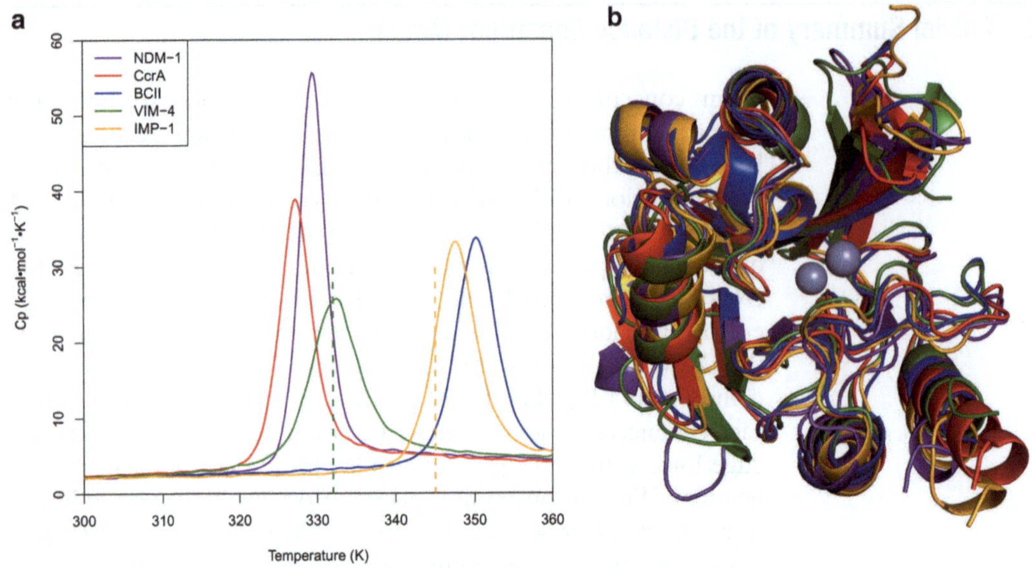

Fig. 1 (**a**) Predicted heat capacity curves of each of the five enzymes. The referenced experimental melting temperatures are marked with *dashed vertical lines*. The T_m of VIM-4 and IMP-1 are, respectively, 332 K and 345 K. (**b**) Superposition of the five metallo-β-lactamase enzymes, which are color-coded the same as in panel (**a**). The average pairwise α-carbon RMSD across the dataset is 2.2 Å (standard deviation = 0.4 Å)

the energy associated with being in a native conformation, and the entropy associated with that native conformation. In the absence of the experimental data to fit to, which is the case for the MBL dataset considered here, the model is parameterized by systematically exploring the parameter space to find $\{u_{sol}, v_{nat}, \delta_{nat}\}$ that is physically reasonable. For our dataset, three different parameter sets produced quantitatively reasonable two-state behavior. While full heat capacity characterizations are not available, the melting temperature of two of the five enzymes is known [28, 29], which are well described using the central set with values $u_{sol} = -2.6$ kcal/mol, $v_{nat} = -0.5$ kcal/mol, and $\delta_{nat} = 1.8$ (cf. Fig. 1a). These model parameter values are well within the expected range established by our prior works across many different globular protein systems.

In addition to the thermodynamic quantities, the mDCM calculates a number of mechanical properties that are appropriately averaged by the thermodynamic ensemble. From the set of QSFR metrics, two are particularly useful. The first, called the Flexibility Index (FI), describes backbone flexibility. Positive FI values quantify the number of DOF within a local region, whereas negative values quantify the number of redundant constraints. When FI = 0, the backbone is said to be *isostatically* rigid, meaning it is marginally rigid (# of DOF = # redundant constraints = 0). The second, called Cooperativity Correlation (CC), depicts the complete set of residue-to-residue couplings. As described below, CC is described by an $N \times N$ matrix, where N is the number of residues in the protein. Each pixel (designating residue pair i,j) is colored based on the correlations therein. When a residue pair is likely to be included in the same rigid cluster, it is colored blue, whereas they are colored red when they are flexibly correlated. White indicates no correlation, but does not necessarily imply rigidity or flexibility. For example, residues i and j might always be rigid, but if their particular rigid clusters never overlap, the CC pixel i,j would be colored white.

3 Metallo-β-Lactamase QSFR

3.1 The Dataset

In this report, we compare five different MBL structures [28, 30–33], all of which correspond to the B1-subfamily [18] (cf. Table 1). A superposition of the five structures is provided in Fig. 1b. Hydrogen atoms were added using H++ [34] with default parameters and the resulting structures were minimized using the AMBER 99 force field [35]. MBLs have two zinc ions per monomer, which are bridged by a well-resolved water molecule. Interestingly, Kim et al. [36] have recently demonstrated that NDM-1 is functionally tolerant of non-zinc metals and a pH-dependent transition from a water molecule to a hydroxide anion. The mDCM is currently not parameterized to deal with metal ions. As such, we employ the same strategy that we used before with CheY [12], where constraints are added across the set of chelating residues to mimic the effect of metal binding within that region. The energy per chelating constraint is set such that the total binding energy is −4.5 kcal/mol per zinc ion, and the bridging water molecule is −3.0 kcal/mol.

3.2 Backbone Flexibility Is Mostly Conserved

A multiple sequence alignment of the five MBLs color-coded by FI is provided in Fig. 2, and structural descriptions are provided in Fig. 3. The backbone flexibility properties are largely defined by

Table 1
Summary of the metallo-β-lactamase dataset

Name	PDBID	Resolution (Å)	R-Free	Structural similarity (Å)[a]
NDM-1	3ZR9	1.91	0.21	–
CcrA	1HLK	2.50	0.23	2.09
BCII	1BVT	1.85	0.27	2.17
VIM-4	2WHG	1.90	0.25	2.33
IMP-1	1DDK	3.10	0.29	3.03

[a]The pairwise α-carbon root mean square distance from each structure is calculated relative to NDM-1

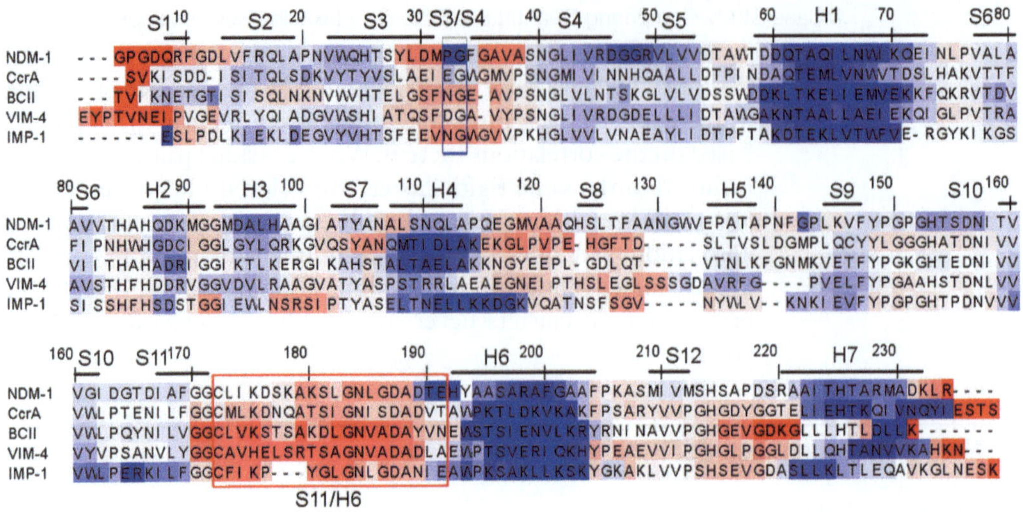

Fig. 2 Multiple sequence alignment of the five enzymes color-coded by backbone flexibility index (FI). *Red* indicates flexibility, whereas *blue* indicates rigidity. *White* corresponds to isostatic regions that are marginally rigid. Secondary structure information is provided above the alignment. In addition, the S3/S4 and S11/H6 loops are respectively indicated with *blue* and *red boxes*

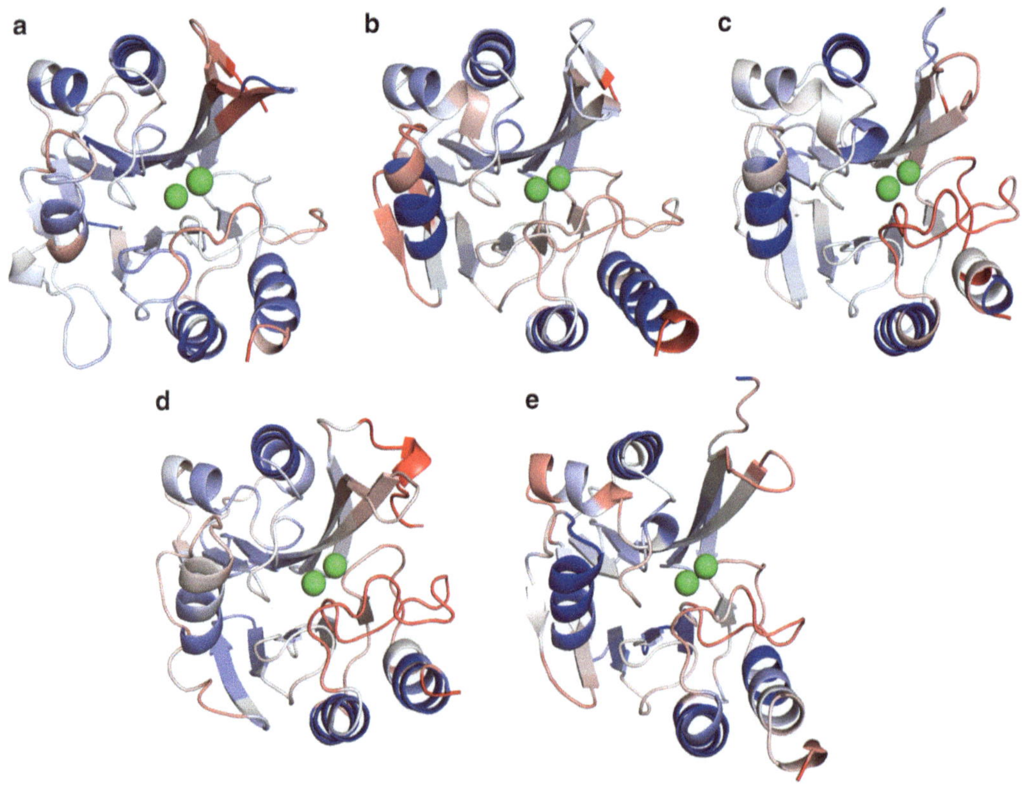

Fig. 3 The five metallo-β-lactamase structures are colored by backbone flexibility index (FI), which are the identical values reported in Fig. 2. *Red* indicates flexibility, whereas *blue* indicates rigidity. *White* corresponds to isostatic regions that are marginally rigid. The structures are centered on the active site region, with the S3/S4 and S11/H6 loops just above and below, respectively, the metal ions. (**a**) NDM-1 (**b**) CcrA (**c**) BCII (**d**) VIM-4 (**e**) IMP-1

secondary structure. For example, the most rigid portions of the protein uniformly correspond to α-helices, whereas the most flexible regions typically correspond to loops. Interestingly, compared to the Class-A BLs, the MBLs tend to be significantly less rigid, especially within the core regions of the protein. In fact, large portions of each structure are very close to isostatic. With the exception of the termini ends, the most consistently flexible portions of the protein correspond to the regions around the S3/S4 and S11/H6 loops [37], which is consistent with our results (cf. Fig. 2). This is particularly interesting because these two regions define the upper and lower borders of the active site. While there are several other structurally conserved loops within MBL, none are so flexibly conserved. Based on their active site proximity, this suggests that flexibility within these regions is critical to MBL function. Similar to our studies on earlier systems, no significant relationship could be found between the (dis)similarities within structure and backbone flexibility. This is an important result because it stresses, while backbone flexibility is in large part defined by secondary structure, it is largely impossible to relate quantitative differences to structural conformation based on the way rigidity propagates through the H-bond network [11].

3.3 Intramolecular Couplings Are Highly Variable

Juxtaposed to the conservation within backbone flexibility, intramolecular couplings within the five MBL structures are visibly different (cf. Fig. 4) owing to the sensitivity within protein dynamics [15, 38]. For example, the extensive blue coloring indicates that the NDM-1 structure is primarily composed of one large rigid cluster, with limited amounts of correlated flexibility. Conversely, much of N-terminal portion the VIM-4 structure is mostly co-rigid, whereas the C-terminal portion is primarily flexibly correlated. The three remaining structures have nearly equal amounts of rigidity and flexibility correlation. The observed variability is quite striking. Note that the large white swaths without any color shading correspond to gaps in the alignment.

Despite the large overall differences, many locally conserved features are observed within the CC plots. For example, the red swaths corresponding to loop L10 centered at alignment position 180 are observed in all five structures. Similarly, there is observable flexibility correlation between S3/S4 and S11/H6 loops in all five structures. This is the only off-diagonal correlated flexibility that is observed in all five structures, which coupled with its active site location suggests that it is critical for enzyme function. Finally, the blue shading is always strongest in the N-terminal portion of the structure, with the region between residues 10 and ~100 being mostly co-rigid (with the exception of S3/S4 loop).

3.4 NDM-1

Taken together, the above results were described in the context of our collective results to represent what we have observed over

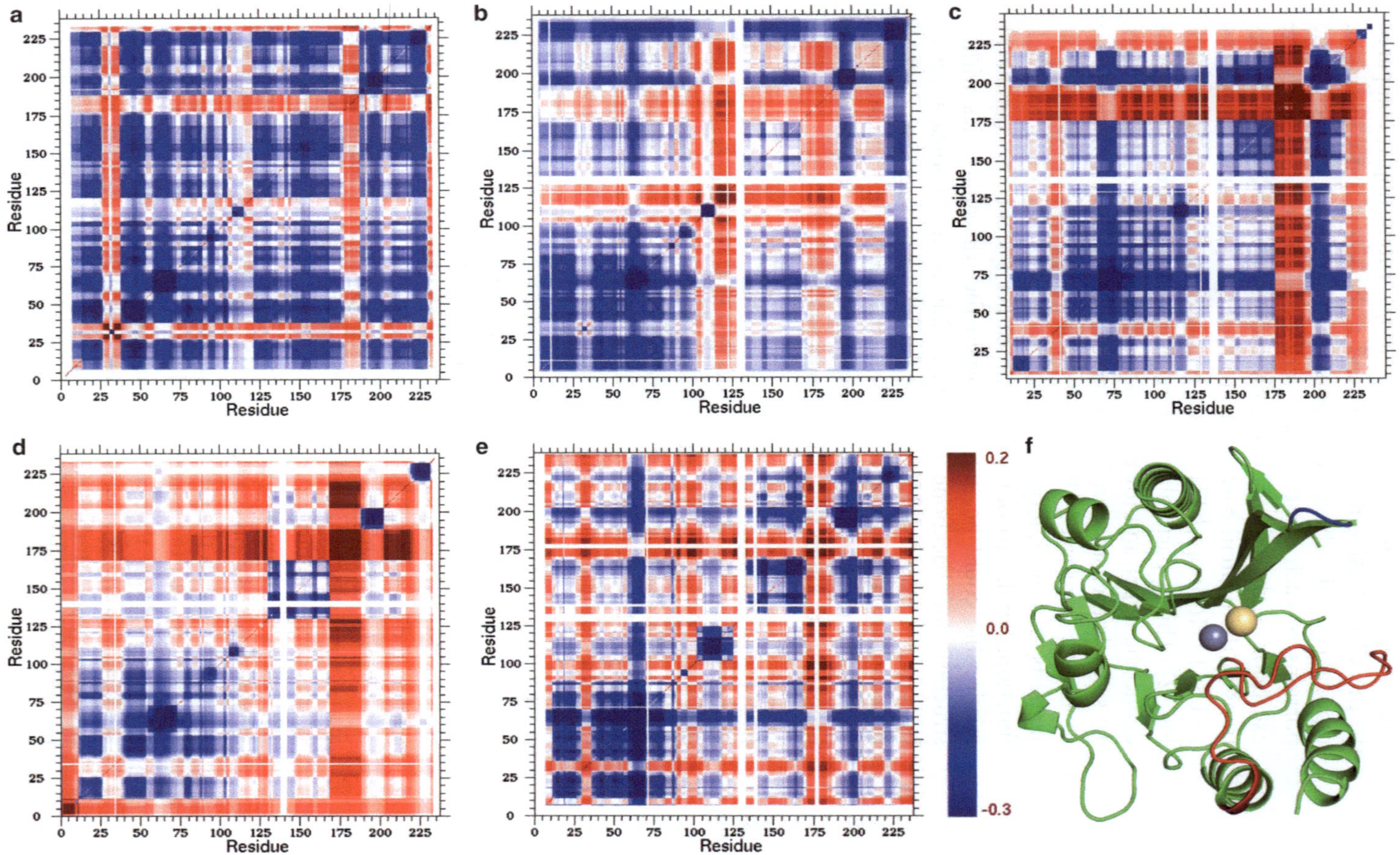

Fig. 4 Cooperativity Correlation (CC) plots reveal pairwise mechanical couplings within structure: (**a**) NDM-1, (**b**) CcrA, (**c**) BCII, (**d**) VIM-4, (**e**) IMP-1. That is, *red* indicates residue pairs that are flexibly correlated, whereas *blue* indicates residue pairs that are rigidly correlated. In each case, the matrices have been aligned so that all pixels are structurally equivalent. *White* indicates either residue pairs that are mechanically decoupled or gaps within the alignment. Notice that in NDM-1 the only appreciable correlated flexibility between loops occurs in the S3/S4 and S11/H6 regions. For reference, the structure of NDM-1 is shown in panel (**f**) with the S3/S4 (*blue*) and S11/H6 (*red*) loops highlighted

several different comparative studies. However, due to the biomedical importance of NDM-1, we focus on what makes it special relative to the other MBLs. Compared to the other MBLs, NDM-1 is unique because it is functionally so promiscuous. It is active against nearly all known β-lactam antibiotics with exceptional efficiency [36]. Moreover, the enzyme is functionally tolerant of substitutions in metal ion cofactors and changes in pH that cause the bridging water molecule to become deprotonated. Nevertheless, as discussed above, there is significantly more co-rigidity within NDM-1 compared to the other structures, and only the S3/S4 and S11/H6 loops are flexibly correlated. The backbone flexibility of the N-terminal half of L10 is significantly reduced in NDM-1 compared to the other structures, whereas the flexibility within the C-terminal portion of L10, which is closest to the active site rim, is similar to the others. Similarly, loop L11, strand β12, and a few other regions are significantly less flexible in NDM-1 as well. Taken together, these results indicate that flexibility and flexibility correlation within NDM-1 tends to be highly focused at the active site region, whereas flexibility and flexibility correlation is more widespread throughout the structure in the other four MBLs.

Another interesting difference is that there are a couple spots of intense rigidity within the NDM-1 structure. The first actually occurs within the β-hairpin composed of strands S3 and S4. The β-turn is rigid, although the flanking strands are flexible. The rigidity occurs due to the optimized structure of the β-turn; it has both a proline in the second position and a glycine in the third, which are hallmarks of a stable β-turn. However, the pronounced rigidity of this region in NDM-1 is mostly attributed to the presence of the proline residue in the second β-turn position, which is unique to NDM-1. While the NDM-1 β-turn is rigid, this does not necessarily imply that it is immobile. Flexibility identifies hinges, which are DOF that control motion elsewhere. In this case, the flexibility surrounding the β-turn allows the rigid body to move through space. This is analogous to a weight swinging from the motion provided by a pendulum. The pivot is flexible and immobile (assuming it is not moving through space), whereas the weight is simultaneously rigid and mobile. The second region of intense rigidity in the NDM-1 structure corresponds to the 3_{10}-capping helix, which is adjacent to the flexible portion of L10, which further underscores the isolation of flexibility within S3/S4 and S11/H6 loops of NDM-1.

It is tempting to speculate on the functional consequences of these observations. The fact that the S3/S4 and S11/H6 loops are flexibly correlated in all structures obviously suggests that it is clearly important to MBL function. No other correlation is strongly conserved across the family, which is not necessarily unexpected considering how different CC can be. Nevertheless, the unique isolation of correlated flexibility within NDM-1 does suggest that it is atypical compared to the other four. Based on this, we suggest that the distilling of the correlated flexibilities to the two active site loops contributes to the enzyme's broad antibiotic resistance activity. However, it remains an open and potentially critical biomedical question of how the pronounced co-rigidity of NDM-1 is reconciled with its functional promiscuity [36].

4 Conclusions

Biology is an inherently comparative science. From Darwin's finches to molecular sequence analysis, the paradigm of comparison and classification has driven biological insight. It follows that comparing the fundamental biophysical properties that govern structure and function within related proteins holds much promise for improved understanding. Unfortunately, however, the bulk of the experimental and computational methods needed to accurately describe these properties are prohibitive to large-scale analyses. To that end, we have developed the DCM in order to allow for robust and comprehensive comparisons of thermodynamic and dynamical properties within proteins. Over the past 7 years, we have established the importance of our comparative QSFR characterizations on several different systems. Without exception, our results establish an intriguing mix of conservation and variation within QSFR properties. Typically, there is obvious similarity within protein structure backbone flexibility across members of the same family. Yet, based on the complexity of the underlying H-bond network, there are simultaneously quantitative differences therein that are not obvious from structure. In contrast, the set of intramolecular couplings are much more sensitive to small perturbations, which can lead to dramatic differences across protein families. This result is entirely consistent with the growing appreciation for the sensitivity of allosteric response to mutation [12, 38–40] and other relatively minor perturbations (i.e., replacing one divalent metal ion for another [41]).

The MBLs characterized here are consistent with these general trends. The five structures are relatively conserved in shape and backbone flexibility. Yet significant differences do occur across the set. Moreover, the similarities (and differences) within backbone flexibility do not simply parallel structural similarity. Manual analysis of the H-bond network can identify the mechanistic origins for some of the differences, but not all due to the long-range nature of rigidity. While there are some conserved features within the CC plots, visually they are overall distinct. This is especially true in NDM-1, which is composed one large rigid cluster with correlated flexibility isolated in the active site S3/S4 and S11/H6 loops. Going forward, it will be important to determine if disruption of this isolated correlated flexibility has an impact on NDM-1's broad-spectrum antibiotic resistance activity.

Acknowledgments

This work has been partially supported by NIH grants R01 GM073082 (to D.J.J. and D.R.L.) and R15 GM101570 (to D.R. L.). Key to the distance constraint model is the use of graph-rigidity algorithms, claimed in US Patent 6,014,449, which has been assigned to the Board of Trustees Michigan State University. Used with permission.

References

1. Feynman R (1963) The Feynman lectures on physics. Addison-Wesley Publishing Company, Reading, MA
2. Henzler-Wildman KA, Lei M, Thai V, Kerns SJ, Karplus M, Kern D (2007) A hierarchy of timescales in protein dynamics is linked to enzyme catalysis. Nature 450:913–916
3. Livesay DR (2010) Protein dynamics: dancing on an ever-changing free energy stage. Curr Opin Pharmacol 10:706–708
4. Jacobs DJ (2006) Predicting protein flexibility and stability using network rigidity: a new modeling paradigm. In: Pandalai SG (ed) Recent research developments in biophysics. Transworld Research Network, Trivandrum, India, pp 71–131
5. Jacobs DJ (2010) Ensemble-based methods for describing protein dynamics. Curr Opin Pharmacol 10:760–769
6. Jacobs DJ, Dallakyan S (2005) Elucidating protein thermodynamics from the three-dimensional structure of the native state using network rigidity. Biophys J 88:903–915
7. Livesay DR, Dallakyan S, Wood GG, Jacobs DJ (2004) A flexible approach for understanding protein stability. FEBS Lett 576:468–476
8. Jacobs DJ, Livesay DR, Hules J, Tasayco ML (2006) Elucidating quantitative stability/flexibility relationships within thioredoxin and its fragments using a distance constraint model. J Mol Biol 358:882–904
9. Jacobs DJ, Livesay DR, Mottonen JM, Vorov OK, Istomin AY, Verma D (2012) Ensemble properties of network rigidity reveal allosteric mechanisms. In: Fenton AW (ed) Methods in molecular biology. Springer, New York, pp 279–304
10. Livesay DR, Jacobs DJ (2006) Conserved quantitative stability/flexibility relationships (QSFR) in an orthologous RNase H pair. Proteins 62:130–143
11. Livesay DR, Huynh DH, Dallakyan S, Jacobs DJ (2008) Hydrogen bond networks determine emergent mechanical and thermodynamic properties across a protein family. Chem Cent J 2:17
12. Mottonen JM, Jacobs DJ, Livesay DR (2010) Allosteric response is both conserved and variable across three CheY orthologs. Biophys J 99:2245–2254
13. Mottonen JM, Xu M, Jacobs DJ, Livesay DR (2009) Unifying mechanical and thermodynamic descriptions across the thioredoxin protein family. Proteins 75:610–627
14. Verma D, Jacobs DJ, Livesay DR (2010) Predicting the melting point of human C-type lysozyme mutants. Curr Protein Pept Sci 11:562–572
15. Verma D, Jacobs DJ, Livesay DR (2012) Changes in lysozyme flexibility upon mutation are frequent, large and long-ranged. PLoS Comput Biol 8:e1002409
16. Savard PY, Gagne SM (2006) Backbone dynamics of TEM-1 determined by NMR: evidence for a highly ordered protein. Biochemistry 45:11414–11424
17. Verma D, Jacobs DJ, Livesay DR (2013) Variations within class-A β-lactamase physiochemical properties reflect evolutionary, but not antibiotic specificity, patterns. PLoS Comp Biol 9:e1003155
18. Cadag E, Vitalis E, Lennox KP, Zhou CL, Zemla AT (2012) Computational analysis of pathogen-borne metallo beta-lactamases reveals discriminating structural features between B1 types. BMC Res Notes 5:96
19. Sharma VK, Guleria R, Mehta V, Sood N, Singh SN (2010) NDM-1 resistance: Fleming's predictions become true. Int J Appl Biol Pharm Technol 1:1244–1251
20. Dill KA (1997) Additivity principles in biochemistry. J Biol Chem 272:701–704

21. Mark AE, van Gunsteren WF (1994) Decomposition of the free energy of a system in terms of specific interactions. Implications for theoretical and experimental studies. J Mol Biol 240:167–176
22. Jacobs DJ, Dallakyan S, Wood GG, Heckathorne A (2003) Network rigidity at finite temperature: relationships between thermodynamic stability, the nonadditivity of entropy, and cooperativity in molecular systems. Phys Rev E Stat Nonlin Soft Matter Phys 68:061109
23. Jacobs DJ, Rader AJ, Kuhn LA, Thorpe MF (2001) Protein flexibility predictions using graph theory. Proteins 44:150–165
24. Jacobs DJ, Thorpe MF (1995) Generic rigidity percolation: the pebble game. Phys Rev Lett 75:4051–4054
25. Katoh N, Tanigawa S (2011) A proof of the molecular conjecture. Discrete Comput Geom 45:647–700
26. Vorov OK, Livesay DR, Jacobs DJ (2009) Helix/coil nucleation: a local response to global demands. Biophys J 97:3000–3009
27. Vorov OK, Livesay DR, Jacobs DJ (2011) Nonadditivity in conformational entropy upon molecular rigidification reveals a universal mechanism affecting folding cooperativity. Biophys J 100:1129–1138
28. Lassaux P, Traore DA, Loisel E, Favier A, Docquier JD, Sohier JS, Laurent C, Bebrone C, Frere JM, Ferrer JL, Galleni M (2011) Biochemical and structural characterization of the subclass B1 metallo-beta-lactamase VIM-4. Antimicrob Agents Chemother 55: 1248–1255
29. Oelschlaeger P, Mayo SL (2005) Hydroxyl groups in the (beta)beta sandwich of metallo-beta-lactamases favor enzyme activity: a computational protein design study. J Mol Biol 350:395–401
30. Carfi A, Duee E, Galleni M, Frere JM, Dideberg O (1998) 1.85 A resolution structure of the zinc (II) beta-lactamase from Bacillus cereus. Acta Crystallogr D Biol Crystallogr 54:313–323
31. Concha NO, Janson CA, Rowling P, Pearson S, Cheever CA, Clarke BP, Lewis C, Galleni M, Frere JM, Payne DJ, Bateson JH, Abdel-Meguid SS (2000) Crystal structure of the IMP-1 metallo beta-lactamase from Pseudomonas aeruginosa and its complex with a mercaptocarboxylate inhibitor: binding determinants of a potent, broad-spectrum inhibitor. Biochemistry 39:4288–4298
32. Payne DJ, Hueso-Rodriguez JA, Boyd H, Concha NO, Janson CA, Gilpin M, Bateson JH, Cheever C, Niconovich NL, Pearson S, Rittenhouse S, Tew D, Diez E, Perez P, De La Fuente J, Rees M, Rivera-Sagredo A (2002) Identification of a series of tricyclic natural products as potent broad-spectrum inhibitors of metallo-beta-lactamases. Antimicrob Agents Chemother 46:1880–1886
33. Green VL, Verma A, Owens RJ, Phillips SE, Carr SB (2011) Structure of New Delhi metallo-beta-lactamase 1 (NDM-1). Acta Crystallogr Sect F Struct Biol Cryst Commun 67:1160–1164
34. Gordon JC, Myers JB, Folta T, Shoja V, Heath LS, Onufriev A (2005) H++: a server for estimating pKas and adding missing hydrogens to macromolecules. Nucleic Acids Res 33: W368–W371
35. Ponder JW, Case DA (2003) Force fields for protein simulations. Adv Protein Chem 66:27–85
36. Kim Y, Cunningham MA, Mire J, Tesar C, Sacchettini J, Joachimiak A (2013) NDM-1, the ultimate promiscuous enzyme: substrate recognition and catalytic mechanism. FASEB J 27 (5):1917–1927. doi:10.1096/fj.12-224014
37. King DT, Worrall LJ, Gruninger R, Strynadka NCJ (2012) New Delhi metallo-beta-lactamase: structural insights into beta-lactam recognition and inhibition. J Am Chem Soc 134:11363–11365
38. Livesay DR, Kreth KE, Fodor AA (2012) A critical evaluation of correlated mutation algorithms and coevolution within allosteric mechanisms. In: Fenton AW (ed) Methods in molecular biology. Springer, New York, pp 385–398
39. Conigrave AD, Franks AH (2003) Allosteric activation of plasma membrane receptors—physiological implications and structural origins. Prog Biophys Mol Biol 81:219–240
40. Forman BM, Umesono K, Chen J, Evans RM (1995) Unique response pathways are established by allosteric interactions among nuclear hormone receptors. Cell 81:541–550
41. Fenton AW, Alontaga AY (2009) The impact of ions on allosteric functions in human liver pyruvate kinase. Methods Enzymol 466: 83–107

Chapter 13

Towards Comprehensive Analysis of Protein Family Quantitative Stability–Flexibility Relationships Using Homology Models

Deeptak Verma, Jun-tao Guo, Donald J. Jacobs, and Dennis R. Livesay

Abstract

The Distance Constraint Model (DCM) is a computational modeling scheme that uniquely integrates thermodynamic and mechanical descriptions of protein structure. As such, quantitative stability–flexibility relationships (QSFR) that describe the interrelationships of thermodynamics and mechanics can be quickly computed. Using comparative QSFR analyses, we have previously investigated these relationships across a small number of protein orthologs, ranging from two to a dozen [1, 2]. However, our ultimate goal is provide a comprehensive analysis of whole protein families, which requires consideration of many more structures. To that end, we have developed homology modeling and assessment protocols so that we can robustly calculate QSFR properties for proteins without experimentally derived structures. The approach, which is presented here, starts from a large ensemble of potential homology models and uses a clustering algorithm to identify the best models, thus paving the way for a comprehensive QSFR analysis across hundreds of proteins in a protein family.

Key words Protein flexibility, Homology modeling, Distance constraint model, Quantitative stability–flexibility relationships

1 Introduction

Compared to sequence space, the relatively small number of quality X-ray crystal structures limits the ability to completely characterize protein flexibility properties across families and superfamilies. In the past we have performed comparative analyses of protein flexibility for a number of protein systems, including bacterial periplasmic binding homologs [1], oxidized thioredoxin [2] and β-lactamase protein families (in review). Thus far, our largest comparative flexibility study has focused on a dataset of 12 ortholog structures; however, a truly comprehensive analysis could require hundreds of structures to describe the flexibility properties of an entire protein family. Unfortunately, there is less than 100 out of 3,900 SCOP families that have 25 or more distinct orthologs with

Dennis R. Livesay (ed.), *Protein Dynamics: Methods and Protocols*, Methods in Molecular Biology, vol. 1084,
DOI 10.1007/978-1-62703-658-0_13, © Springer Science+Business Media New York 2014

experimentally solved structures. As such, a large-scale flexibility analysis on the scale of dozens to 100+ structures will require homology models to fill-in these structure gaps (*see* **Note 1**).

Our previous analyses have shown that mDCM can detect variations in QSFR properties due to subtle structural perturbations introduced by single point mutations [3, 4]. Hence, the key to reproduce accurate QSFR predictions will depend critically on having good homology models so that predicted differences can be trusted to be real. Herein, we present a protocol using human C-type lysozyme as an example that achieves this goal. Starting from 65 human lysozyme homology models constructed from 13 different templates, we use a clustering/filtering algorithm to identify a subset of the models that accurately reproduce the expected flexibility properties. The key difficulty of this problem is that good homology models must be identified a priori without comparisons to the actual structure, which is not available. As a first step toward quantifying homology model quality, we employ a clustering approach that segregates putative structures in terms of QSFR properties. Filtering on the QSFR properties that are most physically reasonable is then applied to screen out poor models, thereby boosting confidence levels in the quality of the remaining models. We test the approach by comparing clustered QSFR properties with those from "held back" real human structures. While a priori identifying the best cluster has not yet been implemented, we show statistically significant results that clearly indicate that homology model structures clustered based on structure similarity, thermodynamic and dynamic properties drastically improve predictions. Moreover, average QSFR quantities calculated over all the identified good homology models successfully reproduced X-ray structures' average QSFR properties. Consequently, this is an important step towards a comprehensive QSFR analysis for hundreds of proteins.

2 Methods

2.1 A Brief Overview of the Distance Constraint Model

The Distance Constraint Model (DCM) is used for simultaneously calculating thermodynamic and mechanical properties of proteins. The DCM is based on a free energy decomposition scheme combined with constraint theory, such that microscopic interactions in the protein are represented as mechanical distance constraints [5, 6]. Each distance constraint is associated with an enthalpic and entropic contribution. The microscopic interactions within the minimal DCM (mDCM) include: covalent bonds, hydrogen bonds and torsional-forces. Covalent bonds are quenched, whereas the other interactions fluctuate. Starting with a native protein structure, an ensemble of conformations is generated from the fluctuating constraints. However, complete enumeration of the partition function is impossible. As such, the mean field free energy of a macrostate, which is defined by the number of H-bond and torsion forces present, is computed using:

$$G(N_{hb}, N_{nt}) = U(N_{hb}) + u(N_{hb}^{max} - N_{hb}) + v(N_{nt}) - T\{S_{conf}(N_{hb}, N_{nt,}|\gamma, \delta_{nat}, \delta_{dis}\} + S_{mix}(N_{hb}, N_{nt})$$

where U is the intramolecular H-bond energy, u is an average H-bond energy to solvent, v is the energy of a native-like torsion angle, $S_{conf}(N_{hb},N_{nt})$ is the conformational entropy and $S_{mix}(N_{hb}, N_{nt})$ is the mixing entropy of the macrostate associated with the number of ways of distributing N_{nt} native-torsions and N_{hb} H-bonds within the protein. As a consequence of integrating mechanical and thermodynamic concepts, accurate flexibility characteristics of a given protein structure is calculated over an ensemble of possible constraint topologies that are appropriately thermodynamically weighted.

2.2 Homology Model Preparation

In this work, we focus on the ability of the mDCM to reproduce QSFR descriptions of human C-type lysozyme models. Starting with 13 different (nonhuman) lysozyme ortholog structures selected from SCOP [7], each is used as a template for the human sequence. The 13 template structures have a wide range of sequence identity to the human lysozyme varying from 37.6 to 77.7 %. MODELLER [8] is used to construct five models per template using otherwise default settings. Hydrogen atoms are added to the model structures and minimized followed by structure minimization using Amber99 force field. To ensure proper ionization, the H++ server [9] is used to add hydrogen atoms to the structures as expected at pH 2.7 based on calculated pK_a values. Other structural details are provided in Table 1. The same structure preparation is applied to seven human crystal structures, which are used to assess the quality of the model predictions.

2.3 Model Parameterization

Model parameter values $\{u, v, \delta_{nat}\}$ are determined by fitting to experimental heat capacity curves from differential scanning calorimetry (DSC) [5, 6]. Once parameterized, the DCM can calculate a number of quantitative stability–flexibility relationship (QSFR) properties, which are thermodynamically averaged over the free energy basin. Each model and human X-ray structure is fit to the same human C_p curve obtained from differential scanning calorimetry [10], which provides empirical constraints that the mDCM leverages. As an example, best-fit curves for the five *rainbow trout* models are shown in Fig. 1. Other model structures exhibit similar heat capacity fit trends, although there are slight differences in parameters (cf. Table 1). Interestingly, the least squares fitting error is not correlated to homology model accuracy, highlighting the importance of other structural features that contribute towards prediction of accurate thermodynamic and mechanical features (*see* **Note 2**).

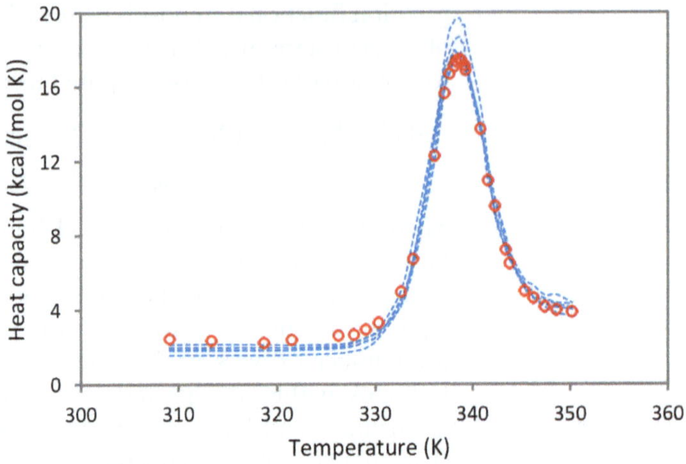

Fig. 1 Five homology models constructed from *rainbow trout* template are fit to human lysozyme heat capacity curve. The *curve* represented by *red dots* describes the data from DSC experiment, and the five *dashed-blue curves* show typical fits

Table 1
Structural template used to construct the human lysozyme homology models

Organism	Template PDB	Resolution (Å)	*R*-value	*u*	*v*	δ_{nat}
Turkey	135L	1.30	0.189	−1.58 ± (13.6 %)	−0.38 ± (69.7 %)	0.86 ± (75.7 %)
Northern bobwhite	1DKJ	2.00	0.177	−2.01 ± (26.2 %)	−0.60 ± (30.0 %)	1.00 ± (48.3 %)
Domestic silkworm	1GD6	2.50	0.181	−1.85 ± (7.6 %)	−0.50 ± (15.2 %)	0.96 ± (29.9 %)
Chicken	1HEL	1.70	0.152	−1.72 ± (13.7 %)	−0.60 ± (18.5 %)	0.50 ± (53.3 %)
Helmeted guineafowl	1HHL	1.90	0.170	−1.69 ± (8.2 %)	−0.40 ± (51.3 %)	0.86 ± (40.8 %)
Tasar silkworm	1IIZ	2.40	0.231	−2.12 ± (13.4 %)	−0.73 ± (18.1 %)	0.96 ± (44.8 %)
House mouse	1IVM	–	–	−2.38 ± (8.4 %)	−0.72 ± (12.8 %)	1.14 ± (11.8 %)
Ring-necked pheasant	1JHL	2.40	0.214	−1.90 ± (24.5 %)	−0.61 ± (19.1 %)	0.75 ± (54.0 %)
Echidna	1JUG	1.90	0.170	−1.85 ± (8.7 %)	−0.33 ± (44.0 %)	1.04 ± (38.7 %)
Rainbow trout	1LMN	1.80	0.174	−1.90 ± (9.7 %)	−0.44 ± (30.0 %)	1.07 ± (29.3 %)
Dog	1QQY	1.85	0.178	−1.92 ± (5.4 %)	−0.50 ± (40.1 %)	0.89 ± (30.2 %)
Horse	2EQL	2.50	0.234	−1.81 ± (18.0 %)	−0.64 ± (24.8 %)	0.82 ± (69.4 %)
Japanese quail	2IHL	1.40	0.165	−1.67 ± (12.6 %)	−0.41 ± (44.9 %)	0.96 ± (59.5 %)

Five models are built from each template. The average and percent variation in $\{u, v, \delta_{nat}\}$ of homology models from each template are also reported

2.4 Assessment

Instead of comparing each model structure to a single human structure that could introduce biases due to a particular conformational state, we compare the models to a background profile established from seven different human X-ray structures. We calculate average properties and define a range of likely values based on the observed fluctuations therein. That is, we are asking the question: "When is an observed QSFR property within the range of expected values, and when is it not?" This approach is the same as we established in an earlier work [4]. Any model QSFR metric within ±1 standard deviation (i.e., $\pm 1\sigma$) of X-ray structures' QSFR baseline is considered to be a "good prediction", at a given residue position. A prediction value falling beyond $\pm 1\sigma$ defines "poor prediction" for that QSFR metric.

3 Results and Discussion

3.1 The Quality of Homology Model QSFR Predictions

For each of the 65 models, Fig. 2a compares the percentage of residues within $\pm 1\sigma$ of the backbone flexibility profile to the sequence identity of the template used. Backbone flexibility is quantified by a Flexibility Index (FI), which quantifies how flexible (or rigid) backbone residues are (*see* **Note 3**). These comparisons show that better agreement between models and X-ray structures result in a better prediction of FI. Despite the overall positive relationship between model and X-ray structure similarity, there remains models that are false-positives and false-negatives. For example, the best human model prediction arises from *rainbow trout* (1LMN) resulting in highest QSFR prediction accuracy of 67 %, while the accuracy of the other four 1LMN models are lower, down to 42 %. Another important observation is the poor FI prediction by models derived from *house mouse* (1IVM), which is the only NMR structure template in our dataset (indicated by the blue dots). The sequence identity of this template is 78 %, but the average prediction accuracy is approximately 30 %, which highlight the large differences between NMR and X-ray structures. Across the whole dataset, only eight models have accuracies better than 60 %, and perhaps more critically, those models come from several different templates.

The situation is similar when comparing the models to the real (held-back) structures. Structures with more similar hydrogen bond networks (*panel b*), TM-Scores (*panel c*), and overall RMSD (*panel d*) also enrich good FI predictions. Unfortunately, the number of structurally similar models giving poor FI predictions is still too large. These results are not unexpected since subtle changes in the model structures can cause drastic differences in hydrogen bond interactions and strength. Note that the upper boundary in Fig. 2 (accuracy = 82 %) is defined by the best result obtained by comparing each of the original X-ray structures to the average

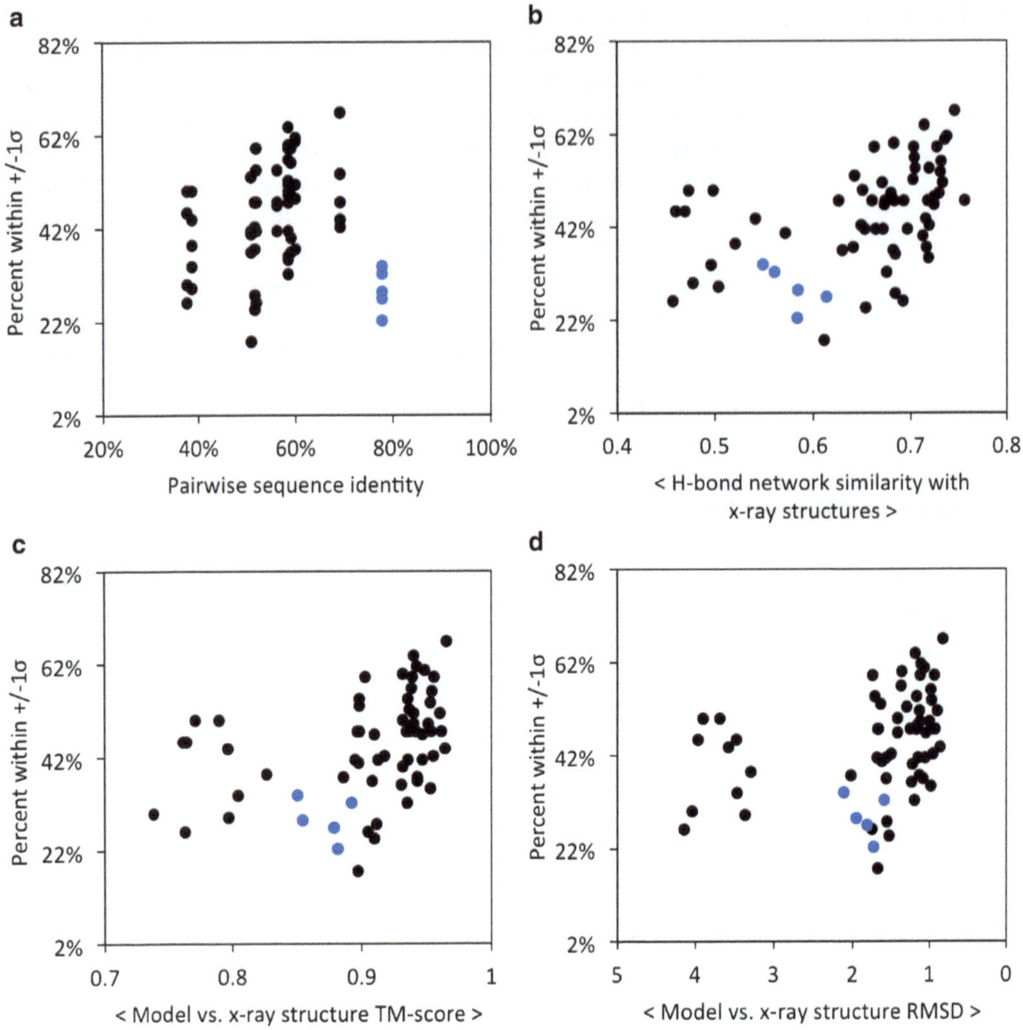

Fig. 2 Comparison of the predicted flexibility index accuracy for all 65 homology models against (**a**) pairwise sequence identity, (**b**) hydrogen bond network similarity, (**c**) TM-scores and (**d**) structure RMSD. The later properties that depend on structure are averaged over seven known X-ray crystal structures. Model structures in close agreement with X-ray structures reproduce the flexibility index with higher accuracy, although false-negatives and false-positives are also present. *Blue data points* represent NMR structures. Note that 82 % is the best score that occurred when comparing a QSFR property associated with each of the original X-ray structures against the average QSFR property taken over all seven structures

X-ray structures' backbone profile. That is, 82 % is the highest similarity between any two of the seven X-ray structures.

The above results clearly suggest that a comprehensive QSFR analysis is possible; however, this is only possible if we can filter out poor models to improve prediction correctness by boosting statistics. For example, we can use model quality assessment score that do not require comparisons to known structures to improve

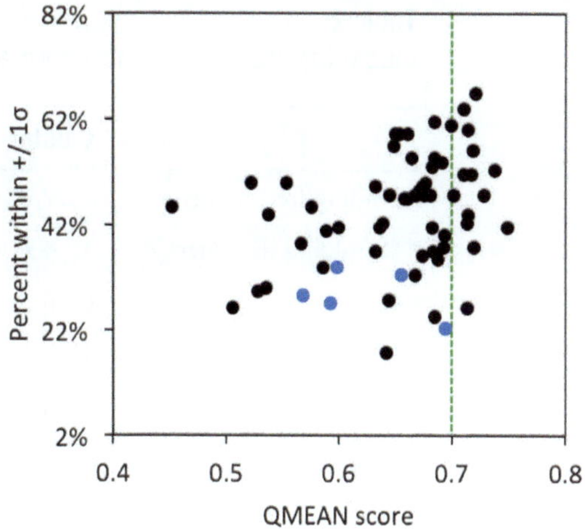

Fig. 3 Comparison of FI percent accuracy prediction. Models with high quality may not necessarily result in higher FI accuracy. *Blue data points* represent NMR structures. The *green dashed line* represents an arbitrary QMEAN threshold value at 0.7

identification of the best models [11]. To that end, we employ QMEAN [12, 13], which evaluates models based on a secondary structure interaction potential, degree of solvent exposure and other structural quantities. Figure 3 plots prediction accuracy versus the QMEAN scores, which indicates further enrichment. The vertical green line identifies an arbitrary threshold of good QMEAN scores. While the best scoring models are above the threshold, there are an unacceptable number of models with poor accuracies as well.

3.2 Expectation Maximization Clustering

The above results indicate that good flexibility predictions using models are possible; however, many models give unsatisfactory results. Moreover, there appears to be no systematic organization of the good models based on model quality or sequence/structure similarity to the target. Therefore, we employ a new clustering/filtering procedure over QSFR data consisting of a large number of heterogeneous metric types. Additionally, structure quality, sequence identity and other thermodynamic information are considered. Taken together, the key assumption of this strategy is that models with similar properties would tend to cluster together.

Clustering is done in two steps, both based on the expectation maximization (EM) algorithm (*see* **Note 4**). The first step focuses on non-QSFR properties discussed above (i.e., percent sequence identity between and QMEAN structure quality score), plus QSFR quantities that characterize the thermodynamic properties related to structure quality (i.e., the free energy barrier height between the native and unfolded basins, and the global flexibility of the native, transition and unfolded states). EM identifies three clusters, whose mean and standard deviation are summarized in Table 2.

Table 2
Clustering results from structure and thermodynamic quantities

	Cluster-1	Cluster-2	Cluster-3
Folding free energy	2.29 ± 0.84	3.57 ± 1.23	2.79 ± 1.05
Unfolding free energy	1.84 ± 0.73	3.12 ± 1.09	2.33 ± 0.96
θ_{nat}	0.75 ± 0.11	0.92 ± 0.14	0.76 ± 0.13
θ_{trans}	1.06 ± 0.15	1.31 ± 0.20	1.10 ± 0.21
θ_{dis}	1.70 ± 0.22	2.07 ± 0.16	1.70 ± 0.30
QMEAN scores	0.67 ± 0.04	0.68 ± 0.03	0.54 ± 0.04
Sequence identity	0.60 ± 0.09	0.58 ± 0.05	0.39 ± 0.01

The values represent average ± standard deviation of data points for the given quantity belonging to a defined cluster. Cluster-2 with best average QMEAN score with least standard deviation is selected. θ_{nat}, θ_{trans}, and θ_{dis} correspond to protein's intrinsic flexibility in native, transition, and disordered states respectively

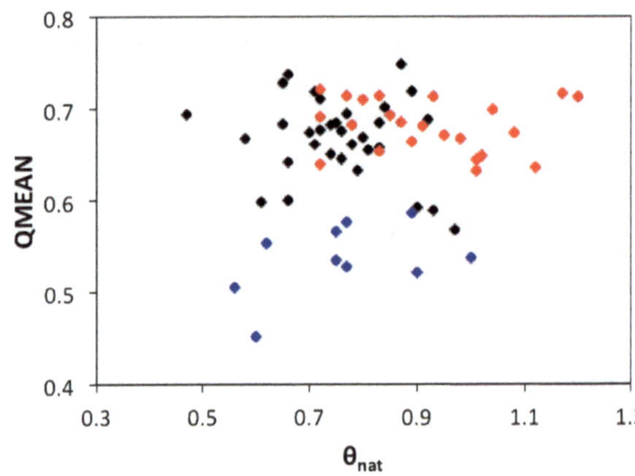

Fig. 4 Clustering using structural and thermodynamic quantities. Clusters are represented by different colors. Models clustered in *red* (cluster-2) are selected for further analyses

Figure 4 plots the QMEAN scores versus θ_{nat}, which describes the global flexibility of the native structure. The three clusters are color-coded. The cluster-2 models (colored red) have the best average QMEAN score and also the smallest variation therein (0.68 ± 0.03). As such, it was selected for further analysis, resulting in 23 models that advanced to the second round of clustering.

The second step starts by calculating an all-to-all correlation for each QSFR metric. For example, the correlation between all FI vectors for all $23 \times 22/2 = 253$ pairs. This process is repeated for all of the other QSFR quantities as well (*see* **Note 5**). The whole set of QSFR correlation coefficients, plus pairwise structure

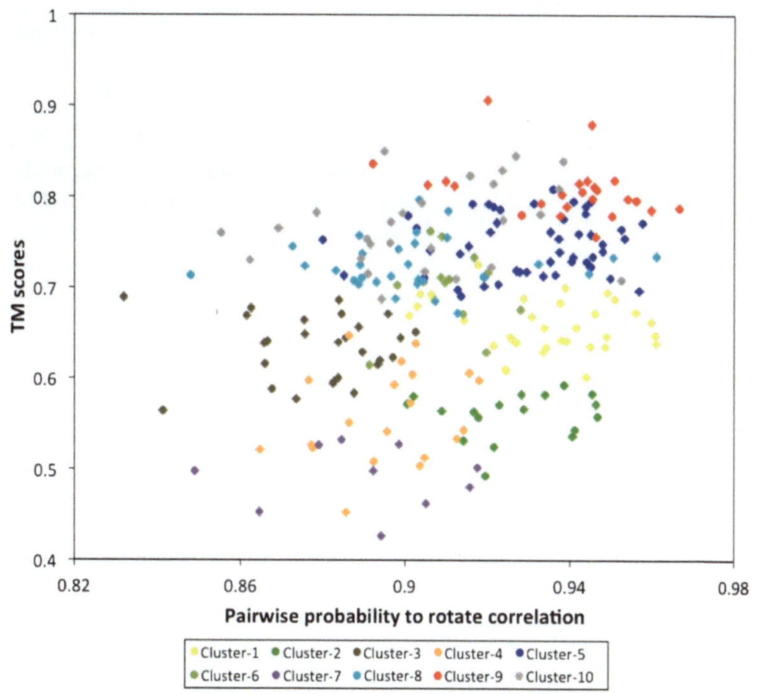

Fig. 5 Results obtained from a second round of EM clustering using structural and mechanical quantities. Clusters are represented by different colors. Model pairs clustered in *red* (cluster-9) are the final filtered models

similarities calculated using RMSD and TM-scores [14], are again clustered using the EM algorithm. Figure 5 shows the resultant clusters from one view of the data, where TM-scores are plotted versus the probability of a backbone torsion angle to rotate. Cluster-9 (shown in red) has the highest average TM-score, and also the QSFR quantities are well conserved therein. As demonstrated next, these 18 models constitute a significant enrichment of models that reproduce the flexibility profiles of the known X-ray structures. Figure 6 summarizes the clustering workflow described above.

3.3 Model Enrichment by Clustering

To compare how well the above cluster of models performs, we compare the average accuracy therein with how well we could have done without clustering. That is, we could have simply used QMEAN to identify the best model, or some set of best models. We compare the accuracy of the QSFR descriptions provided by the EM cluster to: (1) the single best QMEAN model and (2) to the group of five best QMEAN models. As before, accuracy is determined by comparison to the profile developed from the seven human X-ray structures.

Figure 7 plots the results of the three different model sets for eleven different QSFR metrics describing backbone properties.

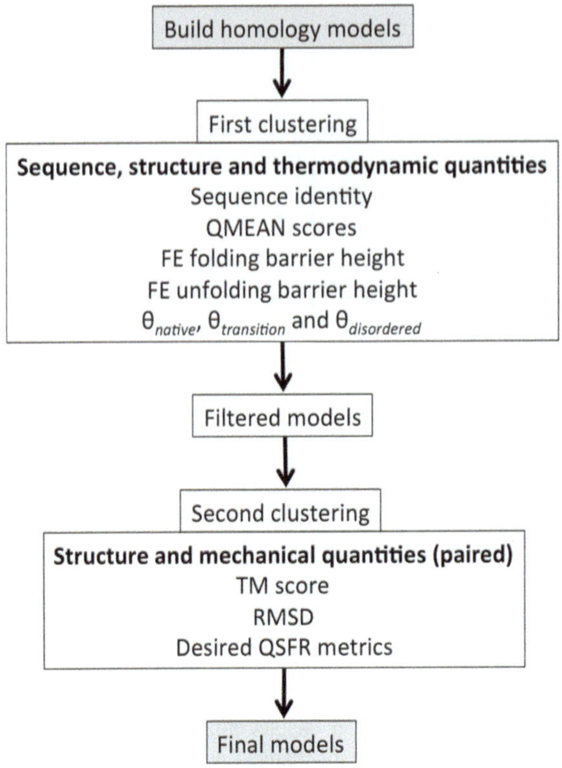

Fig. 6 Expectation maximization clustering workflow to filter best homology models

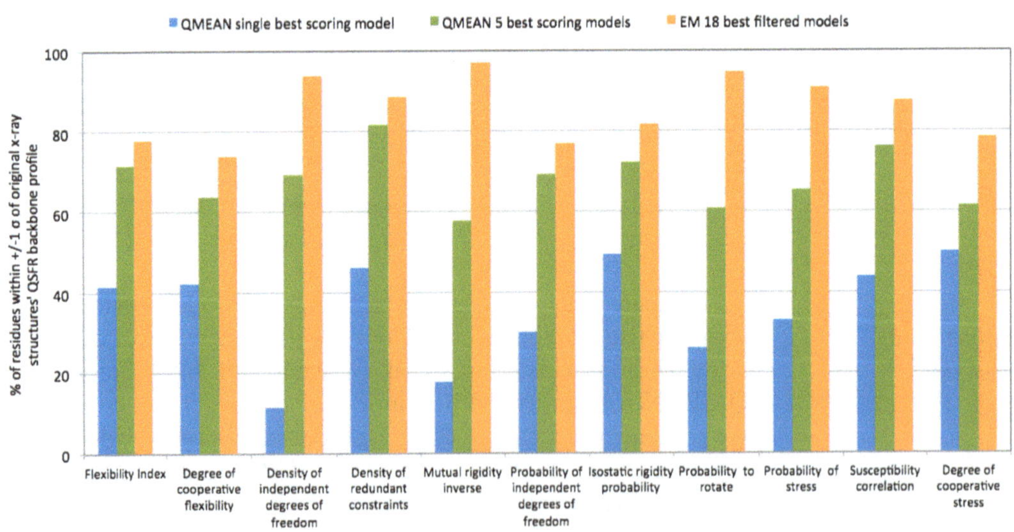

Fig. 7 Comparison of protein backbone QSFR metric accuracy levels for different model sets. Average QSFR quantities obtained from EM 18 best filtered models have higher agreement with X-ray structures' QSFR quantities as compared to QMEAN best and five-best models

Interestingly, the single-best QMEAN model fails to give an accurate prediction in most cases, but averaging over five best QMEAN models significantly improves accuracy. In all cases, averaging over the various EM clusters further improves accuracy, by as much as 20 % in some instances. Taken together, these results suggest that clustering significantly improves the description of backbone flexibility profiles using homology models. Excitingly, note that the EM-set averaged FI accuracy is 78 %, which is very close to the 82 % upper bound identified from the seven X-ray structures (*see* **Note 6**).

Figure 8 compares the five cooperativity correlation plots for QMEAN-5 and EM sets, which are compared to the real plots derived by averaging over the seven X-ray structures. Both sets do a reasonable job of reproducing the experimental structures, but the EM set appears more similar. A key problem with this visual analysis is it neglects variation within real structures. As such, we make profile comparisons, although this time each pixel data point represents a residue pair (versus a single residue). These results are shown in Fig. 9, which confirms EM clustering improves prediction accuracy in all cases. The accuracy in all of the EM clusters is greater than 70 % accuracy, whereas none of the QMEAN-5 sets reach that level. A scatter plot of CC metric values provides insight on correlation distribution (Fig. 10). The EM set is clearly much sharper and more accurate than the QMEAN-5 set.

3.4 Concluding Remarks

The presented method clearly demonstrates that EM clustering based on the model output and structural details leads to an enrichment of models able to reproduce the flexibility profiles of the real structures. The sole remaining point is to determine how to select the final cluster. In the presented work, the second round of clustering identified ten clusters, and the final (Cluster-9) was chosen based on that set best that reproduced the X-ray profile. That is, we knew the correct answer and picked the solution that was closest. While there were some indicators that this cluster was the best (e.g., highest average intra-cluster TM score), we have not yet determined a robust method to identify the best cluster. Nevertheless, the presented method represents a substantial improvement in the ability to use homology models to describe protein flexibility properties, thus making our goal of a comprehensive analysis of a 100+ proteins with a given family that much closer.

4 Notes

1. Within the DCM framework, flexibility and rigidity respectively quantify conformational diversity and regularity. These mechanical origins of flexibility and rigidity is linked to conformational entropy. These thermodynamic and mechanical

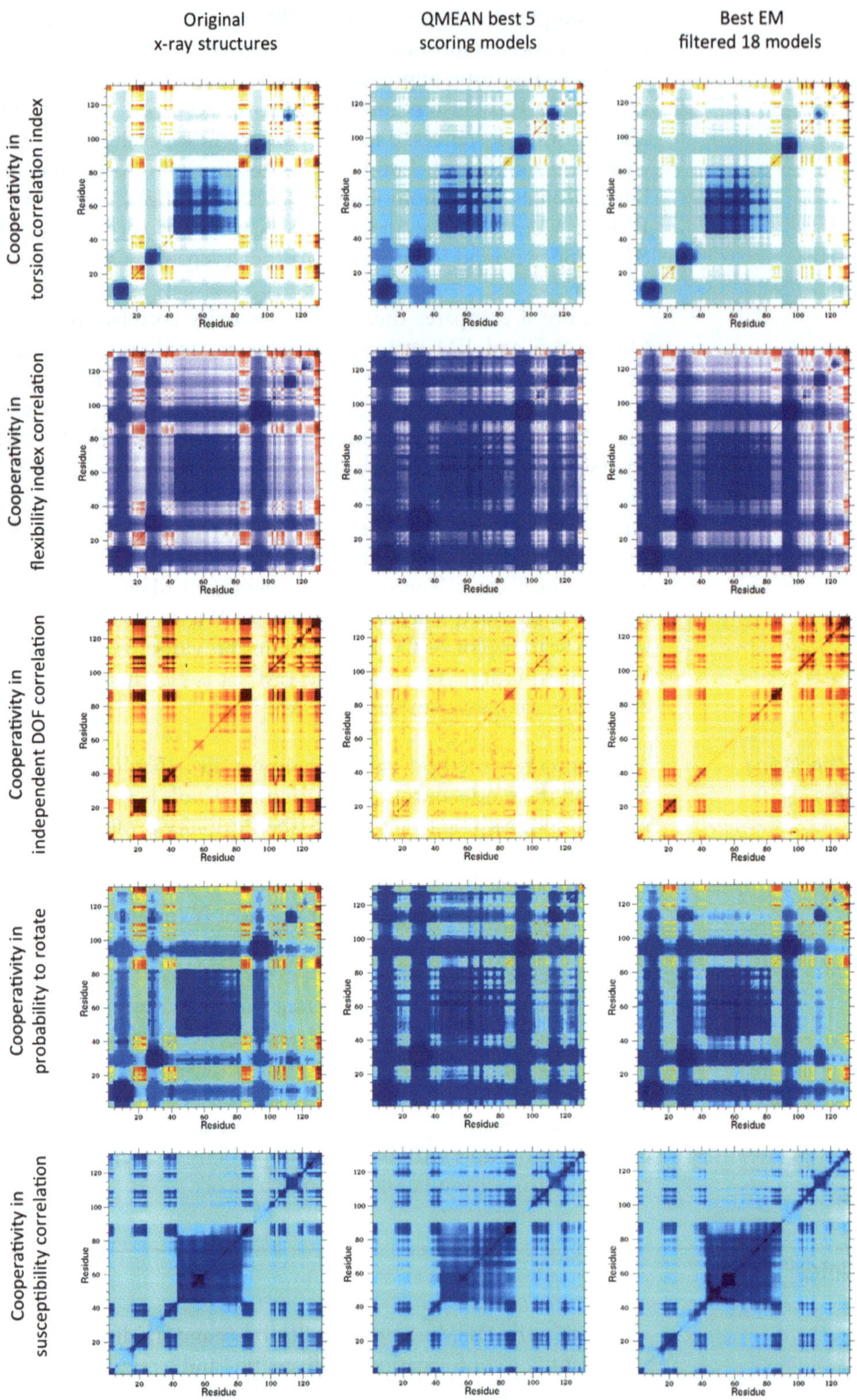

Fig. 8 Comparison of residue–residue coupled QSFR metrics. A good qualitative resemblance is observed between EM 18-best models and original X-ray structures' average QSFR cooperativity metrics

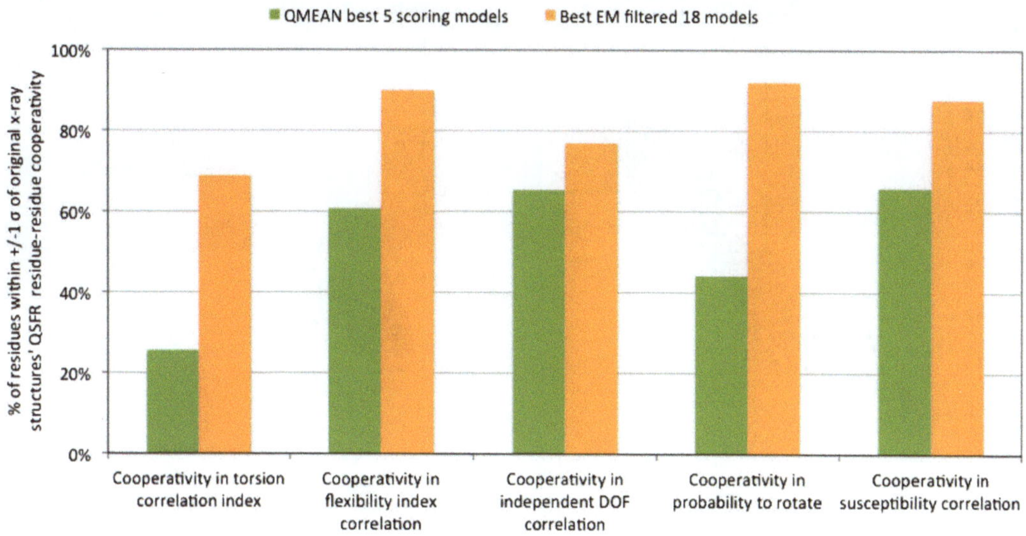

Fig. 9 Comparison of residue–residue coupled QSFR metric accuracy levels for different model sets. Average QSFR quantities obtained from EM 18 best filtered models have better precision levels

measures combine to define QSFR. To be precise, the free energy of the protein can be expressed in terms of a global flexibility order parameter θ. This parameter defines the intrinsic flexibility of the protein and is equal to the average number of disordered torsion constraints divided by the total number of residues in a protein.

2. The definition of a good homology model is difficult to describe, and fairly arbitrary within the field. Many scoring functions assess the quality of homology models based on statistical potentials and physics-based energy calculations [15, 16]. Intrinsic errors that arise in physical properties caused by poor quality model structures are difficult to characterize and control, which are the important but problematic aspects addressed in this work.

3. A frequently used QSFR metric in our analyses is the flexibility index (FI) that quantifies the degree to which a given residue deviates from being isostatic. An isostatic residue is marginally rigid, meaning there are just enough mechanical constraints due to intermolecular interactions to counter act its number of degrees of freedom to keep it rigid. Positive values of the FI represent the number of excess degrees of freedom (DOF) per rotatable dihedral angle within covalent bonds within flexible regions. Negative values of the FI represent the excess number of constraints per covalent bond within a rigid region.

4. Expectation maximization (EM) assigns a probability distribution to every data instance, which defines the probability of it belonging to each of the clusters. The algorithm can create its

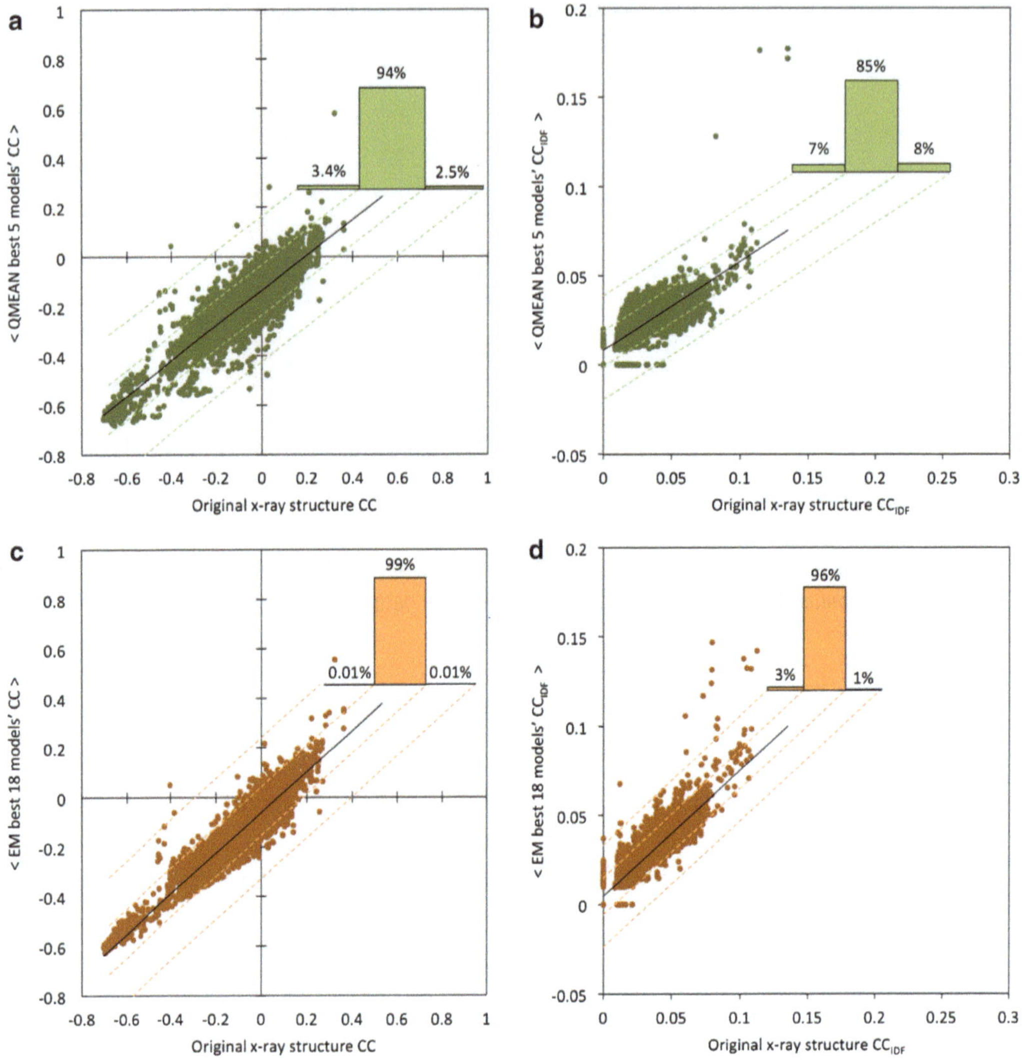

Fig. 10 (**a** and **c**) Flexibility cooperativity correlation (CC) and (**b** and **d**) CC_{IDF} values are compared against original X-ray structures. EM best 18 filtered models can reproduce much more accurate CC and CC_{IDF} properties of original X-ray structures as compared to QMEAN's best five models. QMEAN models' CC comparison with original X-ray structures exhibits wider data distribution, whereas EM models have a narrow distribution. Other metrics show similar trends. The *black line* shown is the best-fit regression across data points. The histograms are constructed by binning data points with equal intervals on *y*-axis on either side of the regression line. The interval size for binning is consistent across both CC and CC_{IDF} plots, respectively

own clusters and does not require a priori information regarding the expected number of clusters. To find the optimum number of clusters the EM algorithm cross validates and calculates the average log-likelihood. Starting with one cluster, the numbers of clusters are increased if the average log-likelihood continuously increases at each step.

5. In addition to the flexibility index, FI, the mDCM calculates a number of other backbone flexibility quantities, including: (a) mechanical susceptibility, which quantifies the fluctuations within a particular residue to be rigid or flexible over the ensemble of constraint topologies; (b) the density of independent DOF; (c) the probability of a backbone torsion angle to rotate (cf. Fig. 7). The mDCM also calculates five Cooperativity Correlation (CC) metrics that quantify different types of residue–residue couplings. For example, the flexibility cooperativity correlation identifies residues pairs that are either co-rigid, flexibly correlated, or have no mechanical coupling. Examples of the five CC metrics are provided in Fig. 8.
6. The X-ray structure upper bound varies for each of the QSFR quantities, ranging from 76 to 98 %.

Acknowledgments

This work has been partially supported by NIH R01 GM073082, S10 SRR026514, and the Department of Bioinformatics and Genomics. Key to the distance constraint model is the use of graph-rigidity algorithms, claimed in US Patent 6,014,449, which has been assigned to the Board of Trustees Michigan State University. Used with permission.

References

1. Livesay DR, Huynh DH, Dallakyan S, Jacobs DJ (2008) Hydrogen bond networks determine emergent mechanical and thermodynamic properties across a protein family. Chem Cent J 2:17
2. Mottonen JM, Xu M, Jacobs DJ, Livesay DR (2009) Unifying mechanical and thermodynamic descriptions across the thioredoxin protein family. Proteins 75(3):610–627
3. Verma D, Jacobs DJ, Livesay DR (2010) Predicting the melting point of human C-type lysozyme mutants. Curr Protein Pept Sci 11 (7):562–572
4. Verma D, Jacobs DJ, Livesay DR (2012) Changes in lysozyme flexibility upon mutation are frequent, large and long-ranged. PLoS Comput Biol 8(3):e1002409
5. Jacobs DJ, Dallakyan S (2005) Elucidating protein thermodynamics from the three-dimensional structure of the native state using network rigidity. Biophys J 88(2):903–915
6. Livesay DR, Dallakyan S, Wood GG, Jacobs DJ (2004) A flexible approach for understanding protein stability. FEBS Lett 576(3):468–476
7. Murzin AG, Brenner SE, Hubbard T, Chothia C (1995) SCOP: a structural classification of proteins database for the investigation of sequences and structures. J Mol Biol 247 (4):536–540
8. Sanchez R, Sali A (1997) Evaluation of comparative protein structure modeling by MODELLER-3. Proteins (Suppl 1):50–58
9. Gordon JC, Myers JB, Folta T, Shoja V, Heath LS, Onufriev A (2005) H++: a server for estimating pKas and adding missing hydrogens to macromolecules. Nucleic Acids Res 33(Web Server issue):W368–W371
10. Takano K, Yamagata Y, Fujii S, Yutani K (1997) Contribution of the hydrophobic effect to the stability of human lysozyme: calorimetric studies and X-ray structural analyses of the nine valine to alanine mutants. Biochemistry 36 (4):688–698
11. Cozzetto D, Kryshtafovych A, Fidelis K, Moult J, Rost B, Tramontano A (2009) Evaluation of template-based models in CASP8 with standard measures. Proteins 77(Suppl 9): 18–28

12. Benkert P, Kunzli M, Schwede T (2009) QMEAN server for protein model quality estimation. Nucleic Acids Res 37(Web Server issue):W510–W514
13. Benkert P, Schwede T, Tosatto SC (2009) QMEANclust: estimation of protein model quality by combining a composite scoring function with structural density information. BMC Struct Biol 9:35
14. Xu J, Zhang Y (2010) How significant is a protein structure similarity with TM-score = 0.5? Bioinformatics (Oxford, England) 26(7): 889–895
15. Gniewek P, Leelananda SP, Kolinski A, Jernigan RL, Kloczkowski A (2011) Multibody coarse-grained potentials for native structure recognition and quality assessment of protein models. Proteins 79(6):1923–1929
16. Gopal SM, Klenin K, Wenzel W (2009) Template-free protein structure prediction and quality assessment with an all-atom free-energy model. Proteins 77(2):330–341

Chapter 14

Using the COREX/BEST Server to Model the Native-State Ensemble

Vincent J. Hilser and Steven T. Whitten

Abstract

Protein structures under normal conditions exist as ensembles of interconverting, transient microstates. A computer algorithm known as COREX/BEST (*B*iology using *E*nsemble-based *S*tructural *T*hermodynamics) was developed to model microstate structures and describe the native ensembles of proteins in statistical thermodynamic terms. This algorithm has been tested extensively and validated through experimental comparisons examining a range of biophysical and functional phenomena, such as structural cooperativity, pH-dependent stability, and cold denaturation. Here, we describe a Web-based implementation of the COREX/BEST algorithm, called the COREX/BEST Server, and demonstrate how to use this online resource to characterize the structural and thermodynamic properties of the native protein ensemble.

Key words Ensemble, Dynamics, Temperature, Electrostatics, pH

1 Introduction

Protein macromolecules often couple structural rearrangements to biological function [1–3] and show significant conformational heterogeneity under normal solution conditions [4–10]. Characterizations of these structural fluctuations are thus needed to understand the physicochemical properties of proteins and the molecular origins to their biological roles. In this chapter, we discuss a statistical thermodynamic model of the protein conformational ensemble known as COREX/BEST [11, 12]. This model uses an algorithm to compute a distribution of states a protein structure may populate under a given set of conditions (e.g., pH and temperature) to provide an ensemble description of protein dynamics. By calculating state probabilities and monitoring quantitatively the population distributions within an ensemble, COREX/BEST has shown an ability to reproduce diverse phenomena such as regional stability variations within protein structures [13–15], long-range intramolecular signaling [16, 17], electrostatic

Dennis R. Livesay (ed.), *Protein Dynamics: Methods and Protocols*, Methods in Molecular Biology, vol. 1084,
DOI 10.1007/978-1-62703-658-0_14, © Springer Science+Business Media New York 2014

contributions to cooperative transitions [18, 19], non-cooperative cold denaturation [20, 21], and functional adaptation [22]. A Web-based implementation of the COREX/BEST algorithm [23] is freely available to the scientific community and may be accessed by visiting http://best.bio.jhu.edu. A demonstration of using the COREX/BEST Server to characterize the structural and energetic properties of a protein conformational ensemble is presented here.

Detailed descriptions of the COREX/BEST algorithm can be found elsewhere [11, 12]. Briefly, this algorithm uses a user-supplied structure (e.g., a PDB file) as a template to generate a large number of partially folded "microstates" that possess a dual structural-thermodynamic character [21]. Each microstate is generated by treating local structural fluctuations as folding-unfolding reactions that occur in an otherwise folded and native-like protein. By combinatorial unfolding of "folding units", which are contiguous blocks of residues that fold and unfold independently, and an incremental shift in the boundaries of the folding units, an exhaustive enumeration of partially folded states is achieved. The relative Gibbs free energy for each microstate i, ΔG_i, is determined by structure-based parameterization of the intrinsic energetics: $\Delta C_{p,i}$, ΔH_i, and ΔS_i (*see* **Note 1**). This parameterization gives temperature-dependent stabilities for the microstates that can be converted to probabilities (P_i) by the Boltzmann relation,

$$P_i = \frac{K_i}{\sum_i K_i}, \tag{1}$$

where the statistical weight of each microstate (K_i) is determined by the relative Gibbs free energy of that microstate ($K_i = e^{-\Delta G_i/RT}$, where R is the gas constant and T is absolute temperature), and the summation is over all ensemble states. Figure 1 shows a representative ensemble calculated for the *staphylococcal* nuclease protein [24].

The statistical formula to calculate state probabilities (Eq. 1) can be expanded to account for additional system perturbations [25], such as proton-binding energies,

$$P(\mathrm{pH})_i = \frac{K_i \cdot \prod_j \left(1 + 10^{\mathrm{p}K_{\mathrm{a},i,j} - \mathrm{pH}}\right)}{\sum_i \left(K_i \cdot \prod_j \left(1 + 10^{\mathrm{p}K_{\mathrm{a},i,j} - \mathrm{pH}}\right)\right)}, \tag{2}$$

where $\mathrm{p}K_{\mathrm{a},i,j}$ is the $\mathrm{p}K_\mathrm{a}$ value of residue j in microstate i (*see* **Note 2**). The addition of the proton-binding energies to the state probabilities yields a structural ensemble sensitive to pH changes and can be used to investigate electrostatic contributions to protein energetics [18, 19]. Currently, the COREX/BEST Server can perform the following tasks: (1) calculate a conformational ensemble based upon an input protein structure, (2) determine the temperature-sensitivity of the computed ensemble, and (3) determine the

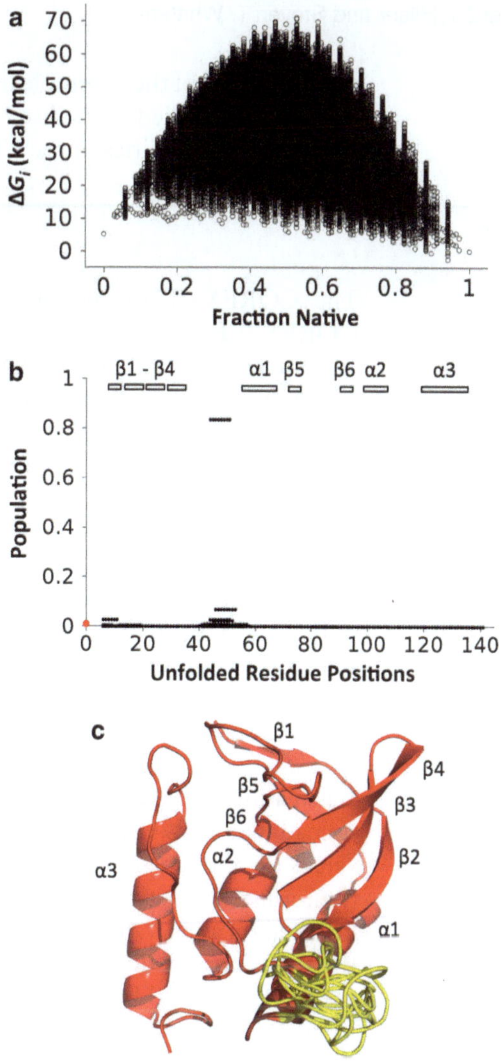

Fig. 1 The nuclease conformational ensemble as calculated by the COREX/BEST Server. Calculations in this figure used 25 °C as the temperature setting and the PDB file 1STN [24]. (**a**) Relative stability (ΔG_i) of each microstate plotted as a function of the fraction native (number of residues in folded segments/total number of residues). Each point corresponds to a particular microstate, and the ΔG values were calculated relative the fully folded state (i.e., the relative stability of the fully folded state in the ensemble was 0). (**b**) Relative populations (P_i) of the 20 microstates with highest computed probability. The *filled black circles* represent the positions of unfolded residues in a microstate. The fully folded native state, which has no unfolded residues, is shown by the *red circle* located at residue position = 0. The *line of black circles* stretching uninterrupted from residue position 6 to residue position 141 represents the fully unfolded state, which had a computed probability of $P = 0.00015$ at this temperature. The locations of secondary structural elements among the residue positions are as indicated. (**c**) Cartoon representation of the most probable microstate of the nuclease ensemble. Segments of protein colored *red* indicate regions that would be folded and regions colored *yellow* would be unfolded (positions 44–51). To illustrate the concept that the unfolded segment freely samples accessible conformational space, this region was modeled in the figure as multiple flexible loops

pH-sensitivity of the ensemble if pK_a values for the ionizable protein groups are supplied. Our discussions in this chapter are limited to the current capabilities of the COREX/BEST Server.

2 Materials

The COREX/BEST Server [23] provides online access to the COREX/BEST algorithm [12]. A computer with a Web browser and internet capabilities is required. The Web site, which can be found at http://best.bio.jhu.edu, was designed to work with standards compliant browsers (e.g., Mozilla's Firefox). Most modern browsers, however, should be sufficient. Adequate data storage space is required to download the results of COREX/BEST calculations to a user's computer.

2.1 Input Structure for Generating an Ensemble

The calculations performed by the COREX/BEST Server require uploading a protein structure file written in plain text and in standard PDB file format [26]. Refer to **Note 3** for common errors encountered by COREX/BEST calculations due to protein structure files that are not properly prepared.

2.2 Native pK_a Values to Calculate pH Sensitivity

Calculating the pH sensitivity of an ensemble using Eq. 2 requires uploading a file written in plain text that contains a list of the pK_a values for the ionizable groups in the uploaded PDB structure file (i.e., a list of "native" pK_a values). Each ionizable group should be listed on a separate line in this file, with each line containing: (1) the three letter code for the amino acid type that has the ionizable group, (2) the residue number identifying the amino acid position in the structure file, and (3) the pK_a value of the ionizable group in the uploaded structure. For example, consider a protein containing a ε-amino group that belongs to a lysine at residue position 25 and ionizes under native conditions with a pK_a value of 11.55. The uploaded file of pK_a values should contain a line specifying "LYS 25 11.55", as well as additional lines of text for each additional ionizable group in the structure file. Native pK_a values for the carboxyl and amino ends of a protein chain should not be listed in the uploaded file; the COREX/BEST algorithm assumes that the C-and N-termini ionize with the same pK_a values as model compounds, (i.e., 3.5 and 7.4, respectively) in all ensemble states and thus the two end groups do not contribute to the pH sensitivity of the ensemble in COREX/BEST calculations.

3 Methods

To perform ensemble calculations using the COREX/BEST Server, a user must first register with the Web site. Registration involves providing a name, a valid e-mail address, and a login password. The e-mail address is required so that users can be contacted when calculations have completed. Following successful registration, a typical user session consists of the following: (1) the user logs into the COREX/BEST Server through their Web browser, (2) the user defines workspaces for each project, (3) the user uploads protein structure file(s) to appropriate workspace(s), (4) the user submits jobs (i.e., calculations) to the COREX/BEST Server, and (5) the user may download all or a subset of the calculated data contained within that user's workspace. Below is the example of performing calculations on the *staphylococcal* nuclease protein (PDB ID 1STN [24]) using the COREX/BEST Server.

3.1 Creating a Workspace

After login to the COREX/BEST Server, a workspace can be created by clicking on the "*Create Workspace*" link on the left side of the webpage under the heading "*WORKSPACE OPTIONS*". The user is required to provide a name for this workspace and will create the workspace by clicking on the "*Create!*" button. Created workspaces can be accessed by clicking on links on the left side of the webpage under the heading "*MY WORKSPACES*". Accessing a newly created workspace, the user is presented with the options to: (1) upload a new PDB file, or (2) delete the workspace. Click on the link "*Upload New PDB*" to upload a plain text PDB file. If the file was uploaded correctly, the server displays a cartoon picture of the structure file in the workspace. Also, the user can click on an "*Upload Report*" link in the workspace that notifies a user of potential errors detected in the PDB file. For example, uploading 1STN gives the following report due to missing the first five and last eight residues in the crystallographic structure:

PDB Upload Report for 1STN on 2012/11/20, 16:55:15:
 * *SEQRES Chain 'A' contains 149 residues*
 * *missing coordinates for residues 1 thru 5, line number 323*
 * *missing coordinates for residues 142 thru 149 (end of sequence)*

Upload Recap:
 * *structure contains 2 atom/residue sequence errors or missing coordinates*
 Upload complete: coordinates for 136 residues with 1092 atoms

Warnings:
 * *2 Discrepancies in uploaded file may cause unexpected results in calculations*
 END PDB Upload Report for 1STN

3.2 Generating a Conformational Ensemble

Next, a user would calculate a conformational ensemble for nuclease based upon the 1STN structure. To do this, click on the "*Perform COREX/BEST Calculations*" link under the "*Protein Options*" heading of the workspace. The user is directed to a webpage titled "*Choosing COREX/BEST Jobs*". On this webpage, click the link "*Run the Ensemble Generator!*". Next, the user is required to specify the size of the folding unit, referred to as the "*Window*

Table 1
Estimates for fully enumerated ensemble generation

Window size	Total states	Estimated time (h:min:s)	MB data
6	37,748,724	1 days 17:22:40	2,208
7	5,242,866	5:44:48	306
8	1,179,632	1:17:34	69
9	393,198	0:25:51	23

Size", which is used to calculate the ensemble. Briefly, the window size is the number of contiguous residues that constitute a folding unit. A window size of 8 means that the protein is folded and unfolded in blocks of eight residues. Smaller window sizes create larger ensembles, since a greater number of folding units can fit along a given chain length for minimally sized folding units. In general, smaller window sizes are good, but larger ensembles require longer run times on the server and COREX/BEST jobs are limited to 24 h each. Any job estimated to run longer than 1 day will not be performed. For reference, the PDB file upload report also contains a table that lists the total number of states (i.e., microstates) that would be generated and estimated run times for ensembles calculated using different window sizes. The table generated for 1STN in its upload report is reproduced in Table 1.

In our experience, COREX/BEST ensembles seem to be able to reproduce experimental results with better accuracy when window sizes of eight to ten residues are used to generate the microstates. For proteins larger than 150 residues in length, using a recommended window size can create ensembles that exceed the run-time limits of the Server. To handle large proteins, a Monte Carlo option is available when generating ensembles [16]. Using this option, ensemble microstates are randomly generated and selected or "sampled" using standard Monte Carlo methods [27] via the computed state statistical weight (Eq. 1). Briefly, microstates with higher probabilities are more likely to be sampled in a Monte Carlo generated ensemble (*see* **Note 4**).

The user is also required to specify a "*Minimum Window Size*" for an ensemble. Since a protein chain isn't necessarily an integer multiple of the user-defined window size, residues at the C-terminal end often don't fit neatly into a standard folding unit. For example, the 1STN structure contains 136 residues. If the user chooses a window size of 9, this would mean that 15 folding units could be overlaid on the first 135 residue positions of the chain, with a C-terminal residue (i.e., the 136th residue) left by itself. For any protein, the C-terminal "left over" residues are given their own folding unit if their number is equal to or exceeds the minimum window size, otherwise those residues are added to the preceding folding unit. This parameter is also needed for the N-terminal end, as the folding unit boundaries are incrementally shifted one residue at a time toward the C-terminal end to exhaustively generate the ensemble microstates, leaving "left over" N-terminal residues [12]. The incremental shifts in folding unit boundaries are used to remove any boundary-related bias in COREX/BEST results.

3.3 Determining Entropy Weighting

The structure-based energy function utilized in COREX/BEST is coarse and provides a rough estimate of the thermodynamic parameters for each microstate (i.e., ΔH_i, $\Delta C_{p,i}$, and ΔS_i). Coarse approximation of the conformational entropy of a microstate ($\Delta S_{\text{conf},i}$), which represents the number of conformational variations that a particular energetic state can occupy, is especially troubling for COREX/BEST simulations. Variations of ±5 % in the computed ΔS_{conf} term can cause shifts in the unfolding temperature by ~20 °C [12]. The conformational entropy term of each microstate is thus normalized using an entropy-weighting factor,

$$\Delta S_i = \Delta S_{\text{solv},i} + W \cdot \Delta S_{\text{conf},i}, \tag{3}$$

where ΔS_{solv} is the solvent entropy (owing to water ordering at the protein surface) and W is the applied weighting factor, typically a value between 0.95 and 1.05. In practice, a common entropy-weighting factor is applied to each microstate of an ensemble. The COREX/BEST algorithm determines an appropriate weighting factor by calculating the value of W that is needed to set the energetic separation between the native and fully unfolded states in the simulation as equal to an experimentally determined free energy difference. For example, the free energy difference between native and unfolded nuclease has been measured to be ~5 kcal/mol at 25 °C via guanidine hydrochloride-induced unfolding [28]. To determine an appropriate weighting factor to use in COREX/BEST calculations with 1STN, the user would click on the link "*Perform COREX/BEST Calculations*" under the "*Protein Options*" heading of the 1STN workspace. The user will be directed to a webpage titled "*Choosing COREX/BEST Jobs*". On this webpage, click on the link "*Determining the Entropy Weighting Factor*". Next, the user is asked to specify the overall free energy difference measured and the temperature at which the measurement was made. For 1STN, a value of 5.0 would be input into the "*Target Delta G*" field, and 25 °C entered for the temperature. Next, click on "*Calc. W*". For 1STN, COREX/BEST calculates that a value of 0.968 should be used for the weighting factor, W.

3.4 Calculating the Temperature-Dependent Stability of an Ensemble

To test the ability of the COREX/BEST algorithm to accurately describe a conformational ensemble, this algorithm was initially designed to reproduce regional stability variations within protein structures as observed by site-specific hydrogen exchange data [13–15]. As demonstrated [11], the propensity of any region in a protein to undergo structural fluctuations can be determined from the microstate probabilities (Eq. 1). Defined as the residue stability constant, k_f, the local stability of each residue for the native fold is calculated as the ratio of the summed probabilities of all microstates in the ensemble in which a particular residue *j* is in a folded conformation, $P_{f,j}$, to the summed probability of all microstates in which residue *j* is in an unfolded (or nonfolded) conformation, $P_{nf,j}$,

$$k_{f,j} = \frac{P_{f,j}}{P_{nf,j}}. \tag{4}$$

Residue stability constants for 1STN can be calculated using the COREX/BEST server by the following: After generating an ensemble for 1STN and determining an appropriate entropy weighting (Subheadings 3.2 and 3.3), click on the link "*Perform COREX/BEST Calculations*" under the "*Protein Options*" heading of the 1STN workspace. The user will be directed to a webpage titled "*Choosing COREX/BEST Jobs*". On this webpage, click on "*Calculate the Residue Specific Stability Constants*". Next, the user will need to specify the entropy weighting factor and the temperature (in °C) to use in the calculation. Also, the user will need to specify which previously generated ensemble should be used. Ensembles are classified by the parameters that were defined in the generation step, namely, window size, minimum window size, and if Monte Carlo sampling was employed. Shown in Fig. 2 are the stability constants calculated for 1STN using a temperature setting of 25 °C, an entropy weighting factor of 0.968, and an ensemble generated with a window size of 8 and a minimum window size of 4. High stability constants identify residues that are folded in the majority of the most probable microstates at this temperature, whereas lower stability constants identify residues that are unfolded in many of those microstates. In nuclease, most residues with higher stability constants were located in β-strands 3–6 and in the α-helices. Lower stability constants were found in residues in β-strands 1 and 2, and, most dramatically, the loop connecting β-strand 4 to α-helix 1. In addition to providing residue stability constants, the COREX/BEST Server also outputs a file listing the 50 most probable microstates, giving data that can be used to generate Fig. 1b.

Notable to the COREX/BEST calculations is the recent suggestion that protein cold denaturation offers experimental access to the native ensemble, allowing for its structural characterization [21]. It's thought that cold temperatures promote noncooperative unfolding of a protein [29, 30] by minimizing the favorable energetics related to the nonspecific burial of hydrophobic groups in aqueous solution [31–33]. As such, the nonspecific hydrophobic packing that favors compact protein structures appears to be minimized at low temperatures in a manner that

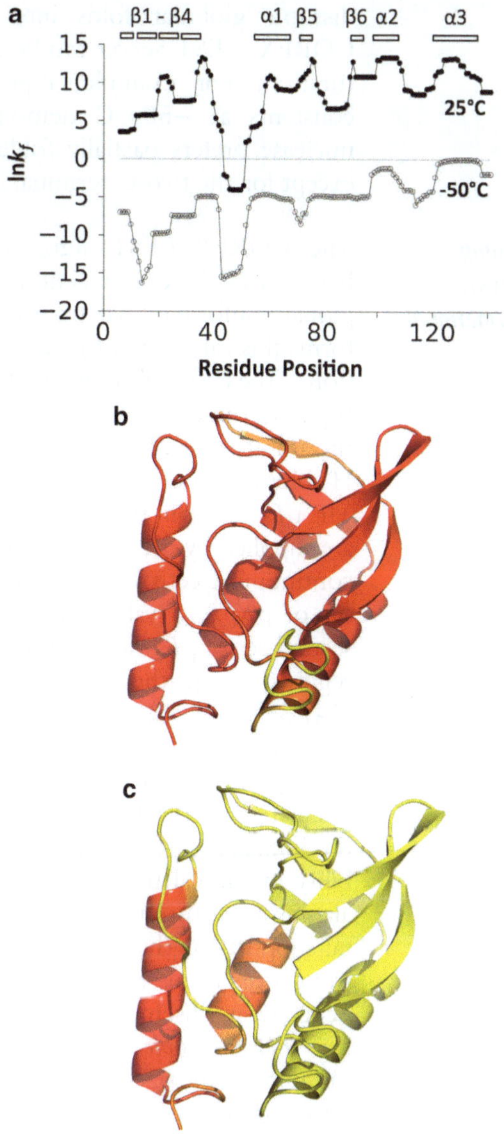

Fig. 2 The effect of temperature on the nuclease ensemble as calculated by the COREX/BEST Server. (**a**) Natural logarithm of the residue stability constants, k_f, for 1STN calculated using the temperature settings of 25 °C (*filled circles*) and −50 °C (*open circles*). The locations of secondary structural elements are as indicated. (**b**) Projection of the residue stability constants onto the nuclease structure where *yellow* represents residue positions with the lowest k_f values and *red* the highest. *Red colored* residue positions were calculated as stable at 25 °C, having large and positive k_f values, whereas *yellow* residues were calculated to have the least stability and lowest k_f values. (**c**) The nuclease structure color-coded the same as in panel **b**, except the stability constants were calculated using a temperature setting of −50 °C

disrupts globular folds into sub-global structural units. The COREX/BEST Server can be used to simulate protein cold denaturation. For example, repeating the calculation of stability constants at −50 °C demonstrates that at low temperatures nuclease prefers partially folded states that are mostly unfolded except for the two C-terminal helices (Fig. 2c).

3.5 Calculating the pH Sensitivity of the Conformational Ensemble

The COREX/BEST algorithm can also model the coupling between H^+ binding reactions, local fluctuations in structure, and global conformational transitions [18, 19]. COREX/BEST performs this calculation by quantifying how the population distribution of the ensemble of microstates is affected by proton binding via Eq. 2. In this calculation, two sets of pK_a values are used to describe the affinity of each H^+ binding site (*see* **Note 2**). One set describes H^+ binding to sites in folded and native-like regions of the protein. The pK_a values in this set are referred to as "native" pK_a values and are supplied by the user (*see* Subheading 2.2). In practice, standard continuum electrostatics methods have been used to generate the set of p$K_{a,\text{native}}$ values for use with COREX/BEST [18]. The second set of pK_a values describes H^+ binding to sites in unfolded regions, using the intrinsic pK_a values of model compounds [34, 35]. p$K_{a,\text{intrinsic}}$ values are supplied by the COREX/BEST algorithm.

Figure 3 shows the pH sensitivity of the 1STN ensemble calculated by the COREX/BEST Server and identifies which of the ionizable residues were responsible for its pH-dependent stability in the calculation. For nuclease, mutagenic tests involving alanine substitutions of the ionizable residues provided experimental support for the COREX/BEST results [18]. To perform this simulation using the COREX/BEST Server, a user should first generate an ensemble by the steps outlined above, but also check the "*pH Ensemble?*" box when specifying window size and minimum window size. The COREX/BEST Server saves structural information needed to assign "native" or "intrinsic" pK_a values to each ionizable group in each microstate only when the "*pH Ensemble?*" box is checked. After the ensemble has been generated, the user should click on the "*Calculate the pH Sensitivity an Ensemble*" on the "*Choosing COREX/BEST Jobs*" page.

4 Notes

1. The relative Gibbs free energy for each microstate, ΔG_i, is determined by structure-based parameterization of $\Delta C_{p,i}$, ΔH_i, and ΔS_i [12]. The heat capacity, $\Delta C_{p,i}$, is known to originate primarily from changes in hydration and has been parameterized in terms of changes in solvent accessible apolar (ΔASA_{apol}) and polar (ΔASA_{pol}) surface area,

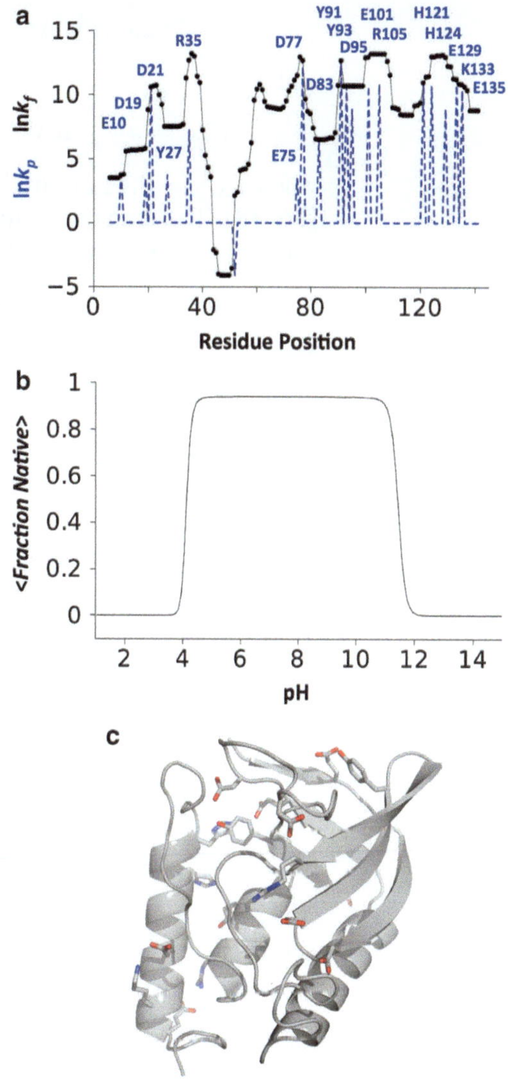

Fig. 3 The effect of pH on the nuclease ensemble as calculated by the COREX/BEST Server. (**a**) Natural logarithm of the stability constants (*black*), k_f, and the residue-specific protection constants (*blue dashed line*), k_p. Protection constants identify which ionizable residues are folded and in a native-like environment in the most probable microstates [19]. In general, ionizable residues with high protection constant values (labeled in the figure) contribute to the pH-dependent population shifts of the ensemble (if $pK_{a,\text{native}} \neq pK_{a,\text{intrinsic}}$), whereas those with low protection constant values do not. (**b**) pH dependent folding of the nuclease ensemble determined by $\langle \text{fraction native} \rangle = \Sigma$ fraction native$_i \bullet P(\text{pH})_i$ where $P(\text{pH})_i$ was calculated by Eq. 2 and the fraction native value of each microstate was calculated as the number of residues in folded segments divided by the total number of residues. These data show that solution changes to acidic or basic conditions promote cooperative unfolding of the 1STN ensemble. (**c**) Cartoon representation of the nuclease structure. The ionizable residues with high k_p values (labeled in **a**) highlighted by showing the positions of their side chain atoms

$$\Delta C_{p,i} = \Delta C_{p,\text{apol},i} + \Delta C_{p,\text{pol},i}, \qquad (5)$$

and

$$\Delta C_{p,i} = a_{\Delta C_p} \cdot \Delta \text{ASA}_{\text{apol},i} + b_{\Delta C_p} \cdot \Delta \text{ASA}_{\text{pol},i}, \qquad (6)$$

where $a_{\Delta Cp} = 0.44$ cal/mol K Å^2 and $b_{\Delta Cp} = 0.26$ cal mol K Å^2 [36–38]. The enthalpy change, ΔH_i, also scales with accessible surface areas and can be written as,

$$\Delta H(60\ ^\circ\text{C})_i = a_{\Delta H} \cdot \Delta \text{ASA}_{\text{apol},i} + b_{\Delta H} \cdot \Delta \text{ASA}_{\text{pol},i}, \qquad (7)$$

and

$$\Delta H(T)_i = \Delta H(60\ ^\circ\text{C})_i + \Delta C_{p,i}(T - 60\ ^\circ\text{C}), \qquad (8)$$

where T is the temperature, $a_{\Delta H} = -8.44$ cal/mol Å^2 and $b_{\Delta H} = 31.4$ cal/mol Å^2 [36, 39]. The entropy change, ΔS_i, includes two contributions, one from changes in solvation and the other from changes in the conformational entropy,

$$\Delta S_i = \Delta S_{\text{solv},i} + \Delta S_{\text{conf},i}, \qquad (9)$$

The solvation contribution can be written in terms of the polar and apolar values of ΔC_p if the temperatures at which $\Delta S_{\text{solv,apol}} = 0$ and $\Delta S_{\text{solv,pol}} = 0$ are known ($T_{\text{s,apol}}$ and $T_{\text{s,pol}}$, respectively),

$$\Delta S(T)_{\text{solv},i} = \Delta C_{p,\text{apol},i} \ln(T/T_{s,\text{apol}}) + \Delta C_{p,\text{pol},i} \ln(T/T_{s,\text{pol}}). \qquad (10)$$

$T_{\text{s,apol}}$ has been shown to be 385 K [31], while $T_{\text{s,pol}}$ has been shown to be 335 K [40]. The conformational entropies are calculated by considering three contributions,

$$\Delta S_{\text{conf},i} = \Delta S_{\text{bu-ex},i} + \Delta S_{\text{ex-un},i} + \Delta S_{\text{bb},i}, \qquad (11)$$

where $\Delta S_{\text{bu-ex},i}$ is the summed entropy change for all side chains that are buried in the fully folded state and become exposed in a microstate, $\Delta S_{\text{ex-un},i}$ is the summed entropy change of solvent-exposed side chains upon unfolding of the peptide backbone, and $\Delta S_{\text{bb},i}$ is the backbone entropy change for residues that become unfolded in a microstate. The magnitudes of the conformational entropy contributions for each amino acid type have been determined from computational analysis of the probabilities of the different dihedral and torsion angles and are reported elsewhere [40, 41]. The temperature-dependent Gibbs free energy for each ensemble state, $\Delta G(T)_i$, is then expressed in terms of the standard thermodynamic equation,

$$\Delta G(T)_i = \Delta H(T)_i - T\left(\Delta S(T)_{\text{solv},i} + \Delta S_{\text{conf},i}\right). \qquad (12)$$

2. For each microstate, each ionizable group can have one of two values: the pK_a value in the native conformation (p$K_{a,native}$, which are user-supplied) or that of the unfolded state (p$K_{a,intrinsic}$, which are supplied by COREX/BEST). The set of p$K_{a,intrinsic}$ values are those of model compounds in water [34, 35]. It's recommended to calculate p$K_{a,native}$ values using continuum electrostatic methods based on solution of the linearized Poisson-Boltzmann equation [42] applied to the high-resolution structure (e.g., 1STN). For each microstate, titratable residues in unfolded regions are assigned the model compound pK_a values (p$K_{a,intrinsic}$). Titratable groups in folded regions are assigned pK_a values based on the solvent accessibility of the titratable atoms. This was done to correct for the case in which a residue may reside in a folded region, however, due to the unfolding of segments of the protein that pack against the titratable site, the ionizable group is exposed to solvent. To apply this correction term in a general manner, a cutoff threshold was determined by comparison of COREX/BEST calculated and experimentally measured proton binding curves [18, 19]. When the averaged solvent accessibility of the ionizable atoms of a titratable group is <31 % (for Glu, Asp, Tyr, Lys, and Arg) or <45 % (for His), these groups are assigned p$K_{a,native}$ values. Otherwise, groups are assigned p$K_{a,intrinsic}$ values. The COREX/BEST Server allows the user to change these two cutoff thresholds when calculating the pH sensitivity of an ensemble.
3. All lines of an uploaded PDB file that begin with the characters "ATOM" are used for COREX/BEST calculations; the algorithm is blindly unaware of the possibility that a PDB file may contain multiple models of the same structure. If a PDB file contains multiple models, either in terms of the whole chain (e.g., NMR model structures) or for the side chain conformations of individual residues, the Server will incorrectly assume that each model represents new atoms. Accordingly, additional atoms that physically aren't present in a structure produce nonsensical COREX/BEST results. Also, PDB files that are missing side chain atoms should be avoided. The energy function used by COREX/BEST (*see* **Note 1**) was parameterized and trained using protein structures that contained side chain atoms for each residue.
4. The COREX/BEST algorithm can use a Metropolis Monte Carlo method of sampling [27] to decrease the run-time required to generate an ensemble by decreasing the number of microstates generated. By design, this method uses the microstate probabilities calculated from the COREX/BEST energy function (*see* **Note 1** and Eq. 1) to bias sampling in favor of the more probable microstates. The COREX/BEST

energy function, however, depends on temperature and pH. The Monte Carlo sampling routine uses a default temperature setting of 25 °C and includes no proton binding energies when calculating microstate probabilities. Thus, Monte Carlo sampled ensembles work best (i.e., produce equivalent results compared to fully enumerated and non-sampled ensembles) for calculations designed to simulate temperatures at or near 25 °C and that include no proton binding energies.

Acknowledgments

This work was supported by NIH grant R01-GM63747 to V.J.H. and the Texas Higher Education Coordinating Board grant 003615-0003-2011 to S.T.W.

References

1. Changeux JP, Edelstein SJ (2005) Allosteric mechanisms of signal transduction. Science 308:1424–1428
2. Karplus M, Kuriyan J (2005) Molecular dynamics and protein function. Proc Natl Acad Sci U S A 102:6679–6685
3. Eisenmesser EZ, Millet O, Labeikovsky W, Korzhnev DM, Wolf-Watz M, Bosco DA, Skalicky JJ, Kay LE, Kern D (2005) Intrinsic dynamics of an enzyme underlies catalysis. Nature 438:117–121
4. Bai YW, Sosnick TR, Mayne L, Englander SW (1995) Protein folding intermediates: native-state hydrogen exchange. Science 269:192–197
5. SwintKruse L, Robertson AD (1996) Temperature and pH dependences of hydrogen exchange and global stability for ovomucoid third domain. Biochemistry 35:171–180
6. Chamberlain AK, Handel TM, Marqusee S (1996) Detection of rare partially folded molecules in equilibrium with the native conformation of RNaseH. Nat Struct Biol 3:782–787
7. Fuentes EJ, Wand AJ (1998) Local dynamics and stability of apocytochrome b(562) examined by hydrogen exchange. Biochemistry 37:3687–3698
8. Itzhaki LS, Neira JL, Fersht AR (1997) Hydrogen exchange in chymotrypsin inhibitor 2 probed by denaturants and temperature. J Mol Biol 270:89–98
9. Yang DW, Kay LE (1996) Contributions to conformational entropy arising from bond vector fluctuations measured from NMR-derived order parameters: application to protein folding. J Mol Biol 263:369–382
10. Li ZG, Raychaudhuri S, Wand AJ (1996) Insights into the local residual entropy of proteins provided by nmr relaxation. Protein Sci 5:2647–2650
11. Hilser VJ, Freire E (1996) Structure-based calculation of the equilibrium folding pathway of proteins. Correlation with hydrogen exchange protection factors. J Mol Biol 262:756–772
12. Hilser VJ (2001) Modeling the native state ensemble. Methods Mol Biol 168:93–116
13. Hilser VJ, Freire E (1997) Predicting the equilibrium protein folding pathway: structure-based analysis of Staphylococcal nuclease. Proteins 27:171–183
14. Hilser VJ, Townsend BD, Freire E (1997) Structure-based statistical thermodynamic analysis of T4 lysozyme mutants: structural mapping of cooperative interactions. Biophys Chem 64:69–79
15. Hilser VJ, Dowdy D, Oas TG, Freire E (1998) The structural distribution of cooperative interactions on proteins: analysis of the native state ensemble. Proc Natl Acad Sci USA 95:9903–9908
16. Pan H, Lee JC, Hilser VJ (2000) Binding sites in Escherichia coli dihydrofolate reductase communicate by modulating the conformational ensemble. Proc Natl Acad Sci USA 97:12020–12025
17. Liu T, Whitten ST, Hilser VJ (2007) Functional residues serve a dominant role in mediating the cooperativity of the protein ensemble. Proc Natl Acad Sci U S A 104:4347–4352
18. Whitten ST, García-Moreno EB, Hilser VJ (2005) Local conformational fluctuations can modulate the coupling between proton binding

and global structural transitions in proteins. Proc Natl Acad Sci U S A 102:4282–4287

19. Whitten ST, García-Moreno EB, Hilser VJ (2008) Ligand effects on the protein ensemble: unifying the descriptions of ligand binding, local conformational fluctuations, and protein stability. Methods Cell Biol 84:871–891
20. Babu CR, Hilser VJ, Wand AJ (2004) Direct access to the cooperative substructure of proteins and the protein ensemble via cold denaturation. Nat Struct Mol Biol 11:352–357
21. Whitten ST, Kurtz AJ, Pometun MS, Wand AJ, Hilser VJ (2006) Revealing the nature of the native state ensemble through cold denaturation. Biochemistry 45:10163–10174
22. Schrank TP, Bolen DW, Hilser VJ (2009) Rational modulation of conformational fluctuations in adenylate kinase reveals a local unfolding mechanism for allostery and functional adaptation in proteins. Proc Natl Acad Sci U S A 106:16984–16989
23. Vertrees J, Barritt P, Whitten ST, Hilser VJ (2005) COREX/BEST server: a web browser-based program that calculates regional stability variations within protein structures. Bioinformatics 21:3318–3319
24. Hynes TR, Fox RO (1991) The crystal structure of staphylococcal nuclease refined at 1.7 A resolution. Proteins 10:92–105
25. Hilser VJ, García-Moreno EB, Oas TG, Kapp G, Whitten ST (2006) A statistical thermodynamic model of the protein ensemble. Chem Rev 106:1545–1558
26. Berman HM, Westbrook J, Feng Z, Gilliland G, Bhat TN, Weissig H, Shindyalov IN, Bourne PE (2000) The protein data bank. Nucleic Acids Res 28:235–242
27. Metropolis N, Rosenbluth AW, Rosenbluth MN, Teller AH, Teller E (1953) Equation of state calculations by fast computing machines. J Chem Phys 21:1087–1092
28. Whitten ST, García-Moreno EB (2000) pH dependence of stability of staphylococcal nuclease: evidence of substantial electrostatic interactions in the denatured state. Biochemistry 39:14292–14304
29. Freire E, Murphy KP, Sanchez-Ruiz JM, Galisteo ML, Privalov PL (1992) The molecular basis of cooperativity in protein folding: thermodynamic dissection of interdomain interactions in phosphoglycerate kinase. Biochemistry 31:250–256
30. Griko YV, Venyaminov SY, Privalov PL (1989) Heat and cold denaturation of phosphoglycerate kinase (interaction of domains). FEBS Lett 244:276–278
31. Baldwin RL (1986) Temperature dependence of the hydrophobic interaction in protein folding. Proc Natl Acad Sci U S A 83:8069–8072
32. Lopez CF, Darst RK, Rossky PJ (2008) Mechanistic elements of protein cold denaturation. J Phys Chem B 112:5961–5967
33. Privalov PL (1990) Cold denaturation of proteins. Crit Rev Biochem Mol Biol 25:281–305
34. Matthew JB, Gurd FR, García-Moreno EB, Flanagan MA, March KL, Shire SJ (1985) pH-dependent processes in proteins. CRC Crit Rev Biochem 18:91–197
35. Schaefer M, van Vlijmen HWT, Karplus M (1998) Electrostatic contributions to molecular free energies in solution. Adv Protein Chem 51:1–57
36. Murphy KP, Freire E (1992) Thermodynamics of structural stability and cooperative folding behavior in proteins. Adv Protein Chem 43: 313–361
37. Gomez J, Hilser VJ, Xie D, Freire E (1995) The heat-capacity of proteins. Proteins 22:404–412
38. Habermann SM, Murphy KP (1996) Energetics of hydrogen bonding in proteins: a model compound study. Protein Sci 5:1229–1239
39. Xie D, Freire E (1994) Structure-based prediction of protein-folding intermediates. J Mol Biol 242:62–80
40. D'Aquino JA, Gomez J, Hilser VJ, Lee KH, Amzel LM, Freire E (1996) The magnitude of the backbone conformational entropy change in protein folding. Proteins 25:143–156
41. Lee KH, Xie D, Freire E, Amzel LM (1994) Estimation of changes in side chain configurational entropy in binding and folding: general methods and application to helix formation. Proteins 20:68–84
42. Fitch CA, Karp DA, Lee KK, Stites WE, Lattman EE, García-Moreno EB (2002) Experimental pKa values of buried residues: analysis with continuum methods and role of water penetration. Biophys J 82:3289–3304

Chapter 15

Morphing Methods to Visualize Coarse-Grained Protein Dynamics

Dahlia R. Weiss and Patrice Koehl

Abstract

Morphing was initially developed as a cinematic effect, where one image is seamlessly transformed into another image. The technique was widely adopted by biologists to visualize the transition between protein conformational states, generating an interpolated pathway from an initial to a final protein structure. Geometric morphing seeks to create visually suggestive movies that illustrate structural changes between conformations but do not necessarily represent a biologically relevant pathway, while minimum energy path (MEP) interpolations aim at describing the true transition state between the crystal structure minima in the energy landscape.

Key words Proteins, Dynamics, Interpolation, Minimum energy path

1 Introduction

In movies and computer graphics, morphing allows an image to be gradually transformed into another in a seamless fashion, where the resulting series of images often depicts an amusing mixture between the two endpoint images. In the biological context, morphing has been adapted to illustrate the conformational changes between two protein structures. A morph movie showing a smooth transition between the two protein conformations is often informative, helping to understand the structural changes taking place. Geometric protein morphing was originally seen as didactic, with the aim of showing graphically pleasing transition, and it is only recently that its aim has changed to predicting the actual trajectory through which the conformational change occurs (*see* Fig. 1).

The first geometric morph of protein structures was created in 1995 by Vonrhein et al. [1], who structurally aligned 17 crystal structures of homologous nucleoside monophosphates to create a picture series from the open to closed conformation. While the initial movie truly represented the experimental data, the authors found that "these movies were rather jerky." To smooth the

Dennis R. Livesay (ed.), *Protein Dynamics: Methods and Protocols*, Methods in Molecular Biology, vol. 1084, DOI 10.1007/978-1-62703-658-0_15, © Springer Science+Business Media New York 2014

transitions, Vonrhein and colleagues used linear interpolation, where each atom moves in a straight line in Cartesian coordinate space from its initial to final conformation, in the spirit of computer graphics morphing. However, unlike computer images, biological macromolecules must adhere to physical realities; atoms cannot move through each other, and bond lengths and bond angles must be preserved. Biological morphs should therefore also be "physically feasible." This is most often achieved by energy minimizing intermediate frames, enforcing bond lengths and angles prescribed by stereochemistry. This can also be achieved by interpolating the internal coordinates, i.e., the torsion angles of the protein main-chain are changed smoothly from their initial to final values, leaving bond length and angles unchanged.

While geometric morphing can be informative for visualizing structural changes, the true pathways that a protein follows and the transition states it passes through are crucial to understanding the biological function of the protein. An accurate description of the transition pathway could theoretically be attained using molecular dynamics (MD) simulation; however, in practice such a computation is infeasible due to the long timescales needed. We therefore introduce a new approach to morphing, which is no longer geometric, but instead finds the minimum energy path (MEP) between two experimentally determined protein conformations, taking into account the underlying energy landscape.

2 Methods

2.1 Geometric Morphing

The initial purpose of morphing was to create movies that would allow the user to visualize the transitions between known protein structures. A selection of online servers and software to create such molecular morphs are summarized in Table 1.

To allow the user to create physically feasible morphs between conformations, in 1998 Gerstein and Krebs created the MolMovDB http://molmovdb.org/ [2, 3]. This remains perhaps the most popular online method to create movies for protein motion visualization, and produces morphs in up to a few hours. The interpolation is linear in the Cartesian coordinates of each atom; however, each frame along the interpolation is energy minimized to ensure correct stereochemistry and valid packing. Again, no attempt is made to predict the actual trajectory or transition state through which the conformational change passes, but structures are no longer distorted. The MolMovDB stores all morphs generated, allowing the user to browse all past movies by motion type, protein name or PDB ID. The MolMovDB server has undergone several updates, and now accepts multi-chain proteins, as well as initial and final conformations from different but related proteins (mismatched residues are mutated to alanine). The input to the

Table 1
Protein morphing software

Program	URL	Algorithm
MolMovDB	http://molmovdb.org/	Linear interpolation in Cartesian coordinates with energy minimization at each step
LSQMAN	http://xray.bmc.uu.se/usf/	Linear interpolation in Cartesian or internal coordinates
FATCAT	http://fatcat.burnham.org/	Rigid body linear interpolation around a minimal number of predetermined hinges
Elastic Network Interpolation	http://bioengineering.skku.ac.kr/kosmos	Nonlinear interpolation of inter-residue distances
Climber	https://simtk.org/home/climber	Nonlinear interpolation of inter-residue distances with self-adjusting spring constants
Pathway	http://pathways.asu.edu	Nonlinear interpolation of initial RMSD to final RMSD of zero
MinActionPath	http://lorentz.dynstr.pasteur.fr/joel/index.php.	Calculates the minimum energy path

server is a PDB ID or a user uploaded PDB file. There is no need to structurally align the input conformations, as this is done by the server. The output comes in the form of a Java movie (which can be viewed and rotated in 3D with appropriate plug-ins) and can be downloaded as a MPEG movie file; PDB files of the intermediates can also be downloaded.

A popular standalone morphing program (not available as an online server) is implemented in the Kleywegt and Jones 1996 program LSQMAN [4]. It is freely available for academic use as part of the DEJAVU package at http://xray.bmc.uu.se/usf/. The program offers several major advantages; most notable is the ability to interpolate in internal coordinates (torsion angles). This avoids the nonphysical deformation of bond lengths and bond angles, so that intermediates no longer need to be energy minimized. Additionally, structures no longer need to be superimposed, since torsion angles are independent of orientation. While using torsion angles offers the advantage of realistic bond lengths and angles in all intermediates, it too has several drawbacks. Firstly, it may be impossible to reach the final conformation using this method. Secondly, in the process of interpolating, atoms can come very close or pass through each other, so that intermediate structures still suffer from unrealistically high internal energies. Finally, since a change in a torsion angle affects all atom positions downstream of that angle, for large conformational changes, the protein may appear to be unfolding on its way to the target structure.

To alleviate these problems to some extent, LSQMAN allows the user flexibility in choosing the atoms to be morphed and the method (Cartesian or internal).

The FATCAT server [5], developed in 2004 by Ye and Godzik and available at http://fatcat.burnham.org/, creates rigid body linear interpolations of proteins. The server simultaneously optimizes the structural alignment of the initial and final conformations, while minimizing the number of rigid body movements around few hinges that are concurrently introduced. This is in contrast to the usual procedure that first aligns two structures and then detects the hinge points that connect the two aligned structures. The optimally aligned rigid protein domains are linearly interpolated in Cartesian space around the detected hinge points and the intermediate structures are energy minimized. The intermediate structures are available for download as series of PDB files.

Many protein conformational changes can be described as linear, for example ribose-binding protein, which binds its ribose ligand with a simple hinging motion of its two domains (Fig. 1). However, many other transitions are nonlinear, as is the case for example with 5′-Nucleotidase (5′-NT), which undergoes a large 96° domain rotation, with an intermediate crystal structure at 43° rotation (PDB ID: 1OID, 1OI8, 1HPU). Nonlinear interpolation methods, such as Elastic Network Interpolation (ENI), Climber and Pathways use a general approach of interpolating inter-residue distances or root-mean-square-distance (RMSD) from the initial to the final configuration, while enforcing an additional set of constraints to ensure physically realistic intermediates [6–8]. ENI, available as an online server at KOSMOS http://bioengineering.skku.ac.kr/kosmos was the first of such nonlinear interpolation methods. In KOSMOS, the inter-residue distances of an initial conformation are incrementally pulled towards the distances in the target conformation using a set of harmonic restraints or springs. The user may determine the number of pairwise restraints by setting a cutoff distance.

Climber is available as a standalone download from https://simtk.org/home/climber. Climber also uses inter-residue distance interpolation, where the harmonic restraints are added to the internal protein energy function, which is minimized at each step. The user can determine the number of steps that will be used, and the spring force-constant is changed so that the conformation reaches the target destination in approximately the user-defined number of steps. This self-adjusting method allows the trajectory to move around high-energy barriers, as long as a sufficient number of steps are allowed, and trajectories differ as the number of steps is varied.

The Pathway Web server is found at http://pathways.asu.edu. Pathway interpolates the initial RMSD to a final RMSD of zero from the target. The program subdivides the amino-acid side chains

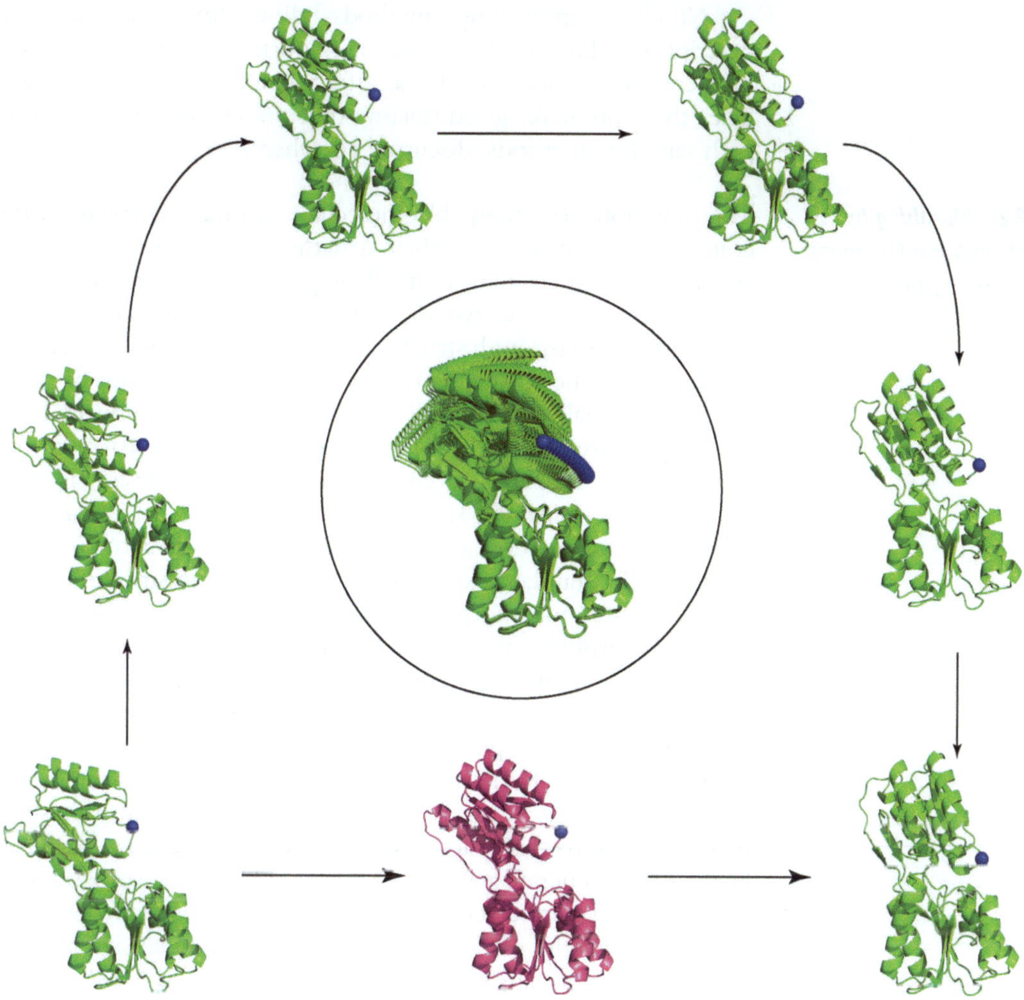

Fig. 1 Morphing between two conformations of ribose binding protein, which binds to its substrate ribose with a hinging motion. The crystal structure of ribose binding protein has been determined in three states, open, intermediate and closed (*bottom row*, PDB code 1BA2 *left* in *green*, 1URP *center* in *magenta*, and 2DRI *right* in *green*). A geometric morph moves the atoms in a linear fashion from the open conformation to the closed conformation (top pathway). The *central figure* shows a superposition of 16 morphed intermediates generated by MolMovDB [2, 3], with the trajectory of one atom highlighted in *blue*. In this case, the geometric morph passes through a transition state close to the actual crystal intermediate, but in general this is not the aim of geometric morphing, which serves to instruct but does not try to predict a biologically relevant trajectory. The figure was prepared with the program Pymol (http:/www.pymol.org)

as rigid units, which can then move subject to constraints, which maintain stereochemistry, hydrogen bonding and hydrophobic packing. Advanced options allow the user to include random motions, to allow for backtracking if the configuration gets "stuck" and to make the protein more or less flexible, thereby producing many possible low energy trajectories.

Nonlinear morphing methods allow for more realistic descriptions of protein motions, and importantly, can move around high energy barriers which linear interpolations must pull through. They therefore make good starting trajectories for more advanced path sampling methods, discussed elsewhere.

2.2 Morphing by Finding the Minimum Energy Path

The functions of many bio-molecules strongly correlate with conformational changes in their structure space, a process usually referred to as their activations. This process for example is very much at the core of enzymatic activity, as an enzyme and its substrate usually go through structural transitions that favor the chemical reaction. The structures of these transition states are of great interest, especially for drug design. Many enzyme inhibitors have been engineered to be transition state analogs, i.e., to resemble the transition state of the enzyme substrate; this design is only possible if the transition state of the enzyme itself is known. This transition state however is very short lived and its structure cannot be studied by standard experimental methods from structural biology. Computational morphing is then a valuable alternative, where the word morphing takes a different meaning than in the geometric morphing techniques described above. Given two conformations for a biomolecule, the problem is a find a plausible path along its energy surface, where plausible usually refers to a path with minimal frustration, also referred to as the Minimum Energy Path (MEP) (*see* Fig. 2). MEP is usually well described within the transition state theory (TST). In this chapter we will briefly discuss how MEP has been used to generate an initial path between two conformations of a molecule. We understand that this description is far from complete and refer the reader to some recent comprehensive reviews [9–12] for a more in-depth presentation of TST and MEP.

In principle, a brute force molecular dynamics (MD) simulation would solve the Minimum Energy Path problem, as it is designed to simulate the dynamics of the system with atomistic details. However, the timescales required for pushing a system over an energy barrier scale exponentially with the barrier height. As a result, traditional MD has difficulty surmounting even small barriers in times that are computationally accessible [13]. Approaches to overcome this limitation involve biasing techniques to focus in relevant regions of conformational space. Probably the most popular of these techniques relevant to morphing is the so-called targeted (or steered) MD [14], which samples conformations using a biased potential that contains information about the target conformation (usually with an additional term the computes the RMS difference between the current and target conformation). Targeted MD has been shown however to introduce a bias in the interplay between local and global motions, leading to a wrong description of the transitions [15, 16].

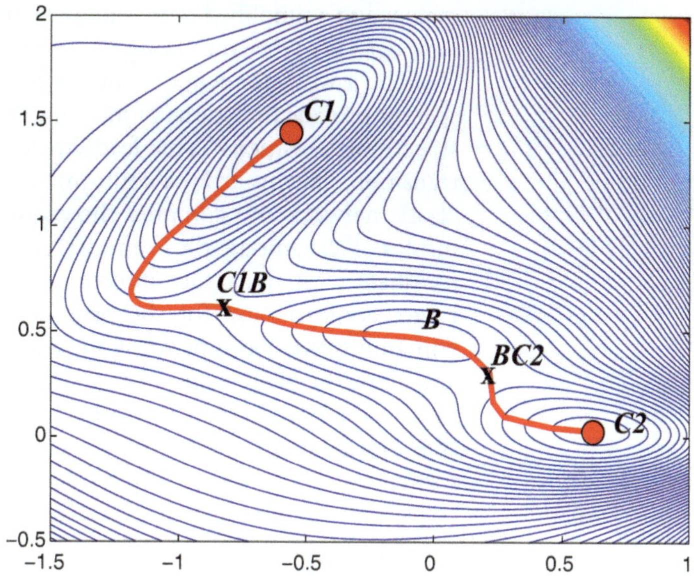

Fig. 2 An example of an energy surface. The Mueller potential [31] shown here was invented as a strict test for algorithms that compute saddle points or minimum energy paths. It has three minima, denoted by C1, B, and C2, and two saddle points, denoted by C1B and BC2. The minimum energy path between C1 and C2 is shown in *red*: it passes through the intermediate minimum B and the two saddle points

The assumption that collective motions govern structural transitions in large biomolecules has led to the development and popularity of the Elastic Network Model [17, 18]. This model represents the molecule by a set of point masses (for example, at the coordinates of the alpha carbons for a protein) that are connected by springs to their neighbors determined by spatial proximity. The actual chemical connectivity is usually ignored. The energy of the system is simply the sum of the Hookean potentials of the springs:

$$E = \sum_{i=1}^{N} \sum_{j \in N_i} \frac{C}{2} \left(r_{ij} - r_{ij}^{0} \right)^2 \qquad (1)$$

where r_{ij} and r_{ij}^0 are the distances between the two points i and j in the current and in the reference structure (usually the X-ray structure), respectively, C is the "bond strength" [17] and N_i is the set of atoms that are within a cutoff distance R_c from i. We note that this energy function is minimal (with a value of 0) at the X-ray reference structure, and that it only depends on two parameters: the bond strength, C, and the cutoff distance R_c. The dynamics of this system is fully characterized by a set of normal modes. The elastic network model provides both a simplified representation of the energy of a system and a simplified representation of its dynamic, both of which have been used for computing transition pathways for biomolecules.

The lowest frequency modes of the corresponding network of springs provide directions of motions that are conjectured to shape structural transitions. This leads to a very simple idea to steer a transition: starting from a conformation A for a protein, derive its elastic network and compute the corresponding lowest frequency modes. Pick among these modes those with large overlap (dot product) with the target direction (i.e., $X_B - X_A$, where B is the target conformation for the protein and X refers to the vector containing the coordinates of all points representing the protein) and allow the protein structure to move along the combined directions corresponding to this mode to a position C. The procedure is then repeated iteratively until C gets very close to the target conformation B. This procedure was successfully applied to study the transitions of the adenylate kinase and hemoglobin [19]. Note that the same idea can be used with any simplified energy model (see for example [20]).

The elastic network model is based on a very simple quadratic potential function; this energy however is only valid in the neighborhood of the conformation from which it is derived. Maragakis and Karplus [21] extended this model to account for two possible conformations C1 and C2 of the same protein. In their approach, referred to as the Plastic Network Model (PNM), if $E_1(X)$ and $E_2(X)$ are the elastic energy functionals for the protein in conformation X computed from the reference conformations C1 and C2, respectively, the total energy $E(X)$ is given by:

$$E(X) = \frac{E_1(X) + E_2(X) - \sqrt{(E_1(X) - E_2(X))^2 + 4\varepsilon^2}}{2} \quad (2)$$

with ε being a constant that is small enough that the energy at C1 and C2 are $E_1(\mathrm{C1})$ and $E_2(\mathrm{C2})$, respectively. For a non-zero value of ε, a saddle point is created at the global minimum of the hypersurface defined by $E_1 = E_2$; this saddle point connects the two energy wells corresponding to C1 and C2 in conformation space. With this (simplified) definition of the energy, it is then possible to build the Minimum Energy Path. Assuming a constant friction for all coordinates, the optimal path satisfies the principle of least resistance [22] and minimizes the functional

$$F = \int_{\mathrm{C1}}^{\mathrm{C2}} \exp(\beta E(X))\mathrm{d}l(X) \quad (3)$$

where $\beta = 1/kT$, with T the temperature, E is given by Eq. 2, and $\mathrm{d}l$ refers to a small arc length on a path on the hypersurface defined by E that connects C1 and C2. Several methods have been developed to minimize the functional F, such as MaxFlux [23], the nudged elastic band method [24], and the conjugate peak refinement algorithm [25] (for a more exhaustive list of methods *see* [11]).

The PNM method described above can be summarized as finding a least resistance path on a single, multidimensional double well potential. Since its publication, several variants have been proposed that either modify the mixing energy or the functional F on the energy surface. For example, Chu and Voth [26] developed the double-well network model (DWNM) in which Eq. 2 is replaced with a network of one-dimensional double well potential:

$$E_{ij} = \frac{E_{1,ij} + E_{2,ij} - \sqrt{\left(E_{1,ij} - E_{2,ij}\right)^2 + 4\varepsilon_{ij}^2}}{2} \tag{4}$$

for each pair of points i, j that are within the cutoff distance R_c. The DWNM was applied successfully on G-actin and adenylate kinase [26].

Equation 3 refers to the principle of least resistance. Alternatively, assuming an overdamped regime and following Onsager and Machlup, it is possible to define a mechanical action for a trajectory between conformations C1 and C2:

$$S = \int_0^{t1} \left(\frac{\mathrm{d}X}{\mathrm{d}t} + \overrightarrow{grad}(E_1(X)) \right)^2 \mathrm{d}t + \int_{t1}^{T} \left(\frac{\mathrm{d}X}{\mathrm{d}t} + \overrightarrow{grad}(E_2(X)) \right)^2 \mathrm{d}t \tag{5}$$

The integrations in Eq. 5 are performed over time, such that at $t = 0$, $t1$, and T the protein is in conformations C1, B, and C2, respectively, where B is the transition state. The optimal path is then defined as the path that leads to a minimum of S as defined in Eq. 5. If E_1 and E_2 are defined according to Eq. 1, minimizing S leads to a system of differential equations that can be solved analytically. This method was implemented in the program MinActionPath [27] and is available online at http://lorentz.dynstr.pasteur.fr/joel/index.php. Figure 3 illustrates the application of MinActionPath on calmodulin.

3 Concluding Remarks

Many questions in biological research are concerned with shapes and their variations. At the macroscopic level for example, Darwin's theory was originally combined with morphometrics (the field of biological shape analysis) to place species within the tree of life. It is well known that the shape of the skull is very important in the study of human and primate evolution. The changes in the shapes of the

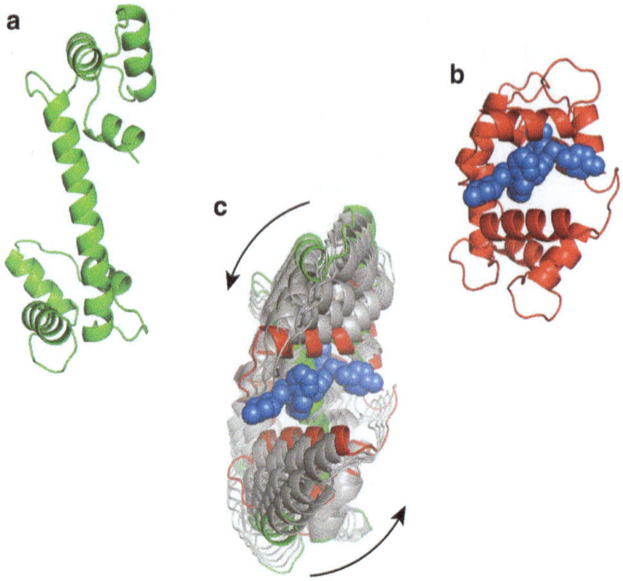

Fig. 3 Morphing between two conformations for calmodulin. Calmodulin is a calcium-binding messenger protein expressed in all eukaryotic cells. The structure of the human calmodulin was determined by X-ray crystallography (PDB code 1CLL) and is shown in panel (**a**). The same calmodulin is shown in panel (**b**), complexed with two trifluoperazine molecules (shown in space filling) (PDB code 1A29). The two conformations differ by 15.2 Å. We have used MinActionPath [27] to generate a trajectory between these two conformations; eight conformations along this trajectory are shown in *grey* in panel (**c**). The figure was prepared with the program Pymol (http:/www.pymol.org)

skulls corresponding to two species define an evolutionary distance between these species; these changes can be quantified by analyzing the transformations required to morph one shape into the other. New morphing techniques that implement and quantify these transformations are seen as key developments for the future of evolutionary morphometrics [28, 29].

The importance of shape and their fluctuations remains valid at the microscopic level. It is widely accepted that the function and structure of a protein are related, and that this relationship comes from the dynamics of the structure [30]. Characterizing biomolecular dynamics has become essential in many aspects of biological research, for instance in the design of therapeutic drugs. Experiments in structural biology are photography-like in that they provide snapshots of (a few) conformations of the molecule of interest. Morphing techniques have proved useful complements as they provide means to generate trajectories between these snapshots. As described in this review, many such techniques have been developed. The field is still new, however, and we can expect many more new algorithms to appear in the future, especially in the domain of finding minimum energy paths.

References

1. Vonrhein C, Schlauderer GJ, Schulz GE (1995) Movie of the structural changes during a catalytic cycle of nucleoside monophosphate kinases. Structure 3:483–490
2. Gerstein M, Krebs W (1998) A database of macromolecular motions. Nucleic Acids Res 26:4280–4290
3. Flores S, Echols N, Milburn D, Hespenheide B, Keating K, Lu J, Wells S, Yu EZ, Thorpe M, Gerstein M (2006) The database of macromolecular motions: new features added at the decade mark. Nucleic Acids Res 34:D296–301
4. Kleywegt GJ (1996) Use of non-crystallographic symmetry in protein structure refinement. Acta Cryst D-Biol Cryst 52:842–857
5. Ye YZ, Godzik A (2004) FATCAT: a web server for flexible structure comparison and structure similarity searching. Nucleic Acids 32:W582–W585
6. Weiss DR, Levitt M (2009) Can morphing methods predict intermediate structures? J Mol Biol 385:665–674
7. Farrell DW, Speranskiy K, Thorpe MF (2010) Generating stereochemically acceptable protein pathways. Proteins 78:2908–2921
8. Kim MK, Jernigan RL, Chirikjian GS (2002) Efficient generation of feasible pathways for protein conformational transitions. Biophys J 83:1620–1630
9. Metzner P, Schutte C, Vanden-Eijnden E (2006) Illustration of transition path theory on a collection of simple examples. J Chem Phys 125:084110
10. Vanden-Eijnden E, Tal FA (2005) Transition state theory: variational formulation, dynamical corrections, and error estimates. J Chem Phys 123:184103
11. Weinan E, Vanden-Eijnden E (2010) Transition path theory and path finding algorithms for the study of rare events. Annu Rev Phys Chem 61:391–420
12. Van Erp TS (2012) Dynamical rare event simulation techniques for equilibrium and nonequilibrium systems. Adv Chem Phys 151:27–58
13. Elber R (2005) Long-timescale simulation methods. Curr Opin Struct Biol 15:151–156
14. Schiltter JM, Engels M, Kruger P, Jacoby E, Wollmer A (1993) Targeted molecular dynamics simulation of conformational change—application to the T-R transition in insulin. Mol Simul 10:291–308
15. Koppole S, Smith JC, Fischer S (2007) The structural coupling between ATPase activation and recovery stroke in the myosin II motor. Structure 15:825–837
16. Noe F, Fischer S (2008) Transition network for modeling the kinetics of conformational changes in macromolecules. Curr Opin Struct Biol 18:154–162
17. Tirion MM (1996) Low-amplitude elastic motions in proteins from a single-parameter atomic analysis. Phys Rev Lett 77:1905–1908
18. Bahar I, Rader AJ (2005) Coarse-grained normal mode analysis in structural biology. Curr Opin Struct Biol 15:586–592
19. Kantarci-Carsibasi N, Haligoglu T, Doruker P (2008) Conformational transition pathways explored by Monte Carlo simulation integrated with collective modes. Biophys J 95:5862–5873
20. Korkut A, Hendrickson WA (2009) Computation of conformational transitions in proteins by virtual atom molecular mechanics as validated in application of adenylate kinase. Proc Natl Acad Sci U S A 106:15673–15678
21. Maragakis P, Karplus M (2005) Large amplitude conformational change in proteins explored with a plastic network model: adenylate kinase. J Mol Biol 352:807–822
22. Berkowitz M, Morgan JD, McCammon JA, Northrup SH (1983) Diffusion-controlled reactions- a variational formula for the optimum reaction coordinates. J Chem Phys 79:5563
23. Huo SH, Straub JE (1997) The MaxFlux algorithm for calculating variationally optimized reaction paths for conformational transitions in many body systems at finite temperature. J Chem Phys 107:5000–5006
24. Jonsson, H., G. Mills, and K. W. Jacobsen. 1998. Nudged elastic band method for finding minimum energy paths of transitions. In Classical and Quantum dynamics in condense phase simulations. World Sci., Singapore. 285-404.
25. Fischer S, Karplus M (1992) Conjugate peak refinement- an algorithm for finding reaction paths and accurate transition states in systems with many degrees of freedom. Chem Phys Lett 194:252–261
26. Chu JW, Voth GA (2007) Coarse-grained free energy functions for studying protein conformational changes: a double well network model. Biophys J 93:3860–3871
27. Franklin J, Koehl P, Doniach S, Delarue M (2007) MinActionPath: maximum likelihood

trajectory for large-scale structural transitions in a coarse-grained locally harmonic energy landscape. Nucl Acids Res 35:W477–W482

28. Wiley, D. F., N. Amenta, D. A. Alcantara, D. Ghosh, Y. J. Kil, E. Delson, W. Harcourt-Smith, F. J. Rolf, K. S. John, and B. Haman. 2005. Evolutionary morphing. In IEEE Viz. 431-438

29. Slice DE (2007) Geometric morphometrics. Annu Rev Anthropol 36:261–281

30. Henzler-Widman KA, Kern D (2007) Dynamic personalities of proteins. Nature 450:964–972

31. Mueller K, Brown LD (1979) Location of saddle points and minimum energy paths by a constrained simplex optimization procedure. Theor Chim Acta (Berl) 53:75–93

Index

Dennis R. Livesay (ed.), *Protein Dynamics: Methods and Protocols*, Methods in Molecular Biology, vol. 1084,
DOI 10.1007/978-1-62703-658-0,

P

Q

R

S

T

U

V

X

MIX
Papier aus verantwortungsvollen Quellen
Paper from responsible sources
FSC® C105338

If you have any concerns about our products,
you can contact us on
ProductSafety@springernature.com

In case Publisher is established outside the EU,
the EU authorized representative is:
Springer Nature Customer Service Center GmbH
Europaplatz 3, 69115 Heidelberg, Germany

Printed by Libri Plureos GmbH
in Hamburg, Germany